WORLD GEOGRAPHICAL ENCYCLOPEDIA

WORLD
GEOGRAPHICAL ENCYCLOPEDIA

VOLUME 1
AFRICA

McGraw-Hill, Inc.

New York San Francisco Washington, D.C. Auckland Bogotá
Caracas Lisbon London Madrid Mexico City Milan
Montreal New Delhi San Juan Singapore
Sydney Tokyo Toronto

English Language Edition

Sybil P. Parker, Editor

Volume 1 (Africa)

Translated by Henry Fischbach and Nicholas Hartmann
Translations Editor: Jeanne De Tar
The Language Service, Inc., Poughkeepsie, New York

Italian Language Edition

Editors in Chief
Virginio Motta and Federico Motta

Managing Editor
Federica Motta

Senior Editor
Guido Bonarelli

Copy Editing
Paolo Marino

Picture Research
Katia Zucchetti

Layout and Design
Beppe Re Fraschini

Art Production
Luigi Senaldi

Cartography and Drawings
L. S. International Cartography; Marco Ghiglieri

Production Manager
Corrado Brizio

Scientific Consultants
Umberto Bonapace, Luigi Bocconi Commercial University
Berardo Cori, University of Pisa
Pedro Cunill Grau, Venezuela Central University
Piero Dagradi, University of Bologna
Ivan Gams, University of Ljubljana
Leszek A. Kosinski, University of Alberta
Lamberto Laureti, University of Pavia
Elio Manzi, University of Pavia
Peter Scott, University of Tasmania
Anne-Marie Seronde Babonaux, University of Caen
Mohammed Shafi, Islamic University of Aligarh
Keiichi Takeuchi, Hitotsubashi University
Alessandro Toniolo, Catholic University of Milan
Eugenio Turri, Italian Geographical Society
Herman Th. Verstappen, Institute for Aerospace Survey and Earth Sciences, Enschede (the Netherlands)
Juan Vilà Valentí, University of Barcelona

Contributors to the present volume
Umberto Bonapace, Mario Casari, Elena dell'Agnese, Lamberto Laureti, Flavio Lucchesi, Alessandro Micheletti, Alessandro Schiavi, Eugenio Turri (*for text material*); Stefano Blaco (*for photo captions*); Giulio Bianchi, Fabio Zucca (*for "Great Routes and Voyages of Discovery"*).

Library of Congress Cataloging-in-Publication Data

Enciclopedia Geografica Universale. English.
World Geographical Encyclopedia / [Sybil P. Parker, editor] — English language ed.
p. cm.
Includes index.
ISBN 0-07-911496-2
1. Geography — Encyclopedias. I. Parker, Sybil P. II. Title.
G63.E5213 1995   910'.3 — dc20   94-29086

The original Italian language edition of this encyclopedia was published as *Enciclopedia Geografica Universale*, copyright © 1994 by Federico Motta Editore, S.p.A. Milan

Typography of the English edition by AB Typesetting, Poughkeepsie, New York
Printing and binding by Arti Grafiche Motta S.p.A., Arese (Milan) – 1994

ISBN 0-07-911496-2

INTRODUCTION

MOTHER AFRICA

"Mother Africa" is how Basil Davidson, speaking as a European and addressing Europe's profound feelings about Africa during the decolonization years, defined the African continent. He thereby acknowledged the maternal role of Africa in the original unity of the human race and, in a certain phase, its culture. This connection had been shattered by the Mediterranean civilizations which historically mediated the discourse with Asia and Europe, relegating Africa to the confines of its own myths. Later, Europe went into Africa and subjected it to its own culture and economics, finding there the space it had since filled on its own soil. There, along with the infancy of the world and the concept of its maternity, it discovered the ethnographic paradise which also served as an anthropological laboratory that enabled it to rediscover humanity. Not the humanity of the Renaissance, which imposed its own order on Nature, or that of Hegelian rationality or the human collectivity that invented the industrial society which transforms or manipulates Nature, but the humanity which enters into a dialogue with Nature and feels inextricably linked to the environment in which it lives—the very foundation of our current "ecological" vision of the world.

There is the presence everywhere in Africa of a Nature so strong and devoid of meekness that it would seem to exclude or limit the human race. Nevertheless our species appeared earlier in Africa than on any other continent, if the recent paleontological findings are to be credited as definitive. Evidence has been unearthed of the presence in eastern Africa (Olduvai Gorge in Tanzania and Lake Rudolf) of hominids capable of chipping stone (*choppers*) a million and a half years or so ago. What road the human species then traveled in Africa is still difficult to say. The impression often gained has been that many of its peoples lived for long stretches of time in a kind of continental enclosure, undergoing very slow cultural mutations. Unlike what happened on other continents, the human presence in Africa has been timid and non-invasive, with few crowded settlements, if we exclude Egypt. Demographic growth has always been, at least until recently, relatively slow.

But the low population density does not adequately explain the meager presence of any anthropic imprint over much of Africa; nor is there evidence of any human overwhelming of Nature. According to the interpretation of some anthropologists (Marvin Harris and others), cultural and technically productive activism is aroused only when existing forms of exploitation no longer succeed in ensuring survival. This would explain the preservation in Africa of activities directly handed down from Neolithic times, such as nomadic herding and hoe cultivation.

Indeed, much of African history has been marked by this balance, which has maintained Africa true to itself, with the same rudimentary methods of Neolithic origin and the same relationship between settlement and land exploitation. Yet there has been no absence of strong groups bent on expansion under the impetus of increased demographic pressures. This was the case of the Bantus and, on a tribal scale, the Mossi in the upper Volta basin. However, following the European penetration and, still more so, after the colonial phase, many of the past balances were upset. Among the factors that impinged most on this was the slave trade, often invoked to justify the static nature or historical depression of Africa in modern times. Although slavery existed before then, energized by the trading between black Africa and the Arab world, it had never reached the dimensions created by Europe's demand for manual labor in the newly colonized American territories. According to some estimates, the slave trade shipped no fewer than 12 million individuals—the youngest and healthiest—across the Atlantic.

EUROPE IN AFRICA

European penetration inland from the coasts was slow, completed only after the difficult exploration of the interior, which was not actually concluded until 1871 with the famous meeting of David Livingstone and Henry M. Stanley in equatorial Africa. But even after this event the delineation of African geography posed complex problems because of the difficulty of surveying vast inland areas where there was no indigenous collaboration.

Already by the end of the 19th century much of Africa was in the hands of the British, who dominated the entire Nile axis, linking it to eastern and southern Africa, where the Boers in the 17th century had found the ideal environmental conditions for establishing settlements and centers of living different from the centers of control and exploitation in the other African regions. British colonies also existed in western Africa (Nigeria, Ghana, Sierra Leone), while large areas were under French rule in the Sahara, equatorial Africa, and on Madagascar. The Germans had occupied Tanganyika and Cameroon, but with their defeat in World War I these territories were divided up between the British and French. The Belgians held the Congo basin and the Portuguese ruled the colonies of Mozambique and Angola. Italy had its colonies in Libya and eastern Africa. Colonial control emanated from the cities, which had been established to fulfill administrative management functions for the subjected territories as well as to spur their economic development.

Exploitation of plantation and mineral resources was the result of a process of conquest which pitted large capitalistic enterprises against each other and which from the very outset was carried out in a most predatory manner, with consequences that have become clearly visible today in the form of extensive deforestation, impoverishment of the soil due to overcultivation, inadequacy of food crops, and the like. Urbanism also developed as a function of this exploitation. The cities built by the Europeans were located mostly along the coasts, from which rail lines and the first modern highways fanned out toward the interior to link up with the inland areas that supplied the products intended for the metropolitan markets and European industry. This structure forced territorial organization into channels which still powerfully determine the conditions of Africa's entire present design. The quality and scope of transportation and communications facilities diminished the further away they were from the coastal cities. The dearth of connections is a factor that contributes to the marginalization of so many areas and to the lack of functional unity of the urban network, which is anemic, unbalanced, and largely devoid of hierarchical structure.

The very political division of the space inherited by the independent states of Africa cannot now be made to fit into the framework of a rational system of modern state management. Needed are financial resources, wise political choices, interventions to restore environmental and territorial balance, defense of cultural assets, and the like—all forms of social and economic direction which Africa has not yet adequately developed. Furthermore, the independence gained as a result of the decolonization process that erupted after World War II led to the creation of states modeled on the constitutional forms of European states, without the necessary cultural, institutional, or economic foundations and especially without the required national cohesion, which is not easy to achieve in a tribally divided Africa.

The advent of independence was both a fortunate and pernicious circumstance because it seemed that Africa, if it were to progress and enter into the conclave of the free world, had to emulate the West. But since the policies were in the grip of the power elites that dominated the new states, they failed to achieve positive results; nor was success to be expected from connivance with the capitalistic former colonial powers, which imposed onerous conditions on the countries from which they were acquiring raw materials, setting themselves up as the trustees of their economic and political interests. Neocolonialism is the word for this more subtle form of control.

One can also speak of neocolonialism in the sense that truly autonomous policies free of all conditions are not easy to apply in a fragile Africa unable to launch development initiatives of its own and internally subject to the domination of urban classes which use the regimes in power, for the most part military, to secure their own privileges. As a result, the process of development is replete with imbalance and inefficiency, being sustained by aid from the developed nations, in regard to which the African countries have an indebtedness that only from a narrowly continental point of view represents a liability for Europe. Such aid is, as it were, a form of historical investment, something which the World Bank's own policy clearly reflects.

TODAY'S DIFFICULT REALITY

In the meantime, Africa is fighting its battles so as not to suffocate completely. But the struggles are endless, involving the very being of the various nations, which are frustrated by a lack of confidence in a traditional world toward which Africans show an insoluble and disconcerting ambiguity.

What will be the fate of the generations to come in a continent whose production has not yet managed to keep pace with its demographic growth? Today, Africa's population increases by an average of 4% a year, one of the highest rates on Earth. This level of growth is due not only to a reduced mortality rate (although infant mortality still reaches levels unknown elsewhere in the world, as high as 2%), but also to a consistently high birth rate. Urban migrations are producing incredibly crowded cities (Cairo, for example, has in excess of 6 million people and its greater metropolitan area over 13 million, Kinshasa has nearly 4 million, and Abidjan 2.5 million). In these cities, with their manifestations of growing urbanism, not even those ethnic quarters survive which, until the recent past, made the city a less alien place for the immigrant. There is no shortage of industries, most of them geared to the production of goods for mass consumption initiated by Western companies or by governments that command the necessary resources and entrepreneurship. But Africa has rich countries and poor ones, countries with mineral resources and countries that are exclusively agricultural, in which the incentive to produce commercial products is adopted to the detriment of food crops.

This, then, is the reality of multifaceted Africa, a continent that is now beginning to shape its own, highly individualized history. Today's crisis, therefore, appears to be one of transition to a new set of conditions, though no one can predict its consequences nor how it will be resolved. In the meantime, the image Europe had of Africa—as the infancy of the world and the land of the Earth's genuineness—is being blurred, and Africa, increasingly disillusioned, is joining the assemblage of history's great areas of production.

Eugenio Turri

GENERAL CONTENTS

Note: All conversions of metric system (SI) units have been rounded off and detailed data may not add up to the totals given. Unless otherwise specified, tons (t) refer to U.S. short tons. In the designation of natural features and place names the local spelling has been retained (or transliterated in Romanized form), except in cases where an English-language conventional spelling is commonly used.

AFRICA

Africa, together with Eurasia, is part of what was known as the Ancient World, but today it is generally regarded as a separate continent. This is due to the fact that, south of the Sahara, Africa was unknown to the peoples of Europe and Asia for thousands of years. Its systematic exploration, started by the Portuguese five centuries ago (Vasco da Gama's first circumnavigation of the globe dates back to 1498), can be said not to have been completed until the end of the last century. The Sahara desert, the largest on Earth, extends from the Atlantic coast to the Red Sea, straddling the Tropic of Cancer. For millennia it has been a sort of diaphragm separating northern Africa, forever linked to the historical events of the Mediterranean countries, from equatorial and southern Africa, which since classical antiquity has been called "black Africa" because of the dark skin of its inhabitants. Only along the Nile Valley did the ancient Mediterranean civilizations have any direct knowledge of the Sudanese and Ethiopian regions.

The unity of the African continent was absolutely unknown to the ancients. The special quality of Africa as a land of mystery, which persisted through the 18th century, particularly with regard to the southern part, is well expressed by Pliny in his famous *Naturalis historia*:

> *Generations of the most varied animals live there… hence the common saying in Greece that Africa always produces something new.*

Even today, notwithstanding the strong geographic unity of this part of the world, a definite cultural and political difference is felt between Mediterranean Africa and Africa south of the Sahara.

NATURAL ENVIRONMENT

If the Mediterranean has always been a smooth passageway between African shores and the other lands facing it, Africa is also physically linked to the Eurasian continental land mass by the Isthmus of Suez and the Straits of Gibraltar, where the ancients situated the mythical pillars of Hercules—the boundary between the known world and the unknown sea Oceanus. In much the same way, the great Sahara desert has been experienced as a hostile space (*Sahara* in Arabic means "emptiness" or "void"), beyond which lands, people, and animals took on the colors of legend.

The unity of Africa is therefore essentially a fact of physical geography. The Mediterranean to the north, the Atlantic Ocean to the west, and the Indian Ocean (with the Red Sea) to the east clearly isolate the African continental mass, which reaches its greatest latitudinal spread around the 12th parallel N between Cape Verde in Senegal and Cape Hafun in Somalia, tapering off south of the Gulf of Guinea and extending into the southern hemisphere in a triangular shape down to Cape Agulhas at 34°50' latitude south. The distances between these extreme points are approximately 5000 mi [8000 km] from north to south and 4600 mi [7400 km] from east to west, covering an area of over 11,657,000 mi^2 [30,200,000 km^2]. There are just over 18,600 mi [30,000 km] of coastline which, in terms of area, give Africa a high degree of continentality, decidedly greater than that of other parts of the world. Africa has, in fact, a stocky shape and its coasts are for the most part compact and devoid of indentations, with few islands. The only large African island (but one characterized by its own environmental features) is Madagascar in the Indian Ocean, off the coast of Mozambique. Africa's mean altitude is 2450 ft [750 m], second only to that of Asia, although Africa has no large and high mountain ranges; this means that, even in plastic relief, there are no marked contrasts and a predominance of squat plateaus and tablelands.

According to Folco Quilici, the overwhelming and fascinating, almost magical nature of Africa has diverted those who have approached this continent from a genuine interest in its people:

> *With its obsessive, abnormal, emotional nature, Africa sums up the most contrasting aspects of geography; from the earliest times to this day, travelers entering the Continent find themselves in front of splendid sights and monstrous offerings of an often hostile and always boundless nature; for centuries, those who came to Africa to cross and explore the Continent, finding themselves facing these natural barriers, had to occupy and even pre-*

occupy themselves above all with how to confront and overcome so many hostile elements in order to survive and complete the undertaking they had set for themselves. Those who succeeded in the undertaking remained so dazzled and devastated by the beauties of the country and its immensity that they were driven "to push on"—neglecting to look around, to gain a more profound knowledge of the people with whom they had come into contact—to reach other marvels beyond the mountains, beyond the forests, beyond the rivers.

And that is certainly one of the wrongs of those who have explored Africa: they have been distracted by its nature, and have overlooked its people in engaging themselves above all with the marvels of its environment. The flocks and packs of wild animals, the baobab, Victoria Falls and the rhinoceros, the tsetse fly and the ferocity of the crocodile, the lure of the rapids, the gems and flowers of the earth, the majestic mouths of solemn rivers, and the luxuriant green forests have often effaced the African in the traveler's eyes. This is why the discovery of Africa has always been partial and incomplete, and still is.

Geological structure and relief. Africa constitutes one of the great masses of ancient Pangaea, of which it probably formed the centermost core. Its composition reveals a great diffusion of compact rocks, predominantly granite and gneiss, dating back to the oldest geological eras, such as the Archeozoic and Paleozoic. These rocks still form the basement complex of the entire continental mass, even if more recent rocks have sometimes been superimposed (as a result of submerging by the sea and major volcanic effusions). In this it is similar to other large continental masses of the Earth, such as Siberia, the Deccan, the Canadian Shield, and the Brazilian plateaus. But Africa is the only part of the world that has not been affected by great tectonic movements which elsewhere have assailed the primitive land masses and raised large mountainous foldings at their edges. The only zones assailed by tectonic paroxysms are north of the Atlas chain, which has experienced recent orogenic movements from the Mediterranean syncline, and the elevated margin of the Cape ranges, displaced above the high continental slope, at the southern tip.

Elsewhere, Africa has undergone mostly vertical movements, which have fractured the rigid crustal mass, causing folding, collapses, and fissures. The most notable tectonic element is the series of fractures which open in the eastern part, from Mozambique to the Ethiopian plateau, circling the tableland of Lake Victoria, and connecting to the Danakil depression. The bottom of these rift valleys are in part filled by lakes (Malawi, Tanganyika, Kivu, and Albert) which clearly outline their course. North of the Danakil Desert, the Great Rift Valley continues into the Red Sea, whose African and Arabian shores exhibit the same geological features, ending in the Syrian and Palestinian grabens of the Dead Sea depression.

These enormous cracks, which formed on the ancient Gondwanaland continent between the Mesozoic and Cenozoic eras and gave rise to great volcanic effusions of lava (the Ethiopia plateau and Mounts Ruwenzori, Kenya, and Kilimanjaro) divide the African land mass lengthwise in two parts which are distinguished by their different altitudes. Southeast of the imaginary line connecting the Gulf of Benguela (in Angola) to Massaua (in Eritrea) lies "high" Africa with a mean elevation of 3300–3900 ft [1000–1200 m]; to the northwest, the rest of the continent has a mean elevation of 1300–2000 ft [400–600 m].

Except for the volcanic cones (Kilimanjaro is the highest peak in Africa, 19,340 ft or 5895 m) and the Atlas Mountains, the continent's relief exhibits tablelands interrupted by clear-cut shelves: grandiose and yet at the same time monotonous landscapes which reflect their ancient geology. Even the lowlands are in most cases the result of sodlike crust depressions shaped like broad basins where the waters of the great rivers collect or settle in vast marshlands (Lake Chad and the Niger's inland delta), or evaporate to form large arheic areas. The edges of these basins, as well as of the large tablelands of southern and eastern Africa, are marked by an arrangement of gridlike reliefs in relation to the meridians and parallels, reflecting in the landscape as well the direction of the fractures which produced the tectonic displacements in the rigid continental mass.

Climate. The almost symmetrical extension of Africa north and south of the equator is the prime determining factor of the continent's peculiar climatic conditions, characterized by a highly tropical climate and a regular succession of wide swaths of climate types arranged in latitudinal direction. These zones are then exposed to symmetrically alternating phenomena (temperatures, pressures, precipitation) in the opposite hemispheres following the fluctuation of the Sun's seasonal culmination, with the northern summer corresponding to the southern winter, and vice versa. With no great orographic obstacles, even the morphology favors such regularity of distribution. Factors of relative diversification, however, are determined by the major influence of the continental masses on northern Africa (which is also much wider) and of the ocean masses on southern Africa.

Because of its location, Africa is where the highest mean temperatures in the world are recorded. Its territory lies almost entirely within the 68°F [20°C] annual isotherms, excluding only the extreme regions of the Atlas Mountains and the Cape of Good Hope. The highest temperatures, combined with high humidity, are obviously found in the equatorial zone, but the heat equator is shifted further north than the geographic equator due to the more extensive continental surfaces; in return, the temperatures tend to diminish more rapidly toward the north while remaining higher toward the south. Likewise, the temperature ranges, much narrower along the equatorial belt, tend to increase the further away one moves from the equator, giving rise to increasingly pronounced seasonal differences.

The high equatorial temperatures create a low-pressure belt which draws masses of air from the tropical regions, predominantly from the north in January and from the south in July. In winter, a high-pressure regimen settles over the Sahara, feeding air currents (of which the *harmattan*, a dry parching breeze charged with sand is typical) to the south; during the northern summer, on the other hand, the torrid heat of the Sahara generates a field of low pressure which draws air from the equator. In southern Africa, by contrast, there are low pressures in January (the southern summer) and a high-pressure zone around the tip of South Africa in July. Monsoon-like conditions prevail along the coasts of eastern Africa and Guinea.

The predominance of low pressures during the year causes extensive rainfall in the equatorial regions, above all in the Gulf of Guinea, the Congo basin, and in the highlands of Ethiopia. In the intertropical zone, peak precipitation occurs when the Sun is at its zenith, which accounts for the fact that the closer the two culminations are (until there is only one solstitial culmination on the tropic lines), the more appreciable the alternation between rainy and drought seasons becomes. Whereas in the equatorial belt there is almost daily precipitation (with one or two peak "sprinkles" and "showers"), in the north there is a sudden transition to the vast drought area of the Sahara caused, as we have seen, by the strong continentality and delimited by the annual isohyet of 3.90 in. [100 mm], whereas to the south the rains decrease more gradually and the 100-mm isohyet circumscribes the other, considerably smaller, African desert of the Kalahari. Other areas of low precipitation are found in the Horn of Africa (Somalia) and the Danakil depression (Ethiopia). In the subtropical latitudes, the two extremities of Africa already mentioned—the Atlas Mountains and the Cape of Good Hope—receive their water from seasonal precipitations (specifically during the winter in the Mediterranean Atlas region, where these rains are also more abundant).

Also contributing to this precipitation regimen, in addition to the winds affected by the general pressure conditions already described, are the ocean currents: the warm currents of the Indian Ocean (the southern equatorial monsoon current of Mozambique) and the cold currents of the Atlantic (from the Canary Islands and the Benguela current), bearers of intense dryness along the coasts.

Hydrography. The geomorphological and climatic conditions of Africa account for the presence of immense basins devoid of runoff to the sea because they actually lack surface water (arheic areas) and also because the running water ends up in lakes and interior swamps (endorheic areas). The entire Sahara Desert west of the Nile has virtually no surface water due to the arid climate and very extensive evaporation; in the depression areas, the rain water collects in large marsh ponds, such as the Lake Chad basin. The same occurs in the Kalahari Desert, where the scarce seasonal rains flow into the marshes of the Okavango. Another large endorheic area is the one which extends from the lakeland plateau south of Lake Victoria through the Ethiopian tablelands to the Danakil depression, where the waters feed a number of lakes (Rukwa, Stefanie, and Turkana) or are absorbed by the soil before reaching the sea (inland Somalia). The endorheic areas with internal drainage cover 13% and the arheic with no drainage as much as 40% of the African continent.

The general inclination of Africa from south to north and from east to west explains why the tributary regions of the Atlantic Ocean (and Mediterranean) are three times greater than those of the Indian Ocean.

The continent has few large rivers. The Congo (2650 mi [4200 km]), with its immense basin, collects 60% of the equatorial precipitation which gives it impressive and constant rates of flow. The sources of the Nile, the longest river in Africa and the world, draw their waters from the equatorial rain belt. In its course of 4187 mi [6671 km], the Nile crosses the eastern section of the Sahara down to the Mediterranean and its flow is strongly affected by both the seasonal precipitation of the Ethiopian highlands and the extensive tropical evaporation.

The Niger (2600 mi [4160 km long]) flows into the Gulf of Guinea with a course that in the north describes a large loop in the pre-Sahara region, where it branches out in a large "inland delta" receiving heavy seasonal evaporations, and then flows southward, replenished by equatorial rains.

Of the major rivers in southern Africa, the Zambesi (1650 mi [2660 km]) and the Limpopo (1100 mi [1900 km]) empty into the Indian Ocean, and a third, the Orange (1300 mi [2140 km]), flows into the Atlantic.

The presence of tablelands with crumbling margins, basins connected by narrows, and plateaus with raised edges explains why the rivers of Africa, although they flow majestically through flat areas, include many rapids and waterfalls which make them difficult to navigate, especially from the coast to the interior. Famous are the Victoria Falls on the Upper Nile, the cataracts of the Middle Nile, the cascades of the Zambesi, and the rapids of the Lower Congo.

Also connected to the geomorphology are the lakes, which dot the eastern highlands in large numbers. Most of them lie in the deep, long rift valleys (Malawi, Tanganyika, Kivu, Albert, and Turkana), outlining clearly the course of the meridian. Lake Victoria, however, the largest in Africa, which collects the waters of the Kagera, furthest headstream of the Nile, is a large reservoir for extensive uplands. But saltwater lakes, whose size varies considerably during the course of the year, lie in the more depressed areas of the endorheic basins, such as Lake Chad in the southern Sahara.

Flora and fauna. More than any other part of the world, Africa has preserved its original flora and fauna. Although human intervention, especially European intrusion, has impoverished and reduced these resources, it has not denatured or irremediably altered them.

The vegetation cover follows fairly regularly the typical climatic zones of Africa and thus contributes to defining them.

The equatorial forest extends from the Gulf of Guinea to the Congo basin, giving way in the east to forests and mountain pastures typical of the eastern African highlands. Extensive rainfall, combined with high temperatures and humidity, promotes the development of the equatorial forest—a dense vegetative area characterized by a continuous plant cycle and a tremendous variety of tree species, typically distributed on several levels, from thick underbrush to gigantic trees 200 ft [60 m] or more tall. This is a most repulsive environment for human life. Its fauna encompasses an extensive variety of insects, reptiles, birds, and monkeys. The large rivers are the most convenient routes for penetrating the interior of the forest and it is along their course that the interior exploration of black Africa has extended.

To the north and south, the rainforest changes into the savanna, an area containing high grasses and shrubs alternating with strips of forest, especially along the watercourses (tropical river forests). The alternating dry seasons, always more prolonged as one proceeds toward the tropics, change the vegetation to plants that are suited to a more arid environment and cause the seasonal migrations of the large mammals which populate the savanna: herbivores (such as gazelles, zebras, antelopes, gnus, and giraffes), which are always seeking fresh pasture land, as well as predatory carnivores (lions, leopards, hyenas, and panthers). The humid areas here are the preferred sites of elephants, hippopotamuses, rhinoceroses, and buffalo. In fact, the savannas are a vast natural zoological garden of the Earth, for the protection of which large natural parks, such as those in Kenya, Uganda, and Tanzania, have been established.

Proceeding toward the tropic of Cancer, the tree savannas are followed by the drier shrub savannas with their broad grass spaces, which are arid during the long summer months and suddenly turn green again during the short rainy season.

In the Northern Hemisphere these areas change into the pre-desert steppe known as the Sahel (meaning shore or beach), which looks out onto the desolate "sea" of sand and rock that is the Sahara. The advancing desert, favored also by the practice of hoe cultivation and cattle raising, and the ever more frequent repetition of year after year without rain make human life especially precarious, with dreadful shortages that have made the Sahel the land of famine and death. The arid steppe is also present in the Horn of Africa, along the coastal strip of Eritrea and Somalia. In the deserts, vegetation is found only where aquifers of ground water emerge to the surface. Typical are the oases, outlined by green palm trees. The fauna of the steppes and deserts is limited and includes species that have adapted to extremely arid climate conditions: reptiles, predatory mammals (among them the hyena and jackal), and birds of prey (vultures). The camel and dromedary, domesticated, are of precious help to the nomadic shepherds.

In the Southern Hemisphere, the tree-lined savanna gives way toward the south to a broad belt of xerophilous shrubs alternating with arid grasslands, which extends to the entire Zambesi basin up to the edges of the eastern highlands. The scrub savanna straddles the Tropic of Capricorn, changing over to the desert steppes of the Kalahari Basin and the desert proper along the desolate coasts of Namibia.

Xerophilous forests and mountain grazing land alternate on the Ethiopian plateau, while interesting vegetation, rich in endemic tree species arranged in altimetric terraces, covers the slopes of the highest peaks (Mounts Kilimanjaro, Ruwenzori, and Kenya).

Completing the picture of Africa's vegetation are the Mediterranean-type formations which, although represented by different species and forms, are found at both ends of the continent, the Cape area and the Atlas range, with oak and Aleppo pine forests.

POPULATION

It now seems to be established that the first *Hominidae* appeared in Central Africa (Ethiopia, Kenya, Tanzania) some six million years ago. From the many remains discovered in this part of the world, it is actually possible to reconstruct some of the evolutionary stages of our species, down to the appearance of *Homo sapiens*, as well as the various cultural periods since the Old Stone Age at the dawn of history. So concludes the anthropologist Yves Coppens in his *Le Singe, l'Afrique et l'homme* [Apes, Africa, and Humans]:

> It is from Asia and Africa that a new hand has been dealt, at a new morphological level, one that is sometimes called ape-like, or the level of the Ape; the road is still difficult, but runs perhaps more smoothly than it has for the past thirty or forty million years; Aegyptopithecus, Proconsul, Kenyapithecus, Australopithecus, and Homo are the milestones, even if all the stops probably are not marked. At any rate, it is increasingly obvious that the essential part of this road is African and that, after a long journey through the forest, the road led into the clearing, the bush and tall elephant grasses of the savanna, and to a vast meadow land in a smiling valley between the wooded banks and the sea. Then, standing erect and attentive, reflecting at times on the strangeness of death, the last born of the family launched out to conquer the world. Humans therefore descend from the Great African Ape.

Furthermore, evidence left by rock carvings (especially those of the Sahara) attests to the changes in climate that have also affected Africa during the past 10,000 years. These changes are no doubt responsible for the human migrations from and to other parts of the world and for the distribution and diversity of the principal racial and ethnic types in Africa today.

Peoples. Also with regard to population groups we must distinguish Mediterranean Africa, equatorial Africa, and southern Africa. Africa north of the Sahara was, in fact, the site of rather ancient settlements of European types of Mediterranean stock, such as the Berbers and Egyptians. (Other subdivisions, classified as "Semites" and "Arabs," fall more properly in the field of cultural anthropology.) The Sahara, whose desertification has become more accentuated during the last millennia, represents a notable diaphragm between these populations and the Negroid peoples of equatorial Africa. The most widespread and numerous among these are the forest people of the large equatorial forests whose physical characteristics are their black skin, pronounced protrusion of the jaw, wide nose, and heavy-set torso.

In contact with them are the two Ethiopic and Nilotic races of Ethiopid and Negroid stock, respectively, who settled the Ethiopian Highlands and the middle and upper valleys of the Nile. They are taller, with less prognathous features and skin which is somewhat lighter among the former and very dark among the latter. It is with these groups that the Egyptians ascending the upper course of the Nile have come in contact since antiquity. Another Negroid race, the Sudanese, inhabits the entire pre-Saharan savanna and steppe area from Senegal to the Sudan.

Surviving in southern Africa, above all in the xerophilous scrub regions, are the Negroid Bushmen (whose name comes from the Afrikaans *boschjesman*, literally "man of the bush") and Hottentots, peoples with less skin pigmentation than those described above, not as pronounced prognathism, and considerable steatopygia (excessive fatness of the buttocks). At one time they lived throughout much of southern Africa, but today they are confined to the Kalahari Desert. The last Negroid people, also confined to the less accessible parts of the equatorial forest, are the Pygmies, short and dark-skinned, who live in isolated groups and are dependent on hunting and the natural products of the rainforest for their food.

Although many questions still remain unanswered concerning the racial characteristics of the African populations (certainly the most studied because the European colonization of Africa coincided with the development of physical anthropology), definitely obsolete in scientific, if not yet political, terms are the racist ideologies based on the supposed supremacy of the white race, which justified the exploitation through slavery and even slaughter of entire African populations.

The slave trade, initiated by the Arabs, was pursued on a large scale by the Europeans after their discovery of America, extensively spreading the presence of Africans throughout that continent. The study and recognition of the original African cultures have been late in coming, and the indigenous peoples became conscious of them only after their subjugation, for good or evil, by the yoke of colonialism.

Emerging from the many African ethnic groups are the

Area and Population

Country (capital)	Area (mi²)	Population	Density (per mi²)	Year	Country (capital)	Area (mi²)	Population	Density (per mi²)	Year
Mediterranean Africa					Congo (Brazzaville)	132,012	2,264,300	17	90
Algeria (Algiers)	919,352	25,360,000	28	90	Equatorial Guinea (Malabo)	10,828	417,000	39	90
Egypt (Cairo)	363,707	53,000,000	146	90	Gabon (Libreville)	103,319	1,299,000	13	78
Libya (Tripoli)	685,343	4,500,000	7	90	Kenya (Nairobi)	224,901	23,882,000	106	89
Morocco (Rabat)	274,388	25,000,000	91	90	Ruanda (Kigali)	10,166	7,100,000	698	91
Tunisia (Tunis)	63,153	8,000,000	127	90	São Tomé & Principe				
					(São Tomé)	386	118,000	306	89
Sahel					Seychelles (Victoria)	175	68,598	392	87
Chad (Ndjamena)	495,624	5,428,000	11	88	Tanzania (Dodoma)	362,635	25,635,000	71	90
Mali (Bamako)	478,714	8,156,000	17	90	Uganda (Kampala)	93,041	16,582,674	178	91
Mauritania (Nouakchott)	397,850	1,803,193	5	87	Zaire (Kinshasa)	905,126	34,138,000	38	90
Niger (Niamey)	457,953	7,450,000	16	90					
Sudan (Khartoum)	967,243	24,485,000	25	88	**Southern Africa**				
					Angola (Luanda)	481,226	9,385,725	20	88
Eastern Africa					Botswana (Gaborone)	224,548	1,255,749	6	89
Djibouti (Djibouti)	8,955	530,000	59	88	Comore (Moroni)	719	466,277	649	90
Eritrea (Asmara)	46,761	3,039,465	65	88	Lesotho (Maseru)	11,717	1,806,000	154	91
Ethiopia (Addis Ababa)	436,234	44,265,539	101	88	Madagascar (Antananarivo)	226,598	11,443,000	50	90
Somalia (Mogadishu)	246,136	5,074,000	21	80	Malawi (Lilongwe)	45,735	8,556,000	187	91
					Mauritius (Port Louis)	787	1,081,669	1,374	89
Western Africa					Mozambique (Maputo)	308,561	14,360,816	47	87
Benin (Porto-Novo)	43,472	4,408,000	99	87	Namibia (Windhoek)	317,734	1,184,000	4	87
Burkina (Ouagadougou)	105,841	7,976,019	75	90	South Africa (Pretoria)	433,565	30,796,000	71	90
Cape Verde (Praia)	1,557	369,000	237	90	–Autonomous Bantu				
Gambia (Banjul)	4,360	875,000	201	90	territories	39,183	6,379,000	163	90
Ghana (Accra)	92,076	14,925,000	162	90	Swaziland (Mbabane)	6,703	768,000	115	90
Guinea (Conakry)	94,901	6,876,000	72	90	Zambia (Lusaka)	290,509	7,818,447	27	90
Guinea-Bissau (Bissau)	13,944	966,000	69	90	Zimbabwe (Harare)	150,833	9,369,373	62	90
Ivory Coast (Yamoussoukro)	124,470	12,100,000	97	90					
Liberia (Monrovia)	42,988	2,436,000	57	88	**Possessions**				
Nigeria (Abuja)	356,574	88,514,501	248	91	British	162	7,000	43	90
Senegal (Dakar)	75,935	6,892,720	91	88	French	1,113	679,000	610	90
Sierra Leone (Freetown)	27,692	4,140,000	150	90	Portuguese	306	274,000	895	90
Togo (Lomé)	21,919	2,970,000	135	84	Spanish	2,887	1,580,000	547	90
					Yemenite	1,400	15,000	11	90
Equatorial Africa									
Burundi (Bujumbura)	10,744	5,382,459	501	90					
Cameroon (Yaoundé)	183,521	11,540,000	63	90					
Central Afr. Rep. (Bangui)	240,260	2,878,253	12	89	**AFRICA**	11,668,545	617,811,000		90

Conversion factor: 1 mi² = 2.59 km²

Bantus, whose widespread distribution throughout much of black Africa and whose singular linguistic and cultural homogeneity testify to a fairly recent expansion. The civilizations of eastern Africa, particularly those of Semitic language in the Ethiopian Highlands and indebted to Asian influences, must be regarded separately, just as the cultures of the peoples of southern Africa (Bushmen and Hottentots).

Ancestor worship, patriarchalism or matriarchalism, prevalence of agriculture or herding, tribalism or the tendency to form state organisms are all aspects that have responded from time to time to environmental influences and the transfer of cultural models, as well as the hostile or friendly encounters with various ethnic groups in the course of an unwritten history whose evolution was surely blocked and original significance distorted by the coming of the Europeans.

The great African cultural heritage has left us original forms of figurative art, religious rites, music, and dance (often passed on by slaves to various far-off cultural environments), an almost symbiotic sense of nature, and forms of social organization still deeply influenced by tribal bonds.

As the Brazilian writer Antonio Olinto has written in his novel *O Rei de Keto* [The King of Keto], steeped in references to African culture:

> For the African in general, politics is probably still a primitive means for satisfying the daily necessities of life, such as food, drink, housing and shelter against bad weather, misfortune, disease, ferocious animals and other humans who are enemies of the community in which they live....
>
> Africans do not dance. They are the dance. Africans do not carve figures in wood. They are the figures they carve....
>
> Not every religion has always served as an instrument for human beings to arrive at their essential nature, to achieve their own wholeness. The difference between the European and the African today is that for the former religion has become a social convention. For us, religion is a part of every moment of our

lives. European religion has rejected pleasures and gaiety as censurable, as sinful. For us, gaiety and pleasures, including those of sex, are part of our worship of the gods....

The foundation of African life is gaiety. Natural and infectious gaiety, something which Europeans, after much theorizing, have long since lost.

Demographic structure and dynamics. At the start of European colonization (1650) the population of Africa was estimated at 100 million people. In the next three centuries, the slave trade led to a dreadful decline, so much so that at the beginning of the last century the number of inhabitants had dropped to 90 million.

The abolition of slavery made possible a rapid demographic rebirth, which has been increasing during this century, especially after the decolonization process of the 1960s.

Thus, from 120 million in 1900, the number more than doubled to 250 million in 1960, and by 1990 it exceeded 600 million people. Propelled by an annual birth rate of 4%, among the highest in the world, Africa's population will surpass one billion in the first few years of the coming century.

This increase is due almost exclusively to natural factors, since immigration from other parts of the world has been minimal. Immigration from Europe, which although it came to a standstill toward the middle of this century, fostered the largest foreign settlements in South Africa (Boers and English), although demographically these remained minorities (1/6 of the population). The immigration from Asia to the coasts of the Indian Ocean is no more notable. Even in the African countries bordering on the Mediterranean the presence of Europeans (French, British, and Italian) has sharply diminished since World War II. Coupled with Africa's high birth rate has been its reduced mortality rate as a result of improvements in sanitary and hygienic conditions. As to the semidesert regions (Sahel, Somalia), afflicted by tragic famines and a high mortality rate, they sustain only a modest, yet ecologically and ethically highly problematical, percentage of the African population.

The distribution and density of Africa's population are quite disparate, and the mean density of 52 inhabitants per mi^2 [20 per km^2] is a highly misleading statistic if we consider that 40% of the continent's land mass consists of desert or subdesert areas that are virtually uninhabited (less than one person per km^2), while some fairly limited areas, like the low valleys of the Nile, have a density of 1000 people per mi^2 [400 per km^2] or more. Of comparatively moderate density (less than 26 inhabitants per mi^2 [10 per km^2]) are the extensive equatorial forest and savanna regions, except for the mouths of the Niger and Congo rivers, the broad southeast coastal belts, and some interior areas (the middle course of the Niger, the eastern highlands), where cattle-raising is combined with plantation agriculture. Among the higher density zones (in addition to the Nile valley mentioned and the interior of the Gulf of Guinea) are the Mediterranean regions of the Atlas Mountains, the mining basins of southern Africa, and the Cape coast. The population distribution clearly reflects the environmental conditions and use of economic resources. This relationship, which in precolonial times was direct and responsible for a better balance between population and environment, changed with the introduction by the colonial powers of new economic activities on an industrial scale (plantations, mines, communications) and the enticements of urban life—fundamentally alien to African culture south of the Sahara—which are still present today in even pathological forms.

Along the Atlantic coast, Dakar, Lagos, Abidjan, and Kinshasa are cities of a million people or more that exercise a strong attraction on the interior populations who end up in suburban agglomerations (shantytowns) typical of the Third World. In South Africa, next to the "white" American-style cities, ghetto cities have been created for the black workers, like the Soweto township near Johannesburg.

Quite different is the urbanism of Mediterranean Africa, anchored in Arab culture to which the Maghreb cities bear great historic testimony. In Egypt, Cairo with its 6 million people (23 million in the Greater Cairo urban agglomeration) is by far the most populous city in Africa and one of the most typical expressions of gigantic urban sprawl characteristic of the developing countries.

ECONOMIC SUMMARY

The economic resources of Africa are closely linked with the particular environmental (geological and especially climatic) characteristics of the continent, whereas the manner in which they are exploited reflects the coexistence of traditional ways of life and the sometimes traumatic innovations introduced by colonialism.

The ancient rock formations are the source of rich mineral deposits, with gold, diamonds, uranium, copper, manganese, tungsten, and zinc particularly widespread in southern Africa, in the eastern highlands, the Congo basin, and the Gulf of Guinea. The large-scale Mesozoic sedimentations which cover the Sahara depressions are rich in phosphates, while hydrocarbon deposits abound (Libya, Algeria).

Another source of wealth is the plant world, especially the equatorial forests, rich in prized essences (ebony, rosewood, teak, iroko, and the like). Moreover, the fertile soil of the tree-lined savannas and plateaus, as well as the areas cleared by deforestation of the rainforest, is propitious for tropical crops (coffee, cacao, sugar cane, bananas, cotton, peanuts, etc.). These are the resources that attracted the Europeans to Africa; they prompted an economy based exclusively on intensive exploitation (mines, plantations) promoted by extremely low labor costs and intended almost completely for the industries and markets of the colonizing countries. And even today these resources predominantly underlie the economies of the independent nations born of the former colonies. Their economic takeoff is severely hampered by single-crop cultures, the interplay of international markets, shortage of technology, and persistent dependence on the foreign supply of equipment and finished products. This state of backwardness is attested to by the fact that Africa's share of the gross world production is rather low (less than 5%).

At the heart of this picture is the persistence of archaic forms of economy, essentially a subsistence economy, such as the cattle-raising extensively practiced by the people of the savannas and steppes who migrate from season to season, ranging far afield in search of pastures, and poverty farming, which often is also dependent on the search for new land to till and which is threatened by desertification in the arid climate belts.

Ancient village farming, characteristic of the tribal communities and focused on the production and trading of products within narrow areas, has collapsed in the face of advancing economic forms based on larger market areas and revolving around major centers. This and the increasing demographic thrust has originated a mass exodus from the villages to the suburbs in search of new and unlikely forms of work and subsistence.

The gross national per capita income in the various states is among the lowest in the world, and only the vast spaces and still low density of the population has made it possible to ensure a minimum threshold of subsistence, but this is betrayed by increasing failures in the less favored areas, introducing the specter of famine and alarming international organizations and world public opinion.

Detached from this picture are the Mediterranean countries, some of which have been granted the role of associate states by the European Economic Community, and the Republic of South Africa, whose economy is firmly in the hands of the white minority. However, the strong demographic surge in the countries of northwest Africa known to the Arabs as the Maghreb is producing a heavy flow of emigration to the countries of Europe, while South Africa is the scene of sharp ethnic tensions.

HISTORY AND CULTURE

For centuries, Europeans have regarded Africa as a world almost devoid of history, limiting their interest to that northern strip of the continent which, facing the *"mare nostrum,"* has since ancient times revolved around the sphere of Mediterranean civilization. A similar field of vision has been quite considerably further narrowed by the prejudices of colonialism, contrary to every reassessment of Africa's indigenous history. It is understandable that such a reassessment has now been in progress for several decades in a climate of conquered independence and the discovery of original cultural roots by African intellectuals, as the ethnologist Giovanna Salvioni has written:

> The "discovery" of African history and of the vital ferments of the entire continent has led to the re-examination of erroneous concepts regarding the life of its people... Although some may have believed that life was played out in the forests and savannas, without any rules, today there is reliable evidence to the contrary... Even if the smaller and more primitive groups have standards and customs, leaving nothing to chance, the more advanced populations of the rich and fertile areas have evolved complex and impressive social forms and institutions... No reason to envy the European societies and their long history; even the kings of Benim, Ghana, Ife, and Mali, just to mention a few, lived in splendor and adulation, contracted alliances, declared wars, centralized power, and laid down rules of succession.

The fact is that the gradual drying up of the Sahara region, where eight thousand years before Christ prehistoric cultures flourished which have left behind telling evidence of rupestral art, has contributed to isolating the Mediterranean strip from the rest of the continent. Egyptian civilization, Phoenician colonization, Greek influence, and Roman domination up to the time of the Arab conquest gave northern Africa a distinctive character, but these were not the only historical realities that flourished on African soil. As early as the 2nd century B.C., the Egyptians were in contact with the Nubian Kingdom of Cush, which was followed by the Axum Kingdom that lasted until the 6th century A.D. in what is now Ethiopia.

The Arab conquest of northern Africa, which gave the Mediterranean strip of the continent the unified face that has characterized the people of the Maghreb down to this day, had to reckon with the presence and resistance of local populations little disposed to letting themselves be subjugated. The Berbers opposed the Arabs since the latter's first expansion in the 7th century. Of symbolic value during this period was the revolt led by Dihia-al-Kahinah who was defeated and killed in 705 after a long resistance. Not even during the centuries that followed, when the cities of Tlemcen, Fez, Marrakech, and Rabat were already among the major cultural and artistic centers of Islam and the Muslim religion had imposed itself throughout the Maghreb, can it be said that the Berbers, although converted, became completely assimilated. From them originated various movements that shook northwest Africa (laying the foundations of the modern states), such as those which established the Rostemid dynasty in Tahert, the Idrisid dynasty in Fez, and the Aghlabite dynasty in Kairouan. Above all, the Almoravide monks in the 11th and 12th centuries breathed life into a vast and powerful empire that reached Spain and, beyond the Sahara, the Sudan.

Beginning in the 11th century, the penetration of Islam south of the Sahara promoted the aggregation of dispersed tribal communities into more organized state structures, with the emergence of the Mandingo empire of Mali, which supplanted the ancient Kingdom of Ghana, followed by the Songhai and Kanem-Bornu empires, and the Hausa Heptarchy. Alongside the states established under the impetus of the Arab presence during the course of the so-called "African Middle Ages," other kingdoms developed in opposition to it or independently, such as the kingdoms of the Congo, Ruanda, Burundi, and so on in the equatorial and subequatorial regions. All these states, which had been nourished in the soil of tribal fragmentation and village particularism, and have been rediscovered by modern African scholars, went down the path of irreversible decadence after the 16th century, with the appearance of the Europeans on African soil. Initially motivated by geographic-commercial incentives, such as finding the route to India, European exploration assumed the character of a military conquest once the resources the continent had to offer were discovered: gold, ivory, and, above all, slaves. Three centuries of the slave trade depopulated entire regions, brought down the ancient kingdoms, and disfigured the very face of the continent, which was even more devastated by the colonization that followed.

As Folco Quilici recalls, it was not only Europe that bore the responsibility for colonialism:

> For almost four centuries the equatorial lands of Africa were scoured by European and Arab slave-traders united in a merciless roundup which led to the death or deportation of millions of human beings. The exact number will never be known and can only be deduced from comparative and analytical studies that are obviously very much at variance... the most plausible number appears to be twelve million slaves in contrast to one hundred million or more mentioned by some African writers. But even if there were not a hundred but "only" twelve million, the number is just as enormous, above all if we consider that those deported were mostly between sixteen and twenty-five years old—the flower of the African people.

The memory of this deportation has been perpetuated down the centuries. Even today, a Banzabi chant repeats:

The waters of the Great River [the Congo–Ed.] are not swollen by the rains but by the tears of the brothers and sisters transported far away into slavery.

Although a presence confined to certain ports and strongholds was considered sufficient by the Spanish, English, French, and Dutch for the trade in slaves and precious tropical products, the African hinterland took on a new appeal when the industrial revolution in Europe created a need for new markets to supply with raw materials and as an outlet for manufactured products. The 1800s were the century of the great explorations (not only the legendary ones of Livingstone and Stanley, but also those of Barth in the Sahara, and Grant, Speke, Baker, and Gessi along the Upper Nile and eastern Africa), followed by the systematic occupation of the territories explored. The "run" on Africa lasted some thirty years at the end of the century and led to the almost total dismemberment of the continent among the great European powers, especially France and Great Britain, which were joined by Germany and Italy, with the latter absorbing the last available territories, namely, Eritrea, Somalia and Benadir, Libya, and finally Ethiopia.

The collapse of European domination in Africa, which was as sudden as its assertion, had its roots in the weakening of the colonial states between the two world wars, the parallel crisis of colonialism in Asia, the revolutionary propaganda after 1917, the spread of "Western" ideas of equality, democracy, and right of self-determination; and, above all, World War II, which involved the continent directly and, calling to arms, as it did, people of different regions and diverse ethnicity, gave birth to the first consciousness of African nationalism. As the poet Paul Niger wrote:

The continent stirs, a race awakens,
A new rhythm infuses the world,
a color never before seen will inhabit the rainbow,
a raised head will trigger the lightning.
Africa is about to speak!

In the 1950s, after the Algerian war, the cultural, political, and social awakening ignited a liberation movement that spread over Africa like wildfire. Having won their independence, the new states found themselves confronting the difficult process of setting themselves free from their colonial past, a process which in so many ways is still far from completion and hindered by "Balkanization" tendencies and neocolonial pressures on the part of huge multinational monopolies, no longer only European.

In the words of the writer Antonio Olinto:

Europeans adore making revolutions in other people's countries, especially in Africa and Latin America. Perhaps they feel that Europe has already reached so advanced a level that it no longer needs immediate reforms....

Even the European revolutionary can be a colonialist when it comes to Africa and other areas of the world where Europe had colonies. At bottom, for any European we are all children in Africa, Latin America, and Asia; we do not know what we are doing, and it is necessary for Europe to show us the way.

Although in material terms the road to African maturity is still in its initial stages and winding its way laboriously in the direction of liberation, economic cooperation, and the search for internal and international balance, strides made in the cultural field are more clear-cut and significant.

The first cultural movement through which Africa became conscious of itself is that of Negritude, launched in the 1930s by the Senegalese poet Léopold Sédar Senghor and the Martinique poet Aimé Césaire. Although successively criticized as inadequate and tendentious because of its propensity to give Africa a mythicized image, this movement has clearly caught some essential points of the African spirit common to the various cultures and traditions living together side by side, superimposed, or settled on the continent itself. Senghor writes:

... the senses of the Negro are open to all contacts, to the slightest stimuli. Negroes "feel" before they see and react immediately to contact with the object, in other words to the waves it emits from the invisible. And it is through this power of emotion that they become conscious of the object.

For the African, every manifestation of the natural world, every social phenomenon, every moment of human life, every form of art is the expression of a reality underlying a visible aspect, and the universe is constituted by a unified whole of symbolic references that can be grasped by means that are not purely rational. These reflections on *Negritude*, fertile because of the very development of Western thought, can serve as an Ariadne's thread through the labyrinth of traditional African civilization and art forms: the ritual dance as much as ancestor worship, carved masks as much as the rhythm of speech, dramatic action as much as music invoke this basic concept of a constant relationship between the human and superhuman, the tangible and the intangible—a relationship which presupposes the "vital force" that circulates among gods, humans, and the world of nature in a continuous flow.

The art of speech, in particular, reveals the rhythms hidden in the universe of which it is a part; language can affect the spirits as well as reality. The budding African literature that has developed in recent decades, extremely diversified as to motive and results, has not cut itself off from its roots in oral tradition: not only does it draw on legendary and epic motifs and premises, but especially gives the word a symbolic and enigmatic value and often imbues literary material with an allusive halo. And this applies not only to poetry, but also to prose, as is so well exemplified by the conclusive passage of Hamidou Kane's novel *L'Aventure ambiguë* [The ambiguous adventure]:

The moment is the bed of the river of my thought. The pulsing of moments has the rhythm of the pulsations of thought; the gust of thought flows into the conduit of the instant. In the sea of time, the moment carries the image of the human profile as does the reflection of the cailcedra tree on the shiny surface of the lagoon. In the strength of the moment you are truly king because your thought is omnipotent when it is there. Where it has passed, the pure blue sky crystallizes into shapes. Life of the moment, ageless life of the moment that lasts; when taking flight from your ardor, you create yourself for all infinity. It is at the heart of the moment that you become immortal because the instant is infinite when it is present. The purity of the instant comes from the absence of time. Life of the instant, ageless life of the instant that reigns; in the luminous arena of your duration you are infinitely extended. The sea! Here is the sea! Hail to you, wisdom rediscovered, my victory! My eyes take in the clarity of your waves. I look at you and you are embodied in the Being. I have no limits. Sea, my eyes take in the clarity of your waves. I look at you and am radiant without limits. I want you for all eternity.

MEDITERRANEAN AFRICA

Physical environment and history are the common denominator of the African lands which front the Mediterranean. This maritime region is distinguished by the uniqueness of its physical characteristics as well as its function as a hinge between the peoples and civilizations of Europe and northern Africa.

In fact, even in classical times the Greeks and Romans regarded the Mediterranean coasts of Africa as a natural extension of Europe's peninsulas and archipelagoes from which they themselves had emerged and their civilizations developed. It was on the shores of the Mediterranean that they founded their colonies and cities which conspicuously still bear witness to them today.

Even in the classical period, the common mold of Africa north of the Sahara was the presence of populations of Hamitic origin, such as the Berbers, Libyans, and Egyptians. The expansion of the Arabs (people of Semitic stock) toward the West in the 7th century A.D., and their rapid conquest and subsequent complete Islamization of all of northern Africa (which they called *al-Maghreb*, "the West"), marked a fundamental turning point in the history and perhaps the development of the environmental conditions of this part of the continent. Africa could then begin to regard itself as truly separate from, but not opposed to, Europe (justifying all the more the name derived from the Semitic root *Afrigah* which the Phoenicians of Tyre coined referring to the city of Carthage they had founded).

It was only at the start of the last century that Europeans began to turn their interest again toward what some called the "fourth shore" of Europe.

Colonization of the old Islamic kingdoms, which had meanwhile come under Ottoman hegemony, even if it did formally end after the last world war as these territories acquired their independence, nevertheless left profound traces, attesting to the process of Westernization according to typically European models which continues to guide the development as well as economic and social transformation of both the countries of the Maghreb and those which extend along the sandbanks and lower course of the Nile.

In strictly physical terms, Mediterranean Africa is composed of two distinctly separate parts: the Maghreb region in the west, constituted by a series of mountain chains and interposed plateaus with Mediterranean-like features such as those found in the southern peninsulas of Europe; and the Libyan-Egyptian region to the east, which coincides with the northernmost portion of the vast Sahara Desert, interrupted in the east by the Nile watershed and the rift valley of the Red Sea.

For a length of 1400 mi [2300 km], from Cape Dra to Cape Bon, the Maghreb Mountain chains, formed by various foldings but especially by the Alpine orogeny during the Tertiary era, succeed each other in an intertwined network variously reaching heights of over 13,000 ft [4000 m], as in the Grand Atlas range, dropping down to large plateaus and closed basins, as in the Algerian section, or branching out into low indented ridges, as in the eastern end of Tunisia. To the south, the two opposing slopes of the Maghreb Mountains overlook the arid expanse of the Sahara Desert from where the hot, dry, tropical winds often blow and, to the north, they incline down to the jagged Mediterranean coastline, interrupted by short alluvial plains or hazardous promontories and dominated by a morphological and bioclimatic landscape that evokes the most typical settings of Mediterranean Europe. One sees the same forests of holm oak and cork, with cypresses and Scotch pines silhouetted against the terraced slopes, and an agricultural panorama reminiscent of such landscapes, with intertwining olive groves, vineyards and citrus orchards forming a variegated mosaic of heterogeneous cultures and land parcels.

The close physical relationship linking the Maghreb to Europe is also evidenced by the structural continuation of some mountain chains, such as Er Rif in relation to the Sistema Penibética in southern Spain and the Tunisian ridge in relation to the Apennines in Sicily.

The extreme complexity of the Maghreb region is contrasted by the clear linearity of the relief and of the climatic conditions of the coastline and hinterland of the Libyan-Egyptian region. Although here there is no change in the ethnolinguistic and cultural substrate that defines the distinctive character of the human landscape, the physical contrasts are rather marked: the desert setting extends directly down to the Mediterranean Sea, even if mitigated by a swath of steppes and often precipitous heights on the coast.

The dominant feature of these vast expanses, however, is dryness (with an annual rainfall that rarely exceeds 4 in. [100 mm]), overcome only by the precious underground water reserves that nurture the rich flowering of the oases. Predominant elsewhere is the desert, the *Sahara* or "desolation" (to use its literal meaning again), with its typical geomorphological varieties: a succession of sand dunes, or *ergs*, the extensive pebble-strewn paths of the *serir*, and the rocky roughness of the *hammada*.

The Sahara Desert, the largest in the world, covering 3.5 million mi^2 [9 million km^2], stretches for over 3000 mi [5000 km] from the Atlantic coast (southern end of Morocco) to the Red Sea coast of Egypt. In fact, beyond this narrow sea basin, this desert swath continues into the heart of Asia.

The only formation other than the oases which breaks the continuity of this region is the course of the Nile, winding its way between a double cornice of vegetation from the depressions of the Sudan to the broad populous delta, extending its close-set fingers into the waters of the Mediterranean. With its more than 50 million people, the long oasis of the Nile is the most densely populated area of Africa, a kind of contraposition to the vast but still thinly inhabited spaces of the Maghreb. This is one of the principal contrasts that characterize the African continent's human geography and represent a counterpoint to the rather sudden variations which distinguish its geomorphological and bioclimatic environments.

Aside from heightening these geographic differences, the Nile can in another sense be regarded as the true gateway to Africa. In fact, although south of the Maghreb region and the Libyan coast itself the arid stretches of the Sahara Desert have always represented a rigid barrier in relation to black Africa (the Sudan and the equatorial and southern regions), whose distant horizons were especially enveloped with an aura of legend and mystery (so well personified in the "*hic sunt leones*" of classical memory), sailing up the waters of the Nile—despite the physical obstacles posed by its frequent cataracts—made possible the gradual establishment of contact with the rest of the continent and, hence, a less traumatic perception of it. Moreover, the river itself has always been the most convenient way to penetrate the highlands of Equatorial Africa, even if the location of its sources was for centuries shrouded in legend and not identified until the second half of the last century following the adventurous travels of Grant, Burton, and Speke, the tragic vicissitudes of Livingstone, and the courageous expeditions of Stanley.

Together, the five African countries which look out on the Mediterranean—Morocco, Algeria, Tunisia, Libya, and Egypt—cover an area of about 2.3 million mi^2 [6 million km^2] and are largely desert land (but not economically negligible despite that fact, due to rich hydrocarbon deposits hidden in their subsoil). Their combined population is slightly more than 110 million. The community of interests that binds these countries of northern Africa was sealed by the stipulation of an agreement (1989) among Algeria, Morocco, Tunisia, and Libya, with the adherence of nearby Mauritania, establishing a Maghreb Union to pursue joint economic and political goals. This is certainly a positive sign in the varied and often tormented geography of the Arab nation.

ALGERIA

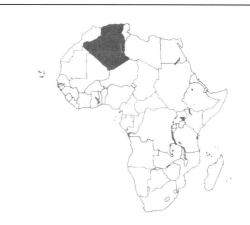

Geopolitical summary

Official name	Al-Jumhuriya al-Jazairiya ad-Dimuqratiya ash-Shabiya
Area	919,352 mi^2 [2,381,741 km^2]
Population	22,971,558 (1987 census); 25,360,000 (1990 estimate)
Form of government	Democratic and popular republic. Legislative power is exercised by the Popular Assembly whose members are designated by the National Liberation Front. The President of the republic is elected for a five-year term by universal suffrage.
Administrative structure	48 prefectures (*wilayat*)
Capital	Algiers (pop. 1,687,579, 1987 census)
International relations	Member of UN, OAU and Arab League
Official language	Arabic; French and various Berber dialects (about 19% of the population) are also spoken
Religion	Islam of the Sunni sect; approximately 60,000 Christians
Currency	Dinar

Natural environment. The second largest country in Africa, Algeria is bounded on the west by Morocco, on the southwest by Mauritania and Mali, on the southeast by Niger, and on the east by Libya and Tunisia.

Geological structure and topography. Like the Sudan, and even more so, the territory of Algeria consists largely of the Sahara, and can be divided into two distinct sections. In the north, the central area of the Maghreb, with the chains and plateaus of the Atlas Mountains, some 560 mi [900 km] in length, bathed by the waters of the Mediterranean (along a coastline of over 700 mi [1100 km]) and with an average width of about 150 mi [250 km], accounts for barely one tenth of the country. The remaining 772,000 mi^2 [2 million km^2] or more is occupied by a vast portion of the Sahara, predominantly geometric in contour and roughly rhomboid in shape. Its outside dimensions vary from approximately 1000 mi [1600 km] in the direction of the meridians to about 1100 mi [1800 km] along the parallels (although more than 1250 mi [2000 km] separate the Mauritanian-Moroccan border from the Libyan-Nigerian border!). Even though limited in size, Mediterranean Algeria represents an important regional link.

Geologically consisting of reliefs formed by the Alpine orogeny during the Tertiary era, like the other chains of the Atlas system which largely constitute the tectonic structure of the neighboring territories of Morocco and Tunisia, Algeria is bounded on the north by a narrow coastal strip hemmed by short plains (including those of Oran, Cheliff, and Annaba) that are separated by rocky promontories, a vestige of ancient Paleozoic massifs, such as the mountains of the Little Kabylia, Collo Kabylia, and Endough. These formations are followed by the Tell Atlas range composed of folded sedimentary rock (sandstone, limestone, and clay). This chain, running almost continuously on a straight line from west to east (with the Tlemcen and Beni-Chougran Mountains and the Ouarsenis Massif), culminates in the Lalla Khedidja at a height of 7570 ft [2308 m], corresponding to the mountains of the Great Kabylia. Stretching south of the Tell Atlas Mountains is a wide swath of semidesert plateaus with a mean elevation of some three thousand feet or a thousand meters, strewn with chotts as well as composed of a sedimentary rock substrate of predominantly tabular structure. Overlooking this depression area toward the south is another series of ranges, the Saharan Atlas (also composed of Mesozoic rock foldings from the Alpine orogeny), running southwest-northeast and culminating at the eastern end in the Aurès Massif (7635 ft [2328 m]). South of the Aurès, by now in the Sahara region, lie the vast depressions of large chott lakes (Chott Melrhir is 102 ft [31 m] below sea level) which continue eastward into Tunisia. At the southern foot of the Saharan Atlas, the surfacing of numerous groundwater tables favored the formation of oases around which inhabited centers sometimes developed, such as Béchar, Laghouat, Biskra, and others. The Algerian Sahara, just like the Libyan Sahara and perhaps even more so, can be considered to epitomize the geological and morphological aspects of the world's largest desert (some 3.5 million mi^2 [9 million km^2]).

Unlike certain stereotypical pictures that regard the desert as only an immense expanse of sand, the Sahara is distinguished by its notable variety of landscapes. The deepest geological substrate consists of the Precambrian crystalline basement, occurring as outcroppings throughout the southeastern sector (Ahaggar Tassili) and the western sector (El Eglab), otherwise being covered either by Paleozoic sedimentary series (Dra Hammada, the Mouydir Mountains, Ajjer Tassili, etc.) or by Mesozoic and Tertiary formations (Tanezkrouft, the Tademait Plateau, etc.). Areas at lower elevations, however, are composed of continental formations dating back to the Tertiary and Quaternary eras as well as of large expanses of sand, such as the Erg Chech, the Great Western Erg, the Great Eastern Erg, and part of the Issaouane Erg. In strictly topographic and geomorphological terms, the hammadas generally consist of vast rocky tablelands together with tassili that surround the Ahaggar region. This region and the neighboring Ajjer Tassili are alive with the residues of numerous volcanic systems derived from fissures in the crystalline basement that were active in the Tertiary and Quaternary eras, with the effusion of acid (trachytic) and basic (basaltic) lava. All these volcanic structures, fashioned by erosion into rather rough but majestic shapes, often rise to over 7000 ft [2000 m] (Adrar Massif, 7400 ft [2254 m]; Garet el Djenoun, 7633 ft [2327 m]; etc.), culminating at 8947 ft [2728 m] at Mount Assekrem and 9856 ft [3005 m] at Mount Tahat.

Hydrography. Its low level of precipitation and the climatic conditions prevent Algeria from being endowed with a sufficiently productive hydrographic network. Perennial and well-developed waterways are available only in the Maghreb region and especially behind the Tell Atlas. The longest of these is the Chéliff, which rises in the interior highlands south of Algiers and then flows westward, running between the Ouarsenis Massif and the coastal Dahra chain, to empty into the Mediterranean slightly north of Mostaganem, after a course of 484 mi [780 km]. Other sizable rivers include the Soumman, which flows into the Gulf of Bejaïa, and the Medjerda, which runs from the region south of Annaba across into Tunisia. Rounding out the hydrographic profile are more or less large salt-lake basins, seasonal in character (*chott, sebkha*) because they dry up during the summer due to their shallowness and the intense evaporation, leaving a characteristic saline crust. These are widespread in the coastal plains, interior highlands, and Sahara depressions. Worth noting among the largest of these lacustrine depressions are the great sebkha of Oran behind that city, the Chott el Hodna on the plateau northeast of Bou-Saâda, and the vast depression of the Chott ech Chergui, as well as the plateau between Mt. Daïa (Tell Atlas) and Mt. Djebel Amour (Saharan Atlas). Finally, there is the deep depression (–102 ft [–31 m]) of Chott Melrhir, already mentioned, which crosses the Tunisian border. Moreover, the oases that dot the desert are fed by springs formed by the surfacing of often abundant groundwater. It is worth remembering, however, that throughout the Sahara region there are still traces, as evidenced by the deep plain gullies (wadis), of an ancient hydrographic network that was already active during the Pleistocene epoch (ancient Quaternary period), when climatic conditions were quite different from today's, with more abundant precipitation and greater vegetation cover.

Climate. Climatic conditions throughout the Sahara are, in fact, characterized by predominant dryness, with annual pluviometric averages of around 4 in. [100 mm]. It is not rare, however, that there is no rainfall at all for several years in a row, such periods of drought alternating at times with sudden and violent, if brief, precipitation. It can sometimes happen that during the winter months the highest peaks of the Ahaggar Mountains are white with snow.

In northern Algeria, however, the climate is typically Mediter-

Climate data				
Location	**Altitude (ft asl)**	**Average temp. (°F) January**	**Average temp. (°F) July**	**Average annual precip. (in.)**
Algiers	33	54	76	30.4
Biskra	397	52	93	6.4
Oran	262	54	76	15.0
Conversion factors: 1 ft = 0.3 m; 1 in. = 25 mm; °C = (°F – 32) × 5/9				

ranean, especially on the coastal strip, with annual precipitation, predominantly in the winter, ranging from 30 in. [760 mm] at Algiers to 15 in. [375 mm] at Oran. On the inland plateaus, characterized by a definitely more continental climate, rainfall is notably reduced, dropping to as low as under 7.8 in. [200 mm] annually, while the annual temperature ranges are more extreme in contrast to along the coast where they are fairly moderate. The mean monthly temperature range is between 52°F [11°C] (at Biskra) and 54°F [12°C] (at Algiers and Oran) in January, while in July it is 93–75°F [34–24°C] (for the same respective locations). Even more pronounced are the temperature ranges in the Sahara areas, where it can reach 122°F [50°C] during the day; at night, the extreme temperatures are often below freezing. The climate is also affected by the wind regime: the wet winds blow predominantly from the west or northwest on the Maghreb regions, whereas the hot Saharan sirocco blows in from the south. Another typical desert wind is the hot and dry *simun*, which blows with great violence, raising veritable sandstorms.

Flora and fauna. The vegetation fully reflects the conditions of climate and morphology. The northern slopes of the coastal chains are covered with dense Mediterranean thickets alternating with the crops typical of this environment (citrus groves, vineyards, and fruit trees), with Scotch pine and cork oak, whereas at the higher elevations, which benefit from sufficient precipitation, Jerusalem pine and evergreen oak prosper, together with splendid cedar forests. Altogether, the forest vegetation, which is annually increased by roughly 100,000 acres [40,000 ha], still covers a rather limited area (about 11.6 million acres [4.7 million ha]) equivalent to about 2% of the country. The cultivation of olives and grain is widespread on the interior plateaus when there is no free space left for steppe meadowland dominated by the culture of alfa and artemisia. Finally, in the sub-Saharan and Saharan areas the vegetation consists of grass steppes and thorny species (*acheb*), as well as palm groves in the oases, whereas in the higher elevations of the Ahaggar it is not unusual to find temperate-zone vegetation, with poplar and fruit trees.

The wild fauna which once characterized Algeria has been considerably reduced, and is now limited to a few species, some of which have been placed under suitable protection, such as the mouflon and the Barbary stag, as well as the gazelle which lives at the edge of the desert, together with predatory species such as the cheetah, hyena, and jackal. This environment is also home to some rodents and many species of reptiles and insects, such as the dangerous scorpion. In the northern areas of the Maghreb, wild boars, hares, and macaques are still widespread; the lynx has become rarer. Birdlife is even more varied and rich, including many migratory species (quail, bustards, partridges, storks, cranes, etc.), in addition to birds of prey, such as falcons and hawks. Widespread among domestic species are the camel and dromedary.

Protected areas include the Chréa National Park, established in 1925 in the elevations south of the city of Blida (Tell Atlas range) and distinguished by majestic cedar forests, with ilexes, yews, and holly, where wild boars and jackals live.

Population. In terms of ethnic profile, the Algerian population is the result of a remarkable mixture of old Berber and Arab elements, although the Arab language and culture absolutely predominate. Preservation of the Berbers' linguistic heritage and

their relevant ethnic and cultural features is still asserted in the Kabylia Mountains (inhabited by the Kabyles, certainly the largest Berber ethnic group), in the Aurès Mountains by the Chaouïa, and by the Mozabites and Tuaregs, who live in the Sahara regions. The latter also include Negroid types descended from slaves (the haratins) deported to the area during the time of the slave trade, but perhaps also from Sudanese populations who settled in the Sahara oases before the Arab invasion.

The European presence was ultimately reduced to several thousand people, temporary residents for reasons of employment as well as economic and cultural cooperation. As a result, the use of French, the old colonial language, has considerably decreased, giving way to Arabic, while the Berber dialects are spoken primarily in the interior highland areas and the Sahara. Although broadly tolerant, Algeria regards Islam as the state religion, which is universally widespread in accordance with Sunnite orthodoxy. There are approximately 60,000 Christians (most of them Catholic, ministered to by the dioceses of Algiers, Oran, Constantine, and Laghouat). Jewish presence has practically disappeared.

Together with Morocco, Algeria is the most populous country in the Maghreb, with a population of over 25 million in 1990, even though its mean density is rather low (about 28 per mi^2 [11 per km^2]) because of the vast uninhabited areas of the Sahara. If the latter is excluded, the density of the inhabited areas reaches a more realistic level of about 207 per mi^2 [80 per km^2]. Toward the middle of the last century, soon after it became a French colony, Algeria had 2 million inhabitants, in addition to the approximately 160,000 Europeans originally from Spain and Italy as well as France. Its favorable environment and climate promoted rapid and intensive colonization with massive immigration from Europe: soon after World War II, the European presence in Algeria exceeded 800,000, in a native population of 7 million. Because of a rate of demographic growth that is among the highest in the world (13% annually), Algeria reached a population of 10 million at the time of its independence in 1962, a figure which doubled over the next two decades. In this connection, it should be noted that about 60% of the population is less than 20 years of age, while only 6% is over 60 years old.

In geographic terms, there is a marked difference, understandable enough, between the Sahara regions (where 6% of the people live) and the Maghreb areas. A phenomenon common to all the Maghreb countries is the increasingly sedentary life of the formerly nomadic population, which tends to settle in the Saharan oases. However, the precarious social conditions in Algeria, where individual incomes are nevertheless fairly high for a developing country (US$2060 a year in 1990), favor a constant demographic exodus from the rural regions of the Maghreb (where the population density in the agricultural areas can exceed 777 per mi^2 [300 per km^2], with a strong imbalance between available resources and nutritional needs) to the country's large urban centers as well as abroad (France and other European countries). At the present time it is estimated that close to one million Algerians are working outside of the country. On the whole, the number of people living in the cities, which at the time of independence totaled about one third of the total population, today accounts for 45%.

There are currently at least a score of cities with more than 100,000 inhabitants, but only Oran, Constantine, and Algiers can

Administrative structure

Administrative unit (prefectures)	Area (mi²)	Population (1987 census)	Administrative unit (prefectures)	Area (mi²)	Population (1987 census)
Adrar	163,087	155,494	M'Sila	6,891	605,578
Aïn Defla	1,759	216,931	Naâma	11,889	112,858
Aïn Témouchent	962	536,205	Oran	816	916,578
Algiers	102	271,454	Ouargla	108,080	286,696
Annaba	462	1,687,579	El Oued	28,255	379,512
Batna	4,679	453,951	Oum al-Bouaghi	2,648	402,683
El Bayadh	30,846	757,059	Relizane	1,936	545,061
Béchar	62,918	183,896	Saïda	2,366	235,240
Bejaïa	1,266	697,669	Sétif	2,566	997,482
Biskra	6,302	429,217	Sidi Bel Abbès	3,574	444,047
Blida	616	704,462	Skikda	1,590	619,094
Bordj-Bou-Arreridj	1,596	429,009	Souk Ahras	1,677	298,236
Bouira	1,765	679,717	Tamanrasset	220,020	409,317
Boumerdes	625	662,330	Et Tarf	1,214	574,786
Ech-Chéliff	1,623	525,460	Tébessa	5,784	94,219
Constantine	830	646,870	Tiaret	7,690	276,836
Djelfa	9,005	490,240	Tindouf	59,058	16,339
Ghardaïa	33,582	215,955	Tipasa	800	615,140
Guelma	1,656	353,329	Tissemsilt	1,342	227,542
Illizi	100,360	19,698	Tizi-Ouzou	1,168	931,501
Jijel	907	471,319	Tlemcen	3,603	707,453
Khenchela	4,090	243,733			
Laghouat	9,806	215,183	**ALGERIA**	919,352	22,971,558
Mascara	2,257	562,806			
Médéa	3,410	650,623	Capital: Algiers, pop. 1,687,579 (1987 census)		
Mila	1,347	511,047	Principal cities (1987 census): Oran, pop. 598,525; Constantine, pop.		
Mostaganem	763	504,124	449,602; Annaba, pop. 227,795; Sidi Bel Abbès, pop. 154,745.		

Conversion factor: 1 mi² = 2.59 km²

boast of being metropolises. Other heavily populated cities, located on the Mediterranean coast, are Annaba, Skikda, Mostaganem, and Bejaïa. There are many urban centers in the basins and valleys of the Tell Atlas, some of notable historic interest, such as Sidi Bel Abbès, Tlemcen, Sétif, Ech-Chéliff, etc. Located in the interior plateau are Batna and Tébessa, and on the southern ridges of the Atlas range, facing the Sahara, are the large settlements of Béchar, Laghouat, and Biskra. But moderately large centers also exist in the Sahara regions, always in relation to the presence of oases, such as El Oued, Ouargla, Ghardaïa, and Tamanrasset, at the foot of the Ahaggar Mountains.

Economy. As did the other countries of the Maghreb, Algeria at the time of its independence (1962) had to face many problems of a logistical and structural nature following the massive exodus of its European residents (in this case, most of them French), who were especially numerous here (11%), such as the drop in consumer goods, flight of capital, decrease in commercial activity and construction, and a worsening of unemployment. Moreover, the removal of a substantially colonial economy was relatively more disruptive in Algeria because of the French occupation lasting for more than a century, and the politically and socially marginal role to which the indigenous population had been relegated, in addition to the fact that at least two thirds of the Europeans were born in Algeria and regarded it as their homeland. Internally, the need to ensure services and fill the positions left

vacant by the Europeans further accelerated the rural exodus, creating a new wave of urban immigration, which likewise fed unemployment and burdened the administrative apparatus.

While relations with the old colonial power to which Algeria owed all of its main infrastructure (roads, railways, ports, dams, electric power stations, land improvement, etc.) were gradually being redefined, the new leaders, after initially attempting to institute a system of joint worker-management control, tended toward a planned, Socialist-type economy, proceeding to fully nationalize the major sources of production. The basis of departure was implementation of the old Constantine Plan launched during the last years of the French colonial government (1958), including

Socioeconomic data

Income (per capita, US$)	2060 (1990)
Population growth rate (% per year)	3 (1980-1989)
Birth rate (annual, per 1,000 pop.)	36 (1989)
Mortality rate (annual, per 1,000 pop.)	8 (1989)
Life expectancy at birth (years)	65 (1989)
Urban population (% of total)	51 (1989)
Economically active population (% of total)	22.5 (1989)
Illiteracy (% of total pop.)	42.5 (1991)
Available nutrition (daily calories per capita)	2726 (1990)
Energy consumption (10^6 tons coal equivalent)	33.3 (1987)

the drilling facilities for extraction of Saharan oil (discovered in 1955) and the projected construction of an iron-and-steel complex at Annaba to supplement the output of existing steel mills, such as the Acilor metallurgical works (in the Oran region), as well the mechanical engineering (automotive assembly), chemical, and rubber (tires) industries run by French corporations.

The principles of self-management were also extended to the agricultural sector by redistributing cultivated land (belonging to the large colonial enterprises or individual absentee owners) to newly established farm cooperatives, while instituting crop reconversion, with the gradual elimination of speculative crops and those marked for export (such as wine grapes) and promoting those destined for domestic consumption (such as grains). The need to expand the latter also prompted various land-reclamation initiatives, including what has come to be known as the "green dam," designed to create a wide uninterrupted swath of vegetation at the edge of the Saharan desert areas from the Moroccan to the Tunisian border.

In the industrial sector, in addition to pursuing the goals of the Constantine Plan, many mixed companies with national and foreign capital were established in order to fully exploit domestic resources, such as the petroleum and methane deposits of the Sahara, connected by pipelines to the ports and refineries on the Mediterranean, and the phosphate deposits of Gebel Onck.

During the 1970s new industrial complexes were built at El-Hadjai (iron and steel), Arzew (chemicals), Constantine (farm machinery), Guelma (motorcycles and bicycles), and Rouiba (industrial vehicles). The 1974–77 Four-Year Plan allowed greater freedom of initiative to private capital, which gave impetus to the development of the mechanical, textile, leather, rubber, and plastics industries.

Demographic pressure and the meager results of the agrarian reform are among the causes which fostered unemployment and the resulting flood of emigration abroad to economically more highly developed European and Arab countries, as well as a certain mood of popular discontent which led to unrest in the fall of 1988, resulting in bloody repression, and to the open conflict between the single government FLN party (the old National Liberation Front) and the new Islamic Salvation Front. This was followed in 1989 by a constitutional reform which, with the renunciation of socialist principles and adherence to a multiparty political system, will certainly affect future developments in the national economy.

Agriculture and livestock. After agriculture had been the main source of income during the colonial period, its importance gradually decreased as exploitation of the Sahara oil deposits got underway. Moreover, failure of the socialist policies in this field, with the creation of large state farms and agricultural villages, became acute due to the terrible drought of 1988 and saw output drop to ever lower levels, forcing the country to import massive quantities of commodities so far paid for by exports of petroleum and methane. The chief crops are wheat and barley, which cover some 30% of the cultivated land and are intended for domestic consumption, but yields are rather low (about 115 bushels per acre). Earmarked for export, on the other hand, is the product of the vineyards (yielding approximately 26 million gal [1 million hl] of wine in 1990), which was fairly extensive in colonial times but has been reduced to occupy only 320,000 acres [130,000 ha]. Also under cultivation are the typical Medi-

terranean crops, from olives to citrus fruit, vegetables (potatoes and tomatoes), and pome fruits. Date palms are widespread in the desert oases. Industrial crops such as tobacco and sugar beets are also found. The forests, covering at least 12.5 million acres [5 million ha], yield valuable wood (Aleppo pine, cedar, cork oak), while the steppe vegetation of the Atlas provides esparto grass and oil palm.

Livestock consists primarily of sheep and goats (total of 17 million head together), as well as 135,000 camels. There are many fishing ports on the Mediterranean with a catch of approximately 110,000 t of fish a year.

Energy resources and industry. Even before the discovery of petroleum, Algeria had considerable mineral resources, although they were not fully exploited, chiefly among them iron ore extracted from the deposits in the Tell Atlas regions (the Oued Chéliff valley, the Medjerda Mountains, etc.) and phosphates (Tébessa Mountains), both predominantly for export. Mining of other minerals, such as lead, zinc, copper, silver, antimony, manganese, and mercury is on a more modest scale. The major oil deposits are located in the Saharan province of Ouargla (Hassi-Messaoud and El Gassi fields), connected by a network of pipelines to the ports of Bejaïa, Skikda, and Arzew (Oran), where refineries are also available.

Other important petroleum deposits are also found on the Libyan border (at Hassi-Mazoula, Zarzaïtine, and elsewhere), linked in turn by a pipeline to the Tunisian port of Sekhira. In recent years, annual production has leveled off at 232 million bbls [37 million metric t], with proven reserves in excess of 8 billion bbls [1.27 billion metric t], which, at the present rate of operation, would ensure active production for another thirty years. Better still are the prospects for the extraction of methane, which is obtained from many deposits in various areas of the Sahara, but especially from the large gas-producing field of Hassi R'Mel (Ghardaïa), connected by pipelines to Mediterranean ports (Arzew, Algiers, Bugia, and Skikda) as well as, through Tunisia, directly to Italy (Sicily) by the gigantic trans-Mediterranean pipeline which has been in operation since 1983. Annual output of gas in 1989 was 1.7 trillion ft^3 [48 billion m^3], but established reserves are 105 trillion ft^3 [3 trillion m^3]. Algerian gas and petroleum are also exported to other Mediterranean and European countries in addition to Italy (in particular France, Spain, Belgium, Greece, Yugoslavia, and Turkey). Natural gas is exported especially after it has been liquefied at special plants, all located at Arzew.

Algeria's industrial apparatus is obviously dominated by the availability of hydrocarbons, which feed many chemical and petrochemical plants that supply the energy needs of other sectors of industry, including traditional ones, such as textiles (cotton yarn and fabrics), metallurgy (Annaba and Oran steel mills), rubber (tires), cement, etc. Almost all of the country's electric energy output (approximately 15 billion kWh) comes from thermal sources. Still in their initial phases are the metal, mechanical (automotive assembly plants) and electrical engineering industries, which avail themselves of appropriate joint-venture arrangements with some of the large European firms. Other industries of moderate size include foodstuffs (wine, sugar, and beer). The arts and crafts industry still produces export articles (carpets, etc.).

Commerce and communications. On the whole, Algeria's trade with foreign countries is substantially in balance, although the overall volume reflects some decline. Its major trading partners include the United States, Germany, and Japan, in addition to the European countries on the Mediterranean.

The communications network, one of the assets inherited from the colonial regime, has been subsequently enhanced, especially in the interior Sahara areas. It currently comprises some 44,600 mi [72,000 km] of roads (1985) and 2500 mi [4000 km] of railways (1987), only a minor portion of which is electrified. The automotive fleet includes over 1.2 million motor vehicles (1989). Moreover, the air transport network connects all the major cities of the coast and the inland regions. The merchant marine can count on a good system of port facilities. The most active are those of Arzew (with total transshipments exceeding 33 million t a year, two thirds of which consist of petroleum products), Algiers, and Oran.

Tourist capacity is still very limited, although the Algerian government is trying to promote a policy of investments in this sector.

Historical and cultural profile. Like all of northern Africa, the area now occupied by Algeria has been inhabited since the earliest times. The graffiti of the Little (Tell) Atlas and the Saharan Atlas in the north and the cave paintings of the Ahaggar massif and Ahaggar Tassili in the south have preserved the record of prehistoric peoples dedicated to hunting and herding. These were the ancestors of those Berbers who, in the 2nd century B.C., established the kingdom of Numidia, which later became a Roman province. Important ruins have remained from the Roman colonization, such as those of Thamugadi (now Timgad), the best preserved city of all Roman Africa; these ruins were to survive all the invasions that followed in the course of the centuries, from those of the Vandals to those of the Byzantines, from the Arab conquest (7th century) to the Spanish and Ottoman empires (16th century).

Arab hegemony. On the political level, this long period of time was characterized by the clash between centralization and centrifugal tendencies, the latter personified by Berber particularism.

On the cultural level, however, Arab hegemony asserted itself irresistibly and was to last down to the present day. Art and literature conformed to Arab canons and even the mosques modeled themselves on the Grand Mosque of Cordoba.

Ibn Khaldun, the greatest of Arab historians, described the grandeur and decadence of the Berbers of Algeria as follows:

> ... *a true people, as so many others of this world, like the Arabs, Persians, Greeks, and Romans. Such was indeed the Berber race; but having declined and lost national consciousness owing to the luxury that developed from the exercise of power, and to the habit of being dominated, it has seen its unity decrease, its patriotism disappear, its group and tribal spirit weaken; to such a degree that its various cores have today become subjects of different dynasties and are bowed down, as slaves, under the weight of taxes....*
>
> *All those virtues which do honor to humankind had become second nature to the Berbers. We might mention their zeal to acquire praiseworthy qualities; the nobility of spirit which elevated them among nations; the deeds which earned them admiration; the courage and readiness to defend guests and patrons; the faithfulness to promises, commitments, treaties....*

A genuine Algerian state was established in 1520 by the Turkish pirate Khàir-ad-din (the famous Barbarossa, as the Europeans called him), who succeeded in imposing himself on disparate local elements and consolidating the nucleus that was to become Algeria. Originally a Turkish colony, this state became a strong base of piracy in the Mediterranean, gradually freeing itself from the sultans of Constantinople. In the 17th century power was in the hands of a local military chieftain, the "dey," while the internal administration of the country was assured by the "beys." From that time onwards, we can speak of Algeria as a country separate from the neighboring states. And we can thus also speak of a strictly Algerian literature, meaning a popular literary tradition.

French occupation. Algeria retained its independence until the first decades of the 19th century. In 1830 the French occupied the major coastal cities of Algiers, Bône (or Annaba), and Oran, from which they launched their conquest of the interior—a conquest resisted by Abd el-Kader, the emir of Mascara, who became a national hero, not surrendering until 1847. Even more difficult was the occupation of the Great Kabylia, where a Berber insurrection broke out led by Mokrani. The new arrivals followed a policy of strict assimilation, governing the country as if it were to all intents and purposes French land, imposing the juridical norms of the mother country (for example, substituting the collective property of the tribes with individual property, fully to the advantage of the colonists), and making French the official language. The reaction to this attitude was, in the cultural sphere, the dissemination of a patriotic literature in the Arabic language and, in general, a nationalist orientation of the intelligentsia.

At the end of World War I, the first campaign for independence was led by the emir Khaled, nephew of Abd el-Kader, but it was unsuccessful. The claims of the independence movement also characterized the programs of the Algerian Communist Party (PCA) and the Algerian Popular Party (PPA), founded in the years that followed. But what really triggered independence was World War II. In 1942, after the American landing, it was actually in Algiers that the Provisional Government of Free France established its seat, and two years later the Committee of National Liberation was constituted in the same city under the alternating presidency of Charles De Gaulle and General Giraud. The participation of Algerians in the war effort and the struggle against Nazism and Fascism prompted broad strata of the population and the moderate Arab middle class itself to adopt nationalist positions for independence. On May 8, 1945, the day of victory, the crowd improvised a great demonstration at Sétif, with incidents against the Europeans. The French army reacted with tough repression resulting in 15,000 victims, the arrest of the Nationalist leaders, and the dissolution of political parties.

Liberation. After the war, however, the need to loosen the reins of colonial government became clear to the French themselves, and some concessions were made to the Algerian people, such as the Statute of 1947 which broadened their voting rights. But the hesitancy and inadequacy of these provisions only contributed to a radicalization of the Nationalist drives, which led to armed conflict. In the summer of 1954, a Revolutionary Committee for Union and Action (CRUA) was established and then transformed into the National Liberation Front (FLN). Insurrection broke out in the night from October 31 to November 1 of that year. At first circumscribed, the revolt soon spread over the entire country, progressively assuming the character of an outright popular war, also by way of reaction to the attitude of the French army, which had unleashed a harsh repression with fierce reprisals against civilians. The war of liberation lasted eight years, cost Algeria eight million lives and France the crisis of the Fourth Republic. In 1958, De Gaulle came to power with the support of the extremists, but even subsequently did not hesitate to use strong-arm measures against the FLN. Gradually, however, he came to recognize the inevitability of Algerian independence, which was finally ratified on July 15, 1962, after a referendum.

Bechir Boumaza, Minister of Labor in the first Algerian government, assessed the events of those years as follows:

> *In Tunisia and Morocco, Nationalist claims were always asserted by the middle class.... But let us not be deceived: at the root are always the people.... Added to this is the great power of suggestion aroused among Tunisians by the idea of pursuing their own independence, since France continued to consider them as a colony. On the international level, however, there was no legitimacy whatsoever in the action of Algeria, which had been duly annexed to France and made a Département. Events developed as follows: the revolt was violent in Algeria because colonization there had sunk deep roots and France was not disposed to relinquish supremacy; if it did yield in Morocco and Tunisia, it was in order to concentrate on Algeria and prevent the struggle for independence from spreading. In the same way, more extremist positions were involved in the acquisition of independence in Algeria as compared to those in the other countries because of the presence of a popular majority movement inspired by the urgency of agrarian reform and nationalization of the land.*

Elected as head of the Republic was Mohammed Ben Bella, one of the historic leaders of the resistance, who committed himself to the establishment of a socialist system based on self-management.

This policy line did not change with the coup d'état of Colonel Boumédienne (1965), who upon his death was succeeded by Chadli Benjadid. The FLN, now in power as the only party, carried out deep-seated reforms during these decades to solve the structural problems inherited from the colonial period and the devastation of war which were made even more acute by the exodus of hundreds of thousands of European settlers, resulting in a general economic crisis with new high levels of unemployment and underemployment that affected 70% of the population. The 1970s were the years of agrarian reform and nationalization of the oil industry, but also of greater dislocation and emigration.

Socialist in its internal policy and anti-imperialist in foreign policy, the FLN favored a gradual process of Arabization in all walks of social life. This trend fostered the advance of Islamic integrationist politics, to such an extent that, in 1991, in the first free elections after thirty years of single-party rule, the FLN was defeated by the Islamic Salvation Front (FIS).

The new Algerian literature. The affirmation of the Arab element also pervaded cultural life. The war of liberation had been accompanied by a literature that expressed the new national consciousness of the Algerian people in essays and novels, although the authors involved (Mouloud Mammeri, Malek Haddad, and Jasin Kateb, to name some of the most significant) preferred to express themselves in French. Among the more recent writers, however, there is the clear intent of building a national literature in the Arabic language in order to achieve full cultural autonomy.

EGYPT

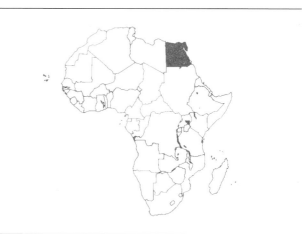

Geopolitical summary

Official name	Al-Jumhuriya Misr al-'Arabiya
Area	386,650 mi² [1,001,449 km²], of which 22,850 mi² [59,202 km²] is in Asia (Sinai Peninsula and Gaza)
Population	48,205,049 (1986 census); 53,000,000 (1990 estimate)
Form of government	Presidential republic. Legislative power is exercised by the National Assembly. The President is elected every six years by universal suffrage.
Administrative structure	26 governorates, 16 provinces, 5 cities, and 5 border districts
Capital	Cairo (pop. 6,052,836, 1986 census)
International relations	Member of UN, OAU and Arab League
Official language	Arabic; English and French are also in common use
Religion	Islam of the Sunni sect; many Christian denominations are also represented (about 7% of the population)
Currency	Egyptian pound

Natural environment. In its present configuration, the territory of Egypt occupies the northeastern corner of the African continent, comparable to a large quadrilateral with sides of approximately 620 mi [1000 km], together with the adjacent Asian peninsula of the Sinai, wedged in the northern end of the Red Sea, between the narrow canals of the Gulf of Suez and the Gulf of Aqaba. The country is bounded by the Mediterranean on the north and by the Red Sea on the east; it borders Israel to the east, the Sudan to the south, and Libya to the west.

Geological structure and relief. The structure of this vast area is fairly simple: most of it does not exceed 1650 ft [500 m] in altitude. The highest elevations are located in the southwest corner of the country, on the Gilf Kebir plateau (3550 ft [1082 m]); along the Red Sea coast, where the elevated edge (up to 7173 ft [2187 m] at the Djebel Shayib el-Banat) of the Red Sea rift valley runs; and in the Sinai Peninsula, where the highest elevations are reached in the Djebel Katherina (8650 ft [2637 m]).

Geologically, Egypt is constituted by a series of sedimentary formations (arenaceous, calcareous, argillaceous, etc.) which have continuously covered the ancient pre-Paleozoic basement rocks of the African continent since the Mesozoic Era. This structure surfaces directly opposite the elevations which dominate the western coasts of the Red Sea where they are often dislocated by successive intrusions of crystalline (granite, diorite, syenite) and volcanic rocks. Some of the ancient basement also emerges in the southern tip of the Sinai Peninsula along the valley of the Nile south of Aswan and at the Libyan-Sudanese border.

Moreover, the geomorphology of the Egyptian territory is completed by the sinuous spreading out of the gully channeled out by the Nile (which separates two distinct portions of the Sahara Desert: the Western Desert to the west, a direct continuation of the Libyan Desert, and to the east the Eastern or Arabian Desert). This led to the wide flat expanse of the Delta, a name derived from its characteristic triangular shape (as the homonymic letter of the Greek alphabet), and an area where fluvial sediments have accumulated at the mouth of the river. Another important structural feature is represented by the frequent depressions disseminated in the Western Desert (generally accompanied by corresponding systems of oases nourished by the outcropping of subterranean water-bearing strata), some of them below sea level (Siwa: –56 ft [–17 m]; Qattara: –436 to –406 ft [–133 to –124 m]; el Faiyum: –144 ft [–44 m]) and others at higher elevations, such as those which include the Farafra, Dakhla, and Kharga oases.

Hydrography. As mentioned, the Nile is the only perennial waterway that flows on Egyptian soil, all the others being valley gullies which cut especially into the topographic profile of the Eastern Desert and the Sinai Peninsula and are simple wadis which fill up with water only after the very short and sporadic winter precipitation. The Nile, which has its source south of the Equator and flows into the Mediterranean after a course of 4135 mi [6670 km], enters Egypt at the Sudanese border and runs for a little more than 930 mi [1500 km] on Egyptian territory. At Aswan, held back by a powerful dam, its waters form a gigantic artificial catchment basin (Lake Nasser) which extends for 300 mi [500 km] with an average width of about 6 mi [10 km], also penetrating Sudanese territory. The Nile has a regimen of the tropical rain type, with floods prevailing in the summer (from July to October), fed essentially by the waters of the Blue Nile which flow down from the Ethiopian Highlands and join the White Nile at Khartoum in the Sudan. The construction of various dams along the entire lower course of the river has led to full control of the floods, which now no longer cause the famous inundations that used to cover the surrounding countryside with fertile mud. In fact, the solids carried down to the mouth of the river have decreased to such an extent that the Delta coastline is already showing signs of receding. On the other hand, especially following construction of Lake Nasser, the possibilities of developing irrigation of areas that are fairly distant from the river's banks have increased. Egypt's hydrographic

resources are supplemented by numerous other lake basins and lagoons distributed both in the Delta region and the Isthmus of Suez (some are even crossed by the canal that was excavated in 1869 to link the Mediterranean to the Red Sea). A lacustrine basin also occupies the bottom of the el Faiyum depression.

Climate. Its geographic location places the territory of Egypt under the direct influence of the dry, hot masses of air coming from the Sahara. In the southern regions of the country, which is crossed by the Tropic of Cancer, a regimen of high pressure predominates which in the spring months sends the tempestuous *khamsin* north. This is a very hot and sand-laden wind which is matched, in the summer months, by the high winds that blow in from the Mediterranean. Egypt's climate, which can be regarded as subtropical on the Mediterranean coasts, cooled by constant marine breezes, is definitely arid and tropical in the rest of the country. Moreover, it is characterized by rather high average monthly temperatures, which in January range from 57–61°F [13–16°C] (at Alexandria and Aswan, respectively) and in July from 79–91°F [26–33°C] at the same localities. Moreover, precipitation is rather scant—more intense along the Mediterranean front (although less than 8 in. [200 mm] annually), an inch or so at Cairo, and virtually nonexistent in the southern regions.

Flora and fauna. Because of Egypt's special climatic conditions, the country has virtually no spontaneous vegetation cover, except along the Nile and in the desert oases where, however, it has been completely transformed by humans in the course of millennia with the planting of specific crops. Date palms in particular prosper in the oases, together with acacia, tamarisk, and mimosa which also characterize the more arid areas. Along the banks of the Nile there was a time when various species of aquatic plants were especially abundant, noteworthy among them lotus and papyrus; the latter is still found in the humid areas of the el Faiyum Depression.

The great wild fauna has also disappeared, but crocodiles still live in the Upper Nile, while some mammals such as hyena, antelope, jackal, and lynx are still seen along the edges of the desert areas. The avifauna is rather varied and includes the ibis, a wader regarded by the ancient Egyptians as sacred.

Population. As to ethnic profile, the people of Egypt can be considered of Hamitic (Berber and Libyan) stock in terms of their more specifically physical characteristics and of Semitic (Arab) stock in terms of their language, culture, and religion. In fact, the overwhelming majority of the population speaks Arabic and practices the Islamic religion (Sunni sect); one fifth still adheres to the Christian faith (Orthodox Copts) and are the guardians of Coptic, the ancient Hamitic language.

Climate data				
Location	Altitude (ft asl)	Average temp. (°F) January	Average temp. (°F) July	Average annual precip. (in.)
Cairo	76	56	83	1.2
Alexandria	16	58	79	7.2
Aswan	344	62	92	–
Ismailiya	16	56	84	1.4
Luxor	246	59	32	0.2
Conversion factors: 1 ft = 0.3 m; 1 in. = 25 mm; °C = (°F – 32) × 5/9				

Administrative structure

Administrative unit (governorates)	Area (mi^2)	Population (1986 census)	Administrative unit (governorates)	Area (mi^2)	Population (1986 census)
Cities			**Gharbiya**	750	2,870,960
			Kafr el-Sheikh	1,327	1,800,129
Alexandria	1,034	2,917,327	**Menufiya**	592	2,227,087
Cairo	83	6,052,836	**Qalyubia**	386	2,514,244
Ismailiya	557	544,427	**Sharqiya**	1,614	3,420,119
Port Said	28	399,793			
Suez	6,888	326,820	**Desert**		
			Matruh	81,897	160,567
Upper Egypt			**New Valley**	145,369	113,838
			Northern Sinai	9,874	171,505
Aswan	262	801,408	**Red Sea**	78,643	90,491
Asyut	599	2,223,034	**Southern Sinai**	12,795	28,988
Beni Suef	510	1,442,981			
Faiyum	705	1,544,047	**EGYPT**	386,559(*)	48,205,049
Giza	408	3,700,054			
Minya	873	2,648,043			
Qena	715	2,252,315			
Sawhaj	597	2,455,134			

(*) Including 22,852 mi^2 in Asia (Sinai Peninsula and Gaza)

Capital: Cairo, pop. 6,052,836 (1986 census)

Principal cities (1986 census): Alexandria, pop. 2,917,327; El Giza, pop. 1,870,508; Shubra al-Khayma, pop. 710,794; El Mahalla el Kubra, pop. 385,300; Port Said, pop. 399,793; Tanta, pop. 374,000; Mansura, pop. 357,800; Hulwan, pop. 352,300; Suez, pop. 326,820; Ismailiya, pop. 236,000; Aswan, pop. 195,700.

Lower Egypt		
Beheira	1,772	3,257,168
Daqahliya	1,340	3,500,470
Dumyat (Damietta)	227	741,264

Conversion factor: 1 mi^2 = 2.59 km^2

At the time of Napoleon's occupation at the end of the 18th century, Egypt had a population of little more than two million or about one fourth of those who lived there during the more florid eras of the Pharaoh dynasties. On the threshold of the year 2000, its population, which actually lives on only 5% of the national territory, will reach 66 million. Its demographic growth rate during the past two decades has fluctuated between 2.5 and 4% annually. The 1976 census recorded approximately 36.6 million inhabitants, which increased to 48.2 million by the next census (1987). In the beginning of the 1990s the population already reached 53 million. Consequently, although the population density of the country as a whole is more or less contained (about 137 per mi^2 [53 per km^2]), it has already reached 2591 per mi^2 [1000 per km^2] in the inhabited regions, with considerably higher peaks in the intensely urbanized areas.

The latter actually comprise some 32% of the population and have mushroomed around Cairo, the capital (an agglomeration of over 10 million people, which includes the city of El Giza, on the opposite bank of the Nile), and Alexandria (where 3 million people live). Other cities of substantial demographic size are Port Said, Suez, and Ismailiya, all on the Suez Canal, and then Aswan, Asyut, El Minya, and Beni Suef on the banks of the Nile south of Cairo. Moreover, there is a considerable urban concentration in the Delta region, with such recently developed cities as Shubra al-Khayma, El Mahalla el Kubra and others established earlier, such as Tanta, Mansura, Zagazig, Dumyat, and Damanhur; to these must be added the city of El Faiyum in the depression of the same name west of the Nile.

The rural population generally lives in large villages scattered in the more intensively cultivated regions. Other than in the Delta and along the Nile, it inhabits the many oases that stud the depressions of the Western Desert. The most populated of these (other than El Faiyum already mentioned) are Bahariya, Farafra, Dakhla, and Kharga, as well as the remotest one of Siwa, beyond the Qattara Depression near the Libyan border.

Economic summary. Although nominally independent, until the monarchy fell (1952) Egypt had developed an essentially colonial type of economy based on agriculture and commerce, while industry, predominantly manufacturing, was in the hands of foreign capital (chiefly British and French). The infrastructure, like the Suez Canal, was managed by essentially foreign corporations. With the advent of the new republican revolutionary government (1953), the country's economy underwent a radical change, reflected by nationalization of the Suez Canal and especially by agrarian reform which noticeably reduced the large landed properties in favor of a greater number of farmers. At the same time, using financial aid from the Soviet Union, construction was started on the new Aswan High Dam on the Nile (the second Aswan dam, completed in 1971) within the framework of a new and more determined energy policy. Other initiatives included various reclamation projects in both the Delta area and the oases of the Western Desert (New Valley); efforts are still being made to encourage industrial development by promoting the inflow of foreign capital. However, a major obstacle to economic development, both under the revolutionary regime of Nasser and under the more liberal Governments of Sadat and Mubarak, has been the constant major demographic growth, which creates ever new population needs, feeds chronic unemployment, and favors urbanization and emigration (there are over 3 million Egyptian workers abroad), thereby contributing to keeping individual income levels low and social conditions static.

Agriculture and livestock. The climatic conditions of Egypt, with its extreme paucity of rainfall, are such that agricultural activity is impossible without the help of irrigation. Consequently, all cultivated areas, totalling some 6.4 million acres [2.6 million ha], are currently irrigated. The numerous dams on the Nile, by eliminating the periodic inundations that covered the soil adjacent to the river with fertile mud, have nevertheless been able to provide permanent irrigation, which has made it possible—but not without the massive support of fertilizer—to harvest as many as three crops a year. In this respect, Egyptian peasants (*fellahin*) distinguish between winter (*scitui*), summer (*sefi*), and autumn (*nili*) crops, consisting of wheat and legumes; cotton, sugar cane, and rice; and corn and millet, respectively. The principal grain crops comprise wheat (the constantly increasing production of which has reached and surpassed 4.8 million t per year), corn (5.8 million t per year), and rice (3.5 million t per year), which are cultivated in the marshy areas of the Delta. Along the Upper Nile, where conditions are more arid, barley and millet are also grown. The range of food crops is rather extensive and includes species grown from seed as well as ligneous species, in addition to potatoes, vegetables, legumes, olives, citrus, and vines. Especially widespread among the tropical crops are peanuts, date palms, and bananas. A prominent position in Egyptian agriculture is occupied by some industrial crops, especially cotton and sugar cane. The latter is grown in the Upper Nile area, with an average annual production of about 11 million t, which yield over 1.1 million t of sugar. Cotton is cultivated in the Delta region, yielding an average of some 330,000 t of fiber and about 550,000 t of seed. Egyptian cotton is especially renowned for the length and resistance of the fiber, and is therefore extensively exported.

Livestock, however, is more modest, above all because of the scarcity of pasture and forage lands. It consists of buffalo, cattle, sheep, goats, and horses, with over 17 million head altogether. Even fishing (along the Nile and in the ponds of the Delta) occupies a marginal position. In the hot salt water of the Red Sea, sponges, coral, and mother-of-pearl are harvested.

Energy resources and industry. Egypt's subsoil is not particularly rich in mineral resources, most of the petroleum production (44 million t a year) coming from the Sinai Peninsula, while modest quantities of methane are extracted in the Delta and Western Desert (El Alamein). There do not seem to be major confirmed reserves of hydrocarbons. Ferrous ore is also extracted around Aswan, while rich deposits of phosphates are found on the Red Sea coasts. Numerous salt works are in production along the Mediterranean coast, and tin, tungsten, talc, titanium, copper, chromium, and gold are extracted from modest mineral deposits present in the elevations of the Eastern Desert. The Sinai Peninsula is also a source of coal and manganese.

Most electric energy production, which has exceeded 39 billion kWh, is of thermal origin, with power plants located in the major cities of the Lower Nile and Delta. The main hydroelectric power plants along the Nile are directly connected to the large dams on the river, such as the two Aswan dams. Moreover, a nuclear power plant is under construction at Sidi Khreis, and the

creation of an artificial basin, with an annual output of 3 billion kWh, is being projected in the Qattara Depression.

The considerable availability of petroleum, in addition to being a source for the production of energy, has favored the development of a petrochemical industry which currently encompasses many refineries as well as ammonia, urea, chlorine, and nitrogenous fertilizer plants. Heavy industry, however, still shows modest development; it comprises iron metallurgy (the Helwan steelworks, and the aluminum plant at Nag Hammadi, for example) and cement works. The metallurgy and machinery sector, on the other hand, is in full expansion, with automotive assembly plants and transportation facilities being operated under suitable joint ventures with other countries (including Japan, Romania, and the United States). The chemical industry also includes plants for the production of superphosphates, soap, sodium hydroxide, and various acids. But the most widespread industries, and those which employ the largest number of people, are food and textiles. The former covers a rather varied range, including sugar refineries, pasta, and edible oil production, as well as mills, canneries, and breweries. The textile industry is chiefly represented by cotton mills which are located in the main cities and whose output is largely exported. Other sectors are also represented, such as the wool, linen, and jute mills, as well as plants producing artificial and synthetic fibers. Still famous are some traditional artisan products, such as hides, carpets, and ceramics.

Commerce and communications. The notable imbalance between domestic resources and the needs of the population accounts for the country's constantly negative trade balance, which shows imports (machinery and transportation equipment, semifinished raw materials, chemicals, fuel, and food) exceeding exports (cotton, petroleum, phosphates, aluminum, shoes, cement, etc.) fully threefold. Egypt's principal trade partners are Italy, Romania, the United States, and Germany.

The transportation network, both rail (the first rail link between Cairo and Alexandria was established in 1851) and highways, follows essentially the direction of the Nile, along which the principal urban centers are located; it is particularly dense in the Delta region and the southern Sinai. Highway connections have also been developed along the Mediterranean coast and the Red Sea, with some penetrating inland, such as the highway that crosses the Western Desert, linking the oases of Bahariya, Dakhla, and El Kharga. All told, the rail network is over 3100 mi [5000 km] long (1986) and the highway network over 19,200 mi [31,000 km]. A modern superhighway now connects Alexandria

Socioeconomic data

Income (per capita, US$)	600 (1990)
Population growth rate (% per year)	2.5 (1980-89)
Birth rate (annual, per 1,000 pop.)	10 (1989)
Mortality rate (annual, per 1,000 pop.)	9.1 (1989)
Life expectancy at birth (years)	60 (1989)
Urban population (% of total)	39 (1989)
Economically active population (% of total)	26.9 (1989)
Illiteracy (% of total pop.)	51.6 (1991)
Available nutrition (daily calories per capita)	3213 (1989)
Energy consumption (10^6 tons coal equivalent)	33.8 (1987)

to Suez through Cairo. The Nile constitutes the principal internal waterway and runs over 1050 mi [1700 km]. Furthermore, Egypt controls one of the main international trade routes represented by the canal which crosses the Isthmus of Suez over a distance of 100 mi [161 km] from Port Said on the Mediterranean to Suez on the Red Sea. The canal was opened in 1869 and managed by a company with predominantly French and British capital. It was nationalized in 1956, but remained closed for 8 years (1967–75) following the conflicts with Israel. Canal traffic averages approximately 20,000 ships a year, displacing a total weight of 350 million gross register tons and carrying about 297 million t of goods. The principal maritime ports of call are Suez, Port Said, and Alexandria. Major airports are located at Alexandria, Cairo, Aswan, Luxor, and Port Said. Tourism (with some 1.8 million visitors annually) is constantly increasing, thereby encouraging development projects for accommodation centers.

Historical and cultural profile. *Ancient Egypt.* The tombs of the Valley of the Kings, the temples at Karnak, Luxor, and Abu Simbel, the Pyramids, and the Great Sphinx have preserved on the surface of their time-polished stone the traces of one of the most luminous civilizations of antiquity that developed in the course of millennia along the Nile and in its Delta—a civilization reflected in that almost 4000-year-old literary masterpiece known as the "Tale of Sanehe."

This fragment of an Egyptian hymn from the year 2000 B.C. addresses the Nile as the Lord, creator of all life:

> *Glory to Thee, O Father of Life,*
> *Secret God emerging from the secret darkness*
> *Inundate the fields the Sun has created,*
> *Quench the thirst of the herds,*
> *Soak the earth.*
> *Heavenly waterway, descend from on high,*
> *Friend of crops, make the ears of wheat grow.*
> *Revealing God, illuminate our abodes.*

After centuries and centuries of economic prosperity and artistic flowering, Egypt lost its stature as a sovereign state a few years before the dawn of the Christian era, when it fell under the sway of Rome. As early as the 4th century B.C., from the time of Alexander the Great's conquest, Egyptian civilization had lost some of its original character, but contamination with the Greek world was not without its positive aspects, as evidenced by the foundation of the city of Alexandria. Under the Ptolemaic Dynasty, there were still great periods of reform and development of the arts, sciences, and letters, when Egypt was one of the main centers of Hellenistic cultural radiation. The burning of the Alexandrian Library during Caesar's military occupation (48 B.C.) is an event that can be read as the point at which an irreversible decline set in. The suicide of Cleopatra VII (30 B.C.) opened the way to the direct dominion of Rome over the ancient land of the Pharaohs, which became progressively impoverished.

The centuries which followed were marked by the spread of Christianity and religious persecution, with ferocious conflicts developing among Christians of various sects after the Edict of Constantine. Having come under Byzantine rule by an "imperial prefect" taking orders from Constantinople, the Egyptians found an ideal guide in the Patriarch of Alexandria, derogatorily known by his enemies as the "new Pharaoh," and followed him

even after the Council of Calcedonia (451) condemned Monophysism. From this schism the Coptic Church was born, with its clear-cut nationalist imprint, and developed its own forms of artistic and literary expression (in the Coptic language) destined, however, to disappear after the country's conquest by the Arabs in the 7th century.

The Arabs and Ottomans. Except in the religious sphere, the assimilation with Arab civilization was then complete. In the year 969, the Fatimids of Tunisian origin who founded Cairo assumed power. With the backing of this dynasty, the city gradually became the most important center for trade with the Orient, the artistic and cultural capital of the Maghreb and the entire Islamic world—to some extent also because of the parallel decline of Baghdad. The Fatimids, for example, also founded the first Islamic university in Cairo, at the al-Azhar mosque. This reborn splendor lasted for centuries, even after power was no longer exercised by the Fatimids, but by the Ayyubids and later the Mamelukes.

Not until the 16th century did Egypt enter a new decline, initiated by a dual order of factors: on the one hand, the Ottoman conquest (1517), and on the other, the shifting of the great trade routes to the Orient, which brought the country to financial ruin. Paradoxically, while the populations of the villages decreased and the people became impoverished, so lively and fascinating a masterpiece as the great *Thousand and One Nights* was anonymously written, and later gained renown in the West.

In 1798, Napoleon occupied Egypt and remained for three years—a short period, but sufficient to bring the country in contact with European civilization. It was actually the commander of the anti-French troops, Mehmet Ali, who won a series of victories over the Turks, obtained the country's independence (1811), and initiated a period of modernization.

European interference. The independence won was rather fragile, as indebtedness of the country to foreign creditors increased from year to year. And so did European interference with Egyptian political life, assuming a special urgency with the opening of the Suez Canal in 1869, when Egypt suddenly acquired a strategic importance of the first order. France and Great Britain's expansionist claims ignited a nationalist revolt headed by Colonel Ahmed Urabi (1882), which was repressed and subjected the country completely to the British government. This subjugation can be said to have become an accomplished formal fact during World War I, when Great Britain declared Egypt a protectorate. In 1922, following antiforeign riots, the protectorate was abolished and independence granted (under the monarchy of Fuad I), but without changing the fundamental subordination of the country to London. This subordination was confirmed at the outbreak of the next world conflict, when Egypt, although neutral, became a British base for war operations in northern Africa.

A 1930 chronicle eloquently expressed the reality of a "Europeanized" country in these words:

> *The train to Cairo started moving. After the daytime desert, we rode through the nocturnal countryside of the Delta, perfumed and fresh. Stars, a myriad of stars, dense clusters of palm trees under the heavens, fields upon fields. Every now and then, the train pulled into a station bathed in light and noise, with Arab men and women in every kind of garb and color of skin. In the large and colorful stations, newsstands of every kind. The*

people of Egypt are covered with newsprint! And not only do they have newspapers, but reproduce in feverish translation into that bottomless well which is the Arabic language everything that Europe sends them, especially stage plays. I have been told that even those of Pirandello are about to be translated and that Egyptian actors of rare sensitivity are all the rage and have every intention of coming to Europe to prove it.

Fifteen minutes after leaving El Kamtara, the train pulls into Cairo.

Cairo: asphalt streets, a city of American layout, resolute demolitions since the advent of independence, intensive construction of large buildings which would be just as much at home in Düsseldorf or some other northern city.

The struggle which in 1952 led to the overthrow of the monarchy and the ruling class associated with Great Britain was not so much an isolated manifestation of the spirit of independence as the fruit of a slow maturing which began in the mid-19th century with the movement of Egyptian "awakening," as it was called: establishment of circles of nationalist intellectuals educated in Europe, dissemination of newspapers and magazines, and assertion of such personalities as Abdallah Nadim (poet and satirical writer) or Nagib ad al Khaddad (playwright and journalist). This made Egypt the most advanced pole of the general autonomy movement of the Arab countries. In our century the password of Egyptian patriots was "tamsir," that is to say, "Egyptianization" of every aspect of culture. All these nationalist drives, which both the masses and intellectual classes claimed as their own, thus found their outlet in the military coup d'état of 1952, which brought to power a group of "free officers," bearers of a reformist appeal within the scope of Arabism and anti-imperialism.

Nasser and Sadat. Asserting itself in this group was the strong personality of Gamal Abdel Nasser who, elected President of the Republic a few years later, introduced a series of radical measures: agrarian reform, which gave the land to the farmers; construction of the Aswan High Dam with Soviet financing; and nationalization of industry, commercial corporations, banks, and even the Suez Canal Company, then still in English hands. Nasser died in 1970, and with his death the revolutionary phase of the Egyptian road to emancipation exhausted itself. Anwar el-Sadat, one of the "free officers" of 1952, assumed the position of head of state; he eliminated the Nasserian left-wing, set aside the socialist measures of the preceding years, and liberalized the economy.

The most serious problem which Sadat inherited from his predecessor, however, was that of relations with Israel, which in 1967 had occupied the Egyptian Sinai, as well as Jordanian and Syrian territories. Following another armed conflict (1973), timid attempts at dialogue were initiated, leading in 1978 to the Camp David agreements and the signing of a formal peace treaty. Egypt urgently needed peace to overcome a state of deep economic depression. In 1981, President Sadat was assassinated by a commando of Muslim extremists, but his political orientation has been substantially maintained by his successor, Hosni Mubarak.

LIBYA

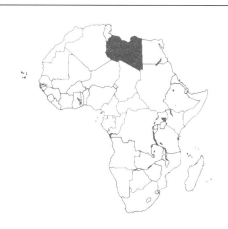

Geopolitical summary

Official name	Al-Jamahiriya al-Arabiya al-Libiya ash-Shabiya al-Ishtirakiya al-Uzma
Area	685,343 mi^2 [1,775,500 km^2]
Population	3,637,488 (1984 census); 4,500,000 (1990 estimate)
Form of government	Socialist People's Republic. Executive power resides in a General Committee and a General Secretariat (whose president assumes the functions of head of state) elected by the General People's Congress.
Administrative structure	24 municipalities (*baladiyat*)
Capital	Tripoli (pop. 989,000, 1982 estimate); the capital is being transferred to Al Jufrah
International relations	Member of UN, OAU, and Arab League
Official language	Arabic; Berber is also spoken locally
Religion	Muslim of the Sunni rite; there are also some 50,000 Christians
Currency	Dinar

Natural environment. Libya includes a vast portion of the central-northern Sahara facing the Mediterranean, with a compact but sinuous coastline marked by the wide Gulf of Sidra and the adjacent squat promontory dominated by the Barqah Plateau. The Libyan coastline stretches almost 1200 mi [1900 km] from the Tunisian border (Abu Kammash) to the Egyptian border (Gulf of Salum) with a broad hinterland extending southward toward the interior in the direction of the rocky highlands or sand-and-pebble depressions which characterize the northern edge of the Sahara. Here the country borders on Niger, Chad, and Sudan.

Geological structure and relief. Geologically, Libya appears to be composed of sedimentary rock formations dating back for the most part to the Mesozoic and Tertiary (limestone, sandstone, marl, and clay). These usually lie subhorizontally or are slightly folded, but not infrequently they are traversed by deep fractures which in some instances prompted the emergence of broadly expanding basaltic magma to form the framework of the central reliefs (Djebel as Sawda and Al Haruj al Aswad), which reach an elevation of as much as 3900 ft [1200 m]. In the southern regions older (Paleozoic) sedimentary outcroppings, partly covered by desert sand, as well as Precambrian crystalline basement remnants also emerge, as at the Egyptian-Sudanese border (Djebel Arkanu, 4706 ft [1435 m]) or the border with Chad (Bette, 7432 ft [2266 m]), where the topography rises corresponding to the northern spurs of the Tibesti volcanic massif, a large part of which is being contested by the two countries.

The geomorphologic structure of Libya appears to be rather differentiated, depending not only on geological conditions, but on topography and climate as well. Moreover, historic and anthropic factors likewise contributed over the years to this regional indentation. The hinterland of the coastal strip includes, from west to east, Tripolitania (an area gravitating around the old capital, which was successively named for the three ancient cities of Sabratah, Leptis Magna, and Oea, and now Tripoli); Syrtica (the area opposite the deep, wide indentation of the Gulf of Sidra with the city of Syrte); Cyrenaica (the squat mountainous area jutting out into the Mediterranean where the ancient city of Cyrene stood), and part of Marmarica which gravitates towards the city of Tubruq and is divided by the Egyptian border. Inland from Tripoli lies the Jefara coastal plain, continuing westward into Tunisia well-irrigated and intensely cultivated and in the south bordering on the edge of the Djebel Nefusah, a rocky tableland which rises over 2600 ft [800 m] and extends still further south to the Hammada al Hamra desert plateau at an average altitude of about 2000 ft [600 m].

The Syrtica region likewise encompasses a low-lying coastal area that gradually rises southward, interspersed by many wadis, to culminate at 2750 ft [840 m] in the Djebel as Sawda which overlooks the Al Jufrah oasis. Towards the southeast, however, the Syrte lowland continues into the wide depression where the Jalu oases crowd together on the threshold of the desolate Calansho Sand Sea. Finally, rising above 2000 ft [600 m] are the Cyrenaica ranges which jut out directly over the sea along a compact and harborless coast. The part of Libya that, strictly speaking, can be considered Saharan consists of two areas: the western Fezzan region, a vast desert that includes extensive stretches of sand, such as the Idehan ergs of Awbari and Marzuq, separated by rocky plateaus, which in turn are criss-crossed by old torrential gullies; and the eastern region, which comprises the

plateaus rising toward the Tibesti Mountains and the expanse of sand and rock of the Libyan Desert (where the Egyptian border crosses at the 25° meridian E of Greenwich), whose desolation is interrupted only by the verdant oasis of Kufra.

Hydrography. Except for a network of wadis, at times fairly dense, there is scant evidence of circulating surface water in Libya. The only stream that can be regarded as perennial is the modest Darnah wadi, which flows down from the heights of Djebel al Akhdar to the coastal city of Darnah in Cyrenaica. Fairly numerous, on the other hand, are the temporary shallow lakes which lie at the bottom of the coastal depressions or inland *(sebkha)* and which are covered during the hot weather by a variegated saline crust created by the intense evaporation. Finally there are the more abundant subsurface aquifers (usually fed by precipitation, but also of fossil origin, formed during the course of geological times) which always surface where there are depressions in the soil, even in the middle of the desert, nourishing the lush oasis vegetation dominated by date palms.

Climate. The climatic conditions are typical of the prevailing dryness that predominates even in the coastal regions, where the average annual precipitation, even if influenced by the presence of the Mediterranean, varies between only 8 and 16 in. [200 and 400 mm] from east to west and is most prevalent during the winter months (in the higher elevations of Cyrenaica, however, the annual rainfall can reach 24 in. [600 mm]). On the plateaus and in the inland zones, further than 90–120 mi [150–200 km] from the coast, precipitation is considerably reduced, even to the point of being almost completely absent for periods of several years. Average monthly temperatures range from about 54°F [12°C] in January to 79°F [26°C] in July on the coast, with no appreciable temperature swings, either during the day or annually. Conditions in the interior and the Sahara regions are quite different; there the daytime temperature may fluctuate by as much as fifty degrees, with extreme temperatures of up to 120°F [50°C]. The Libyan Sahara is often dominated by southeast winds, such as the *ghibli*, which raises violent dust storms and can even blow on the coastal regions.

Flora and fauna. Natural Mediterranean-type vegetation covers the mountains of Cyrenaica with closed formations (thicket), abundant growth of juniper, cypress, wild olive trees, Aleppo pine, and the like. The rest of the coastal strip, where not under cultivation, is covered by steppe-like vegetation, with asphodel, artemisia, and esparto, which thins out toward the interior, giving way to a covering of thorny shrubs and rare small acacia trees.

There are very few animals, consisting mostly of rodents, reptiles, insects, etc. Grazing herds of gazelles can be seen in the steppes, and dromedaries are still relatively widespread.

Population. Ethnically, the Libyan population consists of Arabized Berbers and Arabs inhabiting chiefly the coastal zones. People of Berber race, such as the Tuaregs (in Tripolitania and Fezzan), and of Negro-Sudanese stock, such as the Tebu (Tibesti) live in the inland oases.

The universal language is Arabic, but Berber dialects are commonly spoken in the interior. Still fairly well understood, especially among the upper classes, is Italian as well as English, which is used in business communications with foreign countries. Islam (Malikite rite) is the prevailing religion. The small

Climate data				
Location	**Altitude (ft asl)**	**Average temp. (°F) January**	**Average temp. (°F) July**	**Average annual precip. (in.)**
Tripoli	16	52	81	16.0
Benghazi	50	56	77	10.6
Sabha	1460	54	87	–
Conversion factors: 1 ft = 0.3 m; 1 in. = 25 mm; °C = (°F – 32) × 5/9				

Christian minority (Catholics) falls under the ministration of the Archdioceses of Benghazi, Darnah, Tripoli, and Misratah.

Living in Libya today are several hundred thousand foreign workers primarily from Egypt and other Arab countries in the Mediterranean, as well as a few thousand Europeans (technicians and contractors).

Libya is the least populated country in Mediterranean Africa, both in absolute and relative terms. This is due especially to the low latitude of its coasts and hence the considerable Saharan inflow, which the sea cannot sufficiently stem. At the time of the Italian colonial conquest (1911), after three and a half centuries of Ottoman domination, the Libyan population, which in classical times must have been fairly large to judge by the many ruins of urban settlements (such as Sabratah, Cyrene, Leptis Magna, and others), had declined to little more than half a million. On the eve of World War II, the country had close to 900,000 people, about 13% of them Italian (mostly farmers and artisans).

After the war, as the Italian presence gradually waned and ultimately disappeared, and as economic and social conditions improved following development of the petroleum economy, there was a considerable increase in population growth. In recent years, Libya's annual demographic growth rate stabilized at 4.2%, resulting in a population of 3.6 million according to the 1984 census and 4.5 million in 1990, reflecting a density of approximately 6.6 inhabitants per mi^2 [2.5 per km^2], which is really not very indicative, considering that most of the country is desert and uninhabited.

The populated areas lie in the coastal belt, the plateaus overlooking the coast (Tripoli and Cyrenaic highlands), and the oases scattered throughout the Sahara (Ghadamis, Al Jufrah, Sabha, Kufra, etc.). After the mid-1950s, new population areas were located near the oil fields (Gebel Zelten, Jalu oases, etc.). The highest population density is found around Libya's major cities of Tripoli and Benghazi; other sizable urban centers are Misratah, Az Zawiyah, and Al Khums. Altogether, the urban centers contain about two thirds of the country's total population. The city of Al Bayda was built after the war to become the capital of the Senussite Kingdom. Present plans call for transfer of the capital from Tripoli to Al Jufrah.

Economic Summary. Essentially based on agriculture and grazing, primarily in the coastal belt and oases, the Libyan economy has been completely revolutionized by the discovery (1955) and subsequent exploitation of the rich oil fields south of the Gulf of Sidra. Although coloring their aspirations with a strong ideological taint (vacillating between nationalist socialism and renewed Islamic integralism), the successive governments following the overthrow of the Senussite monarchy left private initiative a free hand, thereby promoting the influx of foreign capital and the creation of mixed corporations, while not disdaining to use the proceeds from oil exports to form partnerships with foreign companies. At the same time, after the former Italian colonizers were excluded, Libya opened its doors to immigration and accepted large contingents of workers, especially from Tunisia and Egypt. The infrastructure created during the period of Italian colonization—for example, the great coastal highway connecting the country's opposite borders over a distance of more than 1100 mi [1800 km]—was subsequently expanded, so that today it is possible to reach even the country's

Administrative structure		
Administrative unit (baladiyat)	Area (mi²)	Population (1984 census)
Ajdabiya	–	100,547
Awbari	–	48,701
Aziziyah	–	85,068
Benghazi	–	485,386
Darnah	–	105,031
Fatah	–	102,763
Ghadamis	–	52,247
Gharyan	–	117,073
Jabal al-Akhdar	–	120,662
Khums	–	149,642
Kufra	–	25,139
Marzuq	–	42,294
Misratah	–	178,295
Niqat al-Khums	–	181,584
Sabha	–	76,171
Sawfajn	–	45,195
Shati	–	46,749
Surt	–	110,996
Tarabulus (Tripoli)	–	990,697
Tarhunah	–	84,640
Tubruq	–	94,006
Yafran	–	73,420
Zawiyah	–	220,075
Zlitan	–	101,107
LIBYA	685,343	3,637,488

Capital: Tripoli, pop. 989,000 (1982 estimate)
Major cities (1982 estimate): Benghazi, pop. 650,000; Misratah, pop. 285,000; al-Khums, pop. 180,000; Zawiyah, pop. 248,000.

Conversion factor: 1 mi² = 2.59 km²

major oases by good roads suitable for vehicular traffic. The country also has an airport.

The improvement in social conditions, undoubtedly bound up with the extensive availability of financial resources and capital goods promoted by the huge oil exports, is not only reflected in absolute terms in the amount of per-capita income, the highest of any country in Africa, but is also due to the advantage of a population still limited in size although substantially increasing. The balance of trade, which has been positive for many years, reveals Libya's marked dependence on imports of capital goods as well as agricultural and food supplies. This is why the government's recent efforts have been geared to expanding land cultivation and increasing industrialization.

Agriculture and livestock. Less than 10% of the country's land area is used for agriculture and grazing, although slightly more than 5 million acres [2 million ha] is actually under cultivation; the rest is predominantly meadows and pastures. The major crops are grain (barley and wheat), vegetables, tobacco, citrus fruits, and other plant species, such as vines, olive trees, and especially date palms, which supply a yearly average of approximately 220 million lb and are grown in small gardens (*suani*) along the coastal strip as well as in the oases. Alfa and esparto are used in the interior steppes to obtain fiber and extract cellulose. A typical culture is also henna, which is used in tanning and hair dyes.

The livestock inventory includes approximately 7 million head of sheep and goats, as well as cattle, horses and camels. Modest fishing activity centers around the port of Zlitan and sponges are harvested on the coasts of Cyrenaica. In order to expand arable land, impressive irrigation projects are being implemented (some are already in operation in the oases of Kufra and Tazirbu) which provide for the construction of what is referred to as the "great artificial river," a system of aqueducts (some 2500 mi [4000 km] long with an average daily output of 1000 MHz) to carry water for the next hundred years from the deep fossil water tables that have been discovered in the subsoil of the southeastern regions to the centers on the Mediterranean coast (Ajdabiya, Benghazi, and Tripoli). The total irrigated area in Libya today covers about 570,000 acres [230,000 ha].

Energy resources and industry. Except for oil, Libya's subsoil does not appear to contain any major resources. Only recently important iron deposits have been discovered (Wadi ash Shati, north of Sabha), the mining of which is intended to supply projected steel works at Misratah. Sodium salts (natron) are extracted from the Awbari lacustrine depression area and saltworks are in operation on the coast.

Almost all major hydrocarbon deposits are situated in the Syrte desert region (Hofra, Bir Zaltan, Sarir, etc.) and are connected by a dense network of pipelines to the coastal terminals of As Sidr, Ras al Unuf, Marsa al Buraiqah, Qaryat az Zuwaytinah, and Tubruq, where the refineries are also located. The production of crude, which in the past had reached as much as 165 million t a year, has now been stabilized at about 55 million t, with proven reserves totaling at least 3.3 billion t. In addition to petroleum, 424 billion ft^3 [12 billion m^3] of methane are also extracted annually. The production of electric energy, amounting to approximately 18 billion kWh, is derived entirely from the combustion of natural gas.

The petrochemical sector is obviously the most highly developed of Libya's industrial activity, with plants at Marsa al Buraiqah (methanol, ammonia, and urea), Ras al Unuf (ethylene and propylene), and Abu Jamash (caustic soda, chlorine, and hydrogen). Other industrial sectors include cement manufacturing, food-processing, and textile and machinery production (car and trailer assembly plants).

Commerce and communications. The entire domestic distribution network is a state monopoly, while foreign trade, which has undergone appreciable fluctuations in recent years, is still

very active, even if adversely affected in large measure by oil price crises. Although Libyan exports consist almost exclusively of hydrocarbons and their derivatives, the range of imports is fairly broad, including machinery, automotive vehicles, food products, capital goods, etc. Libya's major trade partner continues to be Italy (with almost one third of all commercial activity), followed by Germany, Spain, and the United Kingdom. Imports from Japan and South Korea are also substantial.

The communications network is still somewhat disconnected and comprises over 15,500 mi [25,000 km] of roads (1986), some of which also link the oases furthest from the coast, such as those of Ghat and Kufra, which in turn have airports. On the eve of World War II there were 250 mi [400 km] of railroad, which is now being considered for restoration. The merchant marine has at its disposal well-equipped ports (Tripoli, Misratah, Marsa al Buraiqah, Benghazi, and Tubruq) and a fleet of over 834,000 gross register tons. There is also a fairly substantial fleet of cars and trucks with over 770,000 vehicles in circulation (1988).

The tourist industry is negligible, notwithstanding the many centers of special interest, above all archeological, because of internal difficulties and a lack of adequate infrastructure.

Historical and cultural profile. *Libya in ancient and medieval times.* In antiquity, the name Libya referred to northern Africa in general—the only part of the continent then known—as well as to that region of modern Libya inhabited by a people the Romans called *Libii*. These people were of Hamitic stock and were in constant touch with the other peoples of the Mediterranean, especially the Egyptians: *Libii* mercenaries made up entire armies of the Pharaohs, but the Libyans were also invaders who periodically threatened the Nile valley. The name Libya first appeared officially at the time of Diocletian, when the provinces of Upper and Lower Libya were created, but it fell into disuse with the fall of the Roman Empire, not to reappear until the 20th century during the colonial period.

During this period rich architectural remains of the Roman Empire were unearthed in the desert, bringing to light entire cities, such as Sabratah, Leptis Magna, and Cyrene.

This is how the ruins of Sabratah appeared to de Mathuisieulx at the turn of the century:

> *At Zuaga began the series of large oases that end at Zanzur, four hours from Tripoli. Near this first garden and on the shore we find the ruins of Sabratah or Abrotanum.... For a stretch 5000 ft [1500 m] long the shore is covered with cut stones, broken capitals, mosaics, and esplanades which must have been public areas. The rich holiday houses were built on the sea and their imposing foundations can still be seen today.*
> *It is believed that Sabratah was founded by the Phoenicians, who called it "market of grains" in their language. After it was ransacked by the Vandals, it rose again from its ruins and became an often-visited port of call until the end of medieval times. The port was named Old Tripoli by Italian navigators.*

Less extensive appears to be the architectural evidence of Arab domination, which was imposed from the 7th century onward on the regions of Cyrenaica, Fezzan, and Tripolitania without, however, reducing these areas to institutionalized uniformity. This process of Arabization was not as profound or as culturally and artistically successful here as in the rest of the

Socioeconomic data

Income (per capita, US$)	6,060 (1990)
Population growth rate (% per year)	4.2 (1980-89)
Birth rate (annual, per 1,000 pop.)	44 (1989)
Mortality rate (annual, per 1,000 pop.)	9 (1989)
Life expectancy at birth (years)	62 (1989)
Urban population (% of total)	69 (1989)
Economically active population (% of total)	24.9 (1988)
Illiteracy (% of total pop.)	36.2 (1991)
Available nutrition (daily calories per capita)	3,384 (1989)
Energy consumption (10^6 tons coal equivalent)	11.6 (1987)

Maghreb, limiting itself to the simple reassertion of Islamic models. For centuries, the Libyan lands remained dependent—but in fairly imprecise form—on the various dynasties that gradually asserted themselves in northern Africa, from the Fatimid dynasty at Mahdiyah to that of the Banu Matruh, vassals of the Almohads, and the Tunisian dynasty of the Hafsids. Complicating the picture were the recurrent Berber rebellions and the invasion of the Hilalian tribe in the 11th century which had disastrous effects on agriculture.

The modern era and Turkish domination.
In short, the regions which constitute Libya today entered modern times in a state of relative backwardness compared to the other countries of the Maghreb, which further increased when they became provinces of the Ottoman Empire in the 16th century and were subjected to its onerous tax burden. Early in the 18th century the ties to the Sublime Porte were partially loosened when the pasha Ahmed Caramanli founded a dynasty in Tripoli that was substantially independent of Turkish authority and lasted until 1835, when the city was reconquered by the Sultan of Constantinople.

It was the Caramanli dynasty that breathed life into the artistic period of greatest interest, characterized by the great Islamic edifices, such as the Caramanli mosque in Tripoli, with its rich stuccoes and majolica work, the Gurgi mosque with its graceful minaret, and the Bu Ghellar mosque in Benghazi.

Italian occupation.
At the beginning of the 20th century, while the age-long struggle among the European powers for the conquest of colonial possessions was in full force, the regions of Tripolitania and Cyrenaica offered Italy the last chance to carve out for itself an "imperial" space on the opposite shores of the Mediterranean. After waiting for the most favorable international opportunity, the Italian government in 1911 declared war on Turkey, which in the Treaty of Lausanne the following year relinquished to Italy the territory that then took the name of Libya. In fact, Italian control was actually confined to the coastal cities, since the interior populace did not recognize the validity of this treaty. World War I prevented Italy from fulfilling its colonial adventure, which can be said not to have been completed until 1924 to 1931.

Fascism undertook a program of administrative, agricultural, and even architectural modernization in Libya, changing the face of such cities as Benghazi, Misratah, and Tripoli along the lines of lictorian grandiosity.

In her historic memoirs, Isabella Pezzini recalls how the Italian attitude towards Libya was marked by "enlightened" colonialism, so to speak:

> ... precisely in consideration of [Libya's] being the "fourth shore," and therefore sharing the Mediterranean basin, [Italy] officially followed a racial policy of assimilation and not apartheid, as was subsequently to be the case in Ethiopia. This policy was also reflected in its city-planning projects designed to reproduce in the colony the class divisions superimposed on racial divisions in a "Mediterranean" architectural setting. Fascism continued on this road until it granted the Libyans special Italian citizenship in 1934.

During the colonial period a movement of cultural and national rebirth known as "*an-Nachda*" developed in Libya which shook the country out of its lethargy. A literature in Arabic, which had completely fallen into disuse since the time of

the Turkish conquest in the 16th century, was reborn. Achmed asc-Sciarif, Mustafa ben Zichri, and Suleiman al Baruni were profoundly inspired secular poets. Also moved by patriotic sentiments were the lyrics of the poets who gathered around Achmed Rafiq al Machdaui who had come under the influence of Italian culture as well.

Independence.
During World War II Libya was the theater of celebrated military operations of the Italian and German armed forces against Egypt. After the battle of El-Alamein, the country was occupied by the British, the French, and later the Americans. Independence was proclaimed in 1951 and the terms between Italy and the new Libyan state with regard to the transfer of sovereignty were defined in the years that followed. The constitutional monarchy of King Idris proved incapable of solving the problems of a country still so heavily burdened by feudalism. This failure led in September 1969 to a military coup by a group of officers led by Muammar al-Qaddafi. Motivated by socialism, Nasserism, and panarabism, their declared intention was to rediscover a form of direct democracy, as formulated in the *Green Book*, a compendium of the Libyan leader's thoughts:

> Popular Conferences are the only means to achieve popular democracy. Any system of government contrary to this method, the method of Popular Conferences, is undemocratic. All the prevailing systems of government in the world today will remain undemocratic, unless they adopt this method. Popular Conferences are the end of the journey of the masses in quest of democracy.
>
> Popular Conferences and People's Committees are the fruition of the people's struggle for democracy. Popular Conferences and People's Committees are not creations of the imagination; they are the product of thought which has absorbed all human experiments to achieve democracy.
>
> Direct democracy, if put into practice, is indisputably the ideal method of government. Because it is impossible to gather all people, however small the population, in one place so that they can discuss, discern and decide policies, thus nations departed from direct democracy.... It was replaced by various theories of government, such as representative councils, party-coalition and plebiscites, all of which isolated the masses and prevented them from managing their political affairs. These instruments of government ... have plundered the sovereignty of the masses and monopolized politics and authority for themselves.

But his movement—officially defined as the "third way" between capitalism and communism—appeared to be jeopardized, at least at first, by the chronic backwardness of the country, because of the lack of infrastructure and sparse population. To reverse this trend, large-scale exploitation and nationalization of petroleum resources (1971–73) was undertaken. The search for oil had been intensive for over a decade, and Libya soon found itself catapulted to the forefront of the Arab world. This did not prevent episodes of friction with some of the neighboring countries (Tunisia and Egypt) as well as with Italy and the United States, which in 1986 bombarded Tripoli and Benghazi.

In the past forty years a generation of writers has come to the fore in Libya who have rejected the classical forms of Arab poetry and address current events in prose which are related to the Islamic-Socialist ideals that shape today's Libyan republic.

MOROCCO

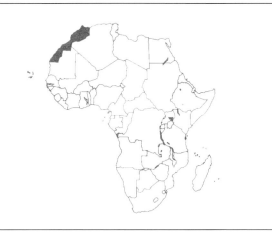

Geopolitical summary

Official name	Al-Mamlaka al-Maghrebia
Area	177,174 mi² [459,000 km²] (not incl. 97,318 mi² [252,120 km²] of former Western Sahara)
Population	20,419,555 (1982 census); 25,000,000 (1990 estimate)
Form of government	Constitutional monarchy. Legislative power is exercised by a Chamber of Deputies of which two thirds are elected by universal suffrage. The King, who is also the paramount religious authority, directly appoints the members of the Government.
Administrative structure	39 provinces and 6 urban prefectures (*wilaya*)
Capital	Rabat (pop. 650,000, 1990 estimate)
International relations	Member of UN and Arab League
Official language	Arabic; Berber (24% of the population) and French are also widespread
Religion	Islam of the Sunni sect; Christians and Jews (approx. 110,000) are also represented
Currency	Dirham

Natural environment. Located on the northwestern end of the African continent, Morocco borders on both the Mediterranean and Atlantic oceans, almost touches Europe from which it is separated by the narrow Straits of Gibraltar, and is bounded on the east by Algeria and on the south by Mauritania.

Geological structure and relief. The topography of Morocco originated geologically from a series of foldings caused by contact between the African and Eurasian continental plates. The elevation of the Atlas ranges, not infrequently accompanied by volcanic manifestations, is attributed to the oldest orogenic phases (Paleozoic), while the Alpine orogeny (Tertiary) accounts for the formation of the mountainous Rif belt which, in turn, is

structurally linked to the Penibético system (in southern Spain) at one end and the Tell Atlas (in northern Algeria) on the other.

With its present borders Morocco appears to have been formed by two major geomorphological domains separated by the Wad Dra fluvial gully: to the southwest, Saharan Morocco, which includes a stretch of the westernmost section of the great north African desert, and to the northeast, historic Morocco, a region of old lands and valleys dominated by the great Atlas Mountains. The Atlas orographic system comprises three major ranges: the Anti-Atlas to the south, a mountain barrier rising above 8200 ft [2500 m] in elevation and extending for about 350 mi [600 km], into which the upper course of the Dra makes a deep cut; the High Atlas chain in the center, delimited by the Dra and the course of the Sous, culminating at 13,665 ft [4165 m] at the Toubkal (the highest peak in Morocco and all of western Africa) and running about 450 mi [700 km] from the Atlantic to the highlands of eastern Morocco; and the Middle Atlas, a shorter and lower mountain range which culminates in the Bou Naceur peak (10,955 ft [3340 m]) and is separated from the High Atlas by the upper valleys of the Moulouya and Oum er Rbia rivers, dropping slightly to the northwest in a series of steplike plateaus of varying elevation down to the coastal plains bathed by the Atlantic and the great Gharb alluvial plain through which runs the winding course of the Sebou. Still further north, overlooking the Gharb plain, is the short but mighty Rif chain which rises above 6500 ft [2000 m] at several points and towers directly over the Mediterranean coast, bordering to the east the last plain of the Moulouya.

Hydrography. The river network of Morocco is strictly conditioned by the relief topography, with the Atlas Mountains constituting the country's main watersheds. Except for the Moulouya (280 mi [450 km] long), which feeds into the Mediterranean near the Algerian border and flows east of the Middle Atlas and the Rif, all the other major waterways are tributaries of the Atlantic, including the Wad Dra which runs along the boundary between the Atlas reliefs and the Sahara Desert for a length of 750 mi [1200 km] but whose bed remains dry for long periods. The basin of the Sebou (280 mi [450 km]) extends from the Rif to the Middle Atlas Mountains, watering the fertile Gharb plain before emptying at Kenitra. The longest water course running down from the Middle Atlas and flowing directly into the ocean is the Oum er Rbia (370 mi [600 km]). Descending from the High Atlas slopes are the Tensift (125 mi [200 km]), which laps at the northern outskirts of Marrakech, and the Sous (110 mi [180 km]), which bathes the Agadir plain. Altogether, the regimen of the Moroccan rivers, often fed by abundant subsurface aquifers, is irregular with sometimes disastrous flooding in the fall and spring. In the words of a Berber poem:

When the river is in flood, little frogs, remain in your hideouts.
When the river is in flood, it is the spirits and not the waters that
  are swept along,
Uprooting the oleander and pomegranate,
Devastating the orchards,
Seeing neither treasured public property nor the holy one's
  garden.

The construction of numerous artificial catchment basins has led to the full use of national water resources. Yet many waterways, especially on the southern Atlas slopes, remain dry during the summer months.

Climate. The climate of Morocco is strongly influenced by three types of air masses: fresh and humid air blowing in from the Atlantic, Mediterranean temperate air, and hot or dry air from the Sahara. An important element of regional contrast is the rains which assault the northwestern slopes of the mountains and fall primarily during the winter semester (from October to May) with an annual accumulated vehemence of as much as 50 in. [1200 mm] in the High and Middle Atlas mountain regions where the snow cover can remain until late spring. In fact, it is not unusual to see sun-bathed beaches on the Atlantic coasts with the snow-capped slopes of the High Atlas in the background. Although the altitude may accentuate certain climatic differences even in the very narrow areas along the coastal belts, the temperature ranges are rather restricted due to the dual effect of the warm waters of the Mediterranean to the north and the cold current of the Canary Islands to the south. As a matter of fact, the average monthly temperatures fluctuate, in the opposite regions of the country, between 50–57°F [10–14°C] in January and 71.5–73.5°F [22–23°C] in July. Very extreme temperatures (near 0°F [–15/–20°C] in the winter and 100–120°F [40–50°C] in the summer) can occur in the desert and high mountain regions.

Flora and fauna. Climate, geomorphology, and the notable differences in local elevation have a considerable effect on Morocco's vegetation which is fairly diversified as well as characterized by a remarkable abundance of flowers. The temperate mixed forests, already strongly depauperated in the past, occupy little more than one tenth of the country and consist primarily of broad-leafed species and conifers in the middle and high altitude zones. Among the most common species are acacia, cork oak, and eucalyptus; the last was imported during the last century and became rapidly acclimated. Splendid cedar forests cover the slopes of the Rif and Middle Atlas, where other conifers, such as arborvitae and Aleppo pine, also grow. Also widespread is the Mediterranean maquis with its typical shrub (lentisk, broom, wild strawberry, etc.) and arboreal species (olive, cypress, cluster pine, etc.). Stretching beyond the upper limit of the woodlands, which rise to 8200–8500 ft [2500–2600 m], one finds meadows of Alpine-type flora. In the more arid regions of the slopes and pre-Saharan lowlands, where rainfall does not exceed 12 in. [300 mm] a year, the vegetation assumes a typical steppe-like character with a prevalence of grasses such as esparto and artemisia. Date palms and the argan tree are typical of the desert oases. Wildlife, which at one time was fairly widespread, has now been reduced by intensive year-round hunting to a few species (the bush pig, mouflon, gazelle, jackal, etc.); there are, however, numerous species of small mammals (mostly rodents), bird species are also quite abundant (including storks, ibis, par-

tridges, ducks, pelicans, and flamingos), and there are insects and reptiles as well.

Today there are two national parks in Morocco, one at Toubkal (High Atlas) and the other at Tazzeka (Middle Atlas).

Population. Like the other regions of the Maghreb, Morocco is inhabited by a population of Berber origin which to some extent retained its own language even after the Arab conquest when the country was otherwise completely Islamized. Even the Ottoman occupation and subsequent European colonization had a minimal influence on Berber ethnic unity. Around the Berber element are various other groups, such as the Masmoudas who live in the more mountainous regions, the Zenatas who had come from Algeria in the Middle Ages, and the Zanagas who settled in the Rif and Middle Atlas. Nomad peoples, including some Tuareg-related tribes (the Reguibate) live in the steppes and desert regions, while populations of Negroid stock, like the Haratin, inhabit the oases. Seminomadic tribes populate the northern regions of the Middle Atlas. Numerous Hebrew groups lived in their own quarters (mellahs) in the Mediterranean coastal cities until the time of European colonization.

Demographic structure and dynamics. At the time of its independence (1956), Morocco had a population of about 8 million people, which has now increased to more than 20 million as a result of an average annual growth rate of 2.5–3.5%. During this same period the number of European residents has dropped from more than half a million to about 50,000. The most densely populated areas are the coastal plains, the interior valleys furrowed by such waterways as the Sebou and the Sous from the Mediterranean coast and inland plateaus to the foot of the High and Middle Atlas mountains. Except for the oases, the steppe and desert regions at the edge of the Sahara and the areas south of the Dra are nearly uninhabited.

As among all Arab peoples, social life and economic activity center around the cities. In Morocco there are fifteen or so with over 100,000 residents. In addition to the two large urban agglomerations of Casablanca (the real economic capital of Morocco) and Rabat, where a fifth of the country's entire population is concentrated, there are the historic cities of Fès and Marrakech which are spread out over fertile inland plateaus, each with about half a million inhabitants. Throughout the country, city dwellers represent slightly less than half of the country's total population. It should be pointed out in this connection that European colonization caused a sometimes abnormal development of the major cities which, even after independence was achieved, grew along Western lines and often haphazardly in terms of patterns and rates of growth. This is evidenced by the proliferation of shanty towns at the outskirts of large urban areas, whose modern skyscrapers now stand in typical contrast, as in the case of Casablanca. The old historical centers have preserved their Muslim urban imprint, however, with their *casbah* (including the often fortified *caid*'s residence and offices), the *medina* (living neighborhoods), and the *suk* (commercial districts). In the rural areas, where concentrated settlements prevail, the typical masonry dwelling is very simple, laid out according to a square floor plan and with a flat or wagon roof according to the Mediterranean custom, or with a slanted roof (as in the mountain areas), but some are more modestly covered with straw or branches. The typical abode of the nomads, however, is the tent. In the southern

Climate data				
Location	**Altitude (ft asl)**	**Average temp. (°F) January**	**Average temp. (°F) July**	**Average annual precip. (in.)**
Rabat	246	55	72	22.6
Agadir	66	57	72	9.0
Fès	1148	50	80	20.8
Marrakech	1689	52	84	9.6
Tangier	49	53	72	35.8
Conversion factors: 1 ft = 0.3 m; 1 in. = 25 mm; °C = (°F – 32) × 5/9				

Administrative structure

Administrative unit	Area (mi²)	Population (1982 census)	Administrative unit	Area (mi²)	Population (1982 census)
Provinces			Tan-Tan	6,676	47,040
			Taounate	2,156	535,972
Agadir	2,281	579,741	Taroudant	6,354	558,501
Azilal	3,879	387,115	Tata	10,007	99,950
Beni Mellal	2,731	668,703	Taza	5,798	613,485
Ben Slimane	1,065	174,464	Tétouan	2,326	704,205
Bon Lemane	5,556	131,470	Tiznit	2,687	313,140
Chechaouèn	1,679	309,024			
Fès	2,084	805,464	**Urban prefectures**		
Figuig	21,612	101,359			
Guelmin	11,098	128,676	Aïn Chok-Hay Hassani	–	298,376
Al-Hoceima	1,370	311,298	Aïn Sebâa-Hay Mohamed	–	421,272
Ifrane	1,278	100,255	Ben Msik-Sidi Othmane	623	639,558
El Jadida	2,316	763,351	Casablanca-Anfa	–	923,630
El Kelâa-Srarhna	3,887	577,595	Mohamedia-Znata	–	153,828
Kénitra	1,832	715,967	Rabat-Salé	492	1,020,001
Khémisset	3,206	405,836			
Khénifra	4,756	363,716	**Saharan provinces**		
Khouribga	1,641	437,002			
Marrakech	5,695	1,266,695	Boujdour	38,646	8481
Meknès	1,542	626,868	Laayoun	15,193	113,411
Nador	2,366	593,255	Oued Ed Dahab	19,640	21,496
Ouarzazate	16,038	533,892	Es Semara	23,839	20,480
Oujda	7,990	780,762			
Er Rachidia	23,000	421,207	**MOROCCO**	274,388(*)	20,419,555
Safi	2,812	706,618			
Es Saouira	2,445	393,683			
Settat	3,764	692,359			
Sidi Kacem	1,567	514,127			
Tangier	461	436,227			

(*) Including 97,318 mi² of the former Western Sahara.
Capital: Rabat, pop. 518,616 (1982 census)
Major cities (1982 census): Casablanca, pop. 2,139,204; Fès, pop. 448,823; Marrakech, pop. 439,728; Meknès, pop. 319,783; Salé, pop. 289,391; Tangier, pop. 266,346; Oujda, pop. 260,082.

Conversion factor: 1 mi² = 2.59 km²

regions one frequently finds fortified villages (*ksar*) protected by high walls.

Social conditions. European colonization, which lasted for about half a century, had a notable influence on Morocco's social structures. After independence was achieved, the country continued the process of Westernization, with the result that traditional life underwent profound changes. In particular, the urbanization trend was intensified, and the emancipation of women gained increasing ground. In the countryside, individual ownership asserted itself as an important economic instrument, replacing traditional forms of land collectivization which characterized the economy of agricultural villages. Appreciable progress has been made in hygiene and sanitation, with a resulting decrease in infant mortality and many endemic diseases. Public education has also advanced considerably, although illiteracy is still very widespread (some 30% of the population). There are actually six universities in Morocco today, at Rabat, Fès, Casablanca, Marrakech, and Oujda. In the rural areas religious instruction, given at Koranic schools, is very widespread.

Economic summary. The economy of the old Moroccan empire, linked to traditional agricultural and pastoral forms, underwent radical changes during the European occupation:

colonial-style agriculture developed rapidly and was promoted by the construction of numerous irrigation systems, while ore prospecting led to the discovery of rich phosphate deposits. At the same time the country was provided with a more efficient road and rail transportation network and dock facilities at the major ports were improved.

The achievement of political independence, which coincided with the mass exodus of the European population, led to the onset of a period of economic crisis caused by a decrease in all commercial and industrial activity, a drop in investments and domestic consumption, as well as to massive unemployment which fed a large flow of emigration abroad. Recovery of the national economy was promoted by the intercession of the World Bank and was carried out under the auspices of a liberal planning policy through gradual land reform and a return of foreign investments. With considerable respect for national identity in corporate management ("Moroccanization"), these investments led the way to gradual industrialization of the large urban centers, with the absorption of a low-cost labor force. A new impulse was simultaneously given to tourism, with an increase in tourist facilities. However, the standard of living of the population remained fairly modest, as is evident from the per-capita gross domestic product, which in 1990 still stood at 950 dollars, the lowest of all the Maghreb countries.

On the whole, the Moroccan economy is still based on farming and livestock in addition to mining, although the industrial and service sectors are rapidly expanding, to such an extent that they now contribute to more than three quarters of the national product.

Agriculture and livestock. The primary sector (agriculture, livestock, and fishing) still monopolizes about 37% of the economically active population, totaling in 1990 almost 8 million.

Cultivated land occupies a little less than one fifth of the total area of the country altogether (excluding the Western Sahara). The chief crops are grain, with a high output of barley and wheat (totaling about 6.4 million t in 1990), in addition to corn and more modest production of rice, rye, millet, sorghum, and oats. In the areas formerly under European agricultural colonization, the presence of efficient irrigation systems has promoted the development of special crops, such as vegetables, legumes, and pome fruits, which are largely exported. There are also extensive exports of industrial crops, such as sugar beets and cane, cotton, peanuts, sunflowers, and tobacco. Especially widespread among tree harvests are olives (on the central plateaus), citrus fruit (with over 1.5 million t annually), and grapes (in the Sebou valley). Dates are typically produced in the oases. The forest cover supplies a small amount of precious wood and cork bark.

Traditionally coupled with agriculture is herding, widespread among the nomadic and semi-nomadic populations and practiced on an extensive scale. Livestock consists of approximately 23 million head of sheep and goats and more than 3.5 million cattle. There has been a considerable decline in the number of other traditional large quadrupeds (donkeys, mules, horses, and camels).

Fishing, a relatively recent development, is done generally along the Atlantic coast where the cold Canary current favors the formation of abundant schools of fish. The chief fishing ports, which are also equipped with canning facilities, are Agadir, Safi, Essaouira, Mohammedia, and Larache. The fish catch, which exceeded 550,000 t in 1989, is constantly increasing.

Mining resources and industry. Phosphates represent Morocco's principal mining resource and are extracted in the inland regions east of Safi or Casablanca (Khouribga, Youssoufia, Meskala) and in the Sahara area (Boukra) in the south. Today, with an average annual extraction of over 22 million t, Morocco is the world's third largest producer of phosphates after the USA and the territory of the former USSR. Other subsoil resources are metal ore (iron, lead, silver, manganese, cobalt, antimony, etc.), flourine, halite, and barite. In addition, there are deposits of coal (Jerada at the southern Algerian border) and modest quantities of hydrocarbons. The as yet unexploited reserves of rock asphalt are, however, considerable. The production of electric energy, about one tenth of which is obtained from water power, had reached 9 billion kWh in 1989.

Still of modest proportions is the industrial sector, which has developed predominantly as a result of foreign investments and is essentially located in the major urban centers (Fès, Sidi Kacem, and Meknès) and port cities (Casablanca, Mohammedia, Tétouan, Kenitra, and Safi). The basic industries include metallurgical and steel plants, petroleum refineries, cement works, and chemical production facilities (plastics, acids, phosphate fertilizers, etc.). Also very active are the food and textile industries. The former includes sugar refineries, oil and other mills, canneries,

breweries, etc.; while the latter, which has been developed in recent decades, includes cotton and wool mills, rug production—heir to a traditional artisan activity—as well as the processing of leather goods. Other industries are paper, tobacco, tires, and automotive assembly plants, and also electromechanical capital goods as well. All told, a little more than one fourth of the economically active population works in industry.

Commerce and communications. Considerable demographic pressure, which tends to swell domestic consumption, and the as yet inadequate industrial and agricultural output are the major causes of the country's heavily negative trade balance. Exports are limited to farm and fish products, raw materials of mineral origin, and carpets, while imports consist primarily of capital goods and consumer commodities in addition to foodstuffs. Morocco's major trade partners are France and the European nations, the Arab countries, the USA, and the former Soviet Union.

The transportation network, largely built during the protectorate and often governed by the ruggedness of the terrain, consists of over 36,500 mi [59,000 km] of road (1987) and 1200 mi [1900 km] of rail. The automobile fleet is still small, but air transportation, with its network of a dozen or so airports, is more efficient. Maritime traffic, on the other hand, has access to a system of fairly developed ports only on the Atlantic coast. The principal port of call is Casablanca, which handles about 17.6 million t of merchandise (1990).

There is a fairly large tourist industry, which already exceeds 2 million visitors a year, mostly from France, Germany, and Spain.

Historical and cultural profile. *Origin of the Maghreb.* It was not until the Middle Ages that Morocco became a state with its own identity, after its territory had experienced a succession of wars and dominations, revolts and conflicts. Visited by the Phoenicians as early as the first century B.C., conquered by the Romans and Vandals, and reconquered by the Byzantines, the land of Morocco was subjected to Arab conquest and converted to the Muslim religion, as was all of northern Africa, in the 7th century. The adherence to Islam did not snuff out the spirit of independence and natural pride of the local Berber tribes, whose revolt in the year 740 gave birth to the western Maghreb, which severed all ties with the Arab Caliphate. Although Islamized, these Berber people maintained a strong autonomy in matters of tradition, social models, and cultural forms. Moroccan civilization was established over the centuries on the foundations of this original Berber culture to which the

Socioeconomic data	
Income (per capita, US$)	950 (1990)
Population growth rate (% per year)	2.6 (1980-89)
Birth rate (annual, per 1,000 pop.)	36 (1989)
Mortality rate (annual, per 1,000 pop.)	9 (1989)
Life expectancy at birth (years)	61 (1989)
Urban population (% of total)	47 (1989)
Economically active population (% of total)	30 (1989)
Illiteracy (% of total pop.)	29.3 (1991)
Available nutrition (daily calories per capita)	2820 (1989)
Energy consumption (10^6 tons coal equivalent)	7.8 (1987)

architecture, orally transmitted popular literature, arts and crafts, music, and dance still bear witness.

The Berbers held power between the 7th and 10th centuries with the Idrisid dynasty, which was followed by the Almoravids and the Almohads, linked to a rigorous and puritanical concept of Islam, a religion which they had grafted on the trunk of their original tribal warlike spirit. It was the Almoravids who founded Marrakech (1062) and the Almohads who extended their dominion over the entire Maghreb, reaching all the way to Tripolitania and Cyrenaica and even overflowing onto the Iberian peninsula.

This era of conquests and intense trade exchanges proved culturally and artistically most fertile, and saw the austerity of the nomad populations tempered in the hearth of Arab civilization and highly refined by Andalusian influences. Almoravid art reconciled the Islamic east and west, as can be seen in the Qarawiyyin mosque in Fès, with its reminders of Cordoba and Persia. Literature and science flourished together with the arts; philosophy asserted its autonomy through the meditations of Ibn Rushd (1126-98), who lived at the Almohad court and was known in Europe by the name Averroës.

Decadence. The collapse of the Almohad empire in the 13th century led to the tripartition of the Maghreb corresponding substantially to the modern distinction among Tunisia, Algeria, and Morocco. Several dynasties succeeded each other in Morocco, which managed for centuries to maintain a state authority over the tribal centrifugal forces.

Under the Merinids, who ruled until 1471, Hispano-Moorish art reached levels which would never again be approached; it flourished primarily at Fès and its most prominent figure was the architect Abdou el-Hassan. Stylistic models which would become normative were established in the mosques, governing palaces, and Koranic schools of this period. Under the Merinids, Fès was a lively cultural as well as artistic center; in addition to the wealth of religious literature, a secular literature of a courtly nature also flourished there, but was destined to die out in later centuries.

These were in fact centuries of regression, if not decadence. The *reconquest* of Arabized Spain encouraged by the Catholic kings changed in the 16th century into a direct attack on the key points of Morocco's Mediterranean coast, at a time when the country was simultaneously being threatened in the east by the advancing Ottoman empire. There were indeed times of prosperity, as under the Sadi caliphs Abd el-Malike and Ahmed el-Mansur, who even undertook the ambitious project of expansion toward black Africa with the ephemeral conquest of the Songhai empire. And times of magnificence, when the court of the Hasanids, the dynasty which supplanted the Sadi dynasty in the 18th century, claimed direct descendance from the Prophet and is still in power today. But the turn of events combined to bring Morocco to a state of isolation and relative weakness at the threshold of the 19th century, when it was exposed to the ever-pressing penetration by the European powers, whose interest it was to control a strategically important territory. Especially fateful was the French occupation, starting in 1830, of Algeria, which then became the base for a series of military actions (such as the shelling of Tangier in 1844), debilitating interventions, and schemes of political influence, which in 1912 brought the country under the French protectorate, with Spain being conceded control over certain areas (Rif, Ifni, Tarfaya).

From French domination to independence. The ability to maintain its independence almost since the end of World War I depended on the solid framework of state union that had gradually established itself—albeit on a backward social fabric—under the Hassanid monarchy, which could count on fairly efficient administrative and military structures. Although these structures were largely preserved during the initial years of the protectorate, France did not hesitate to turn to ever more direct control after the first revolts, such as the Berber revolt (1924–25) led by Abd el-Krim.

The Franco-Moroccan clash finally led to a rupture after World War II. A major unity party, the Istiqlal (Independence), had formed and in 1943 promulgated a definite program of opposition to the protectorate. In that same year, following a meeting with President Roosevelt in Casablanca, Sultan Mohammed V convinced himself that he could count on United States support. And, on a broader basis the crisis of colonialism was accompanied by the political awakening of the Islamic world with the establishment of the Arab League.

In 1956, France and Spain signed two treaties which recognized the independence of Morocco, with the exception of the city of Tangier (which until that time had been under international administration and returned to Moroccan jurisdiction in 1960) and the territories of the Spanish Sahara, which were to become the object of prolonged international controversy in subsequent years. After independence, Morocco was confronted with a phase of delicate adjustment, the impetus to change clashing with court interests and feudal vestiges. The prestigious figure of Mohammed V served as an element of mediation and cohesion, but upon his death (1961) the succession of Hassan II marked the predominance of authoritarian tendencies among the ruling classes and conservative forces in society. These tendencies and forces have characterized recent Moroccan history, both in matters of economic choice (designed not to undermine old privileges by industrial development) and in matters of internal and foreign policy, where the emphasis on nationalism and the role of the armed forces represents an important factor of stability for the crown itself.

Closely related to twentieth-century historical events of Morocco is the cultural evolution of the country, tending above all to end the long isolation and fairly sensitive to "the temptation of the West." The novelists Ahmed Sefrioui and Driss Chraibi, the poets Khair Eddin and Mustafa Nissaboury, and the essayists Aziz Lahabi and Abdallah Laroui wrote in French. A work denouncing the foreign protectorate was composed by a group of nationalist writers, such as Muhammad Hadir ar-Raysuni and Abd al-Karim Gallab. Later, after the achievement of independence, many progressive and Marxist intellectuals asserted themselves, sometimes persecuted by the monarchical regime. A special place is occupied today by Tahar Ben Jelloun, an author intent upon grasping a new and agonizing reality confronting those of his fellow Moroccans forced to leave the country and embark on the life of an emigrant.

TUNISIA

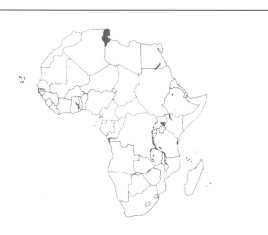

Geopolitical summary

Official name	Al-Jumhuriya at-Tunisiya
Area	63,153 mi² [163,610 km²]
Population	6,966,173 (1984 census); 8,000,000 (1990 estimate)
Form of government	Presidential republic. Legislative power is exercised by the National Assembly, elected every 5 years by universal suffrage together with the President of the Republic.
Administrative structure	23 governorates
Capital	Tunis (pop. 596,654, 1984 census)
International relations	Member of UN, OAU, and Arab League
Official language	Arabic; French is very widespread
Religion	Muslim of the Sunni rite; there are also some 40,000 Christians and Jews
Currency	Dinar

Natural environment. Located at the eastern end of the Maghreb Mountains and facing the Mediterranean with a rather indented coastline, with deep gulfs and protruding peninsulas which come fairly close to the southern shores of Europe (Cape Bon is a mere 87 mi [140 km] from Sicily), Tunisia also extends southward into a small portion of the Sahara region, bordering Algeria to the west and Libya to the southeast.

Geological structure and relief. The territory exhibits pronounced geomorphological and regional differentiation, seeming to be divided in half by the deep depression latitudinally aligned inside the Gulf of Gabès. This depression lies 75 ft [23 m] below sea level and includes a number of chotts which also extend into Algeria. North of this depression is historic Tunisia, which in Roman times was known as "Africa" and as *Ifriqiya* to the Arabs, a region of rather complex morphology

modeled on sedimentary rocks (clay, sandstone, and limestone) of varying age and affected by Alpine foldings at the end of the Tertiary.

The orographic features are closely dependent on the Atlas mountain system, as evidenced by the end chains which consist of direct alignments running predominantly from southwest to northeast and rising to an elevation of 4600 ft [1400 m]; these are the Medjerda Mountains which continue toward the northeast with the Tell region, a complex of highlands which contain a fairly irregular network of waterways and drop slightly in the direction of the Gulf of Tunis. Extending south of this alignment is the Medjerda valley, also of an irregular pattern, beyond which lies the compact barrier of the Teboursouk Mountains, which are followed, after the tectonic depression of the High Atlas, by the broad massif of the Great Dorsal, which is a direct continuation northeastward of the Tebessa Mountains that are bisected by the Algerian border. The Dorsal, whose highest peaks vary between 4300 ft [1300 m] and 5000 ft [1500 m], constitutes the orohydrographic element which characterizes the topography of all of northern Tunisia. Toward the east it drops rapidly down to the coastal plain (which here is named the "Sahel") and fronts on the gulfs of Hammamet and Gabès. To the south, beyond the depression of the Chott el Djerid and el Fedjadj, is Saharan Tunisia, characterized by a tabular relief with Mesozoic sediments and interrupted toward the east by the Djebel Matmata and Djebel Dahar hills delimiting the fertile Jifarah plain which is crossed by the Libyan border.

In this travel log, Grevin describes the beauties of southern Tunisia, where natural and human elements are in perfect harmony:

Pink undulating desert, sand and rocks; garhas, that is mountains in the shape of truncated cones, such as those found at the confines of the Ahaggar: here we find the strange aspect of the first spurs of this mysterious mountain chain.... One has the profound feeling of true mountain Sahara (and this only 90 miles [150 km] from Gabès). Color reigns supreme and absorbs everything: the pink and red of the first plains, the mauve in the distance. One is lost in a sea of gray-green.

And how impressive the picture upon arriving all of a sudden in front of Douirat, this eagle-nest village clinging to the peak of a mountain and the first sight of which is only a white mosque halfway up the cliff because the houses are no more than holes in the pink rock which encircle a ksar with sheer walls.... The cars stop at the foot of this fantastic human rock. ... Douirat is one of the most alluring places, more "south" than the Tunisian South.

Hydrography. The system of waterways is fairly well developed throughout the country, including the Sahara region, where the old valley gullies (*wadi*) still subsist, attesting to active sliding in the early Quaternary and currently running seasonally. But the only perennial watercourse is the Medjerda, which rises in the Atlas Mountains in Algeria and runs through Tunisia for 167 mi [270 km] (of its 230-mi [370-km] total course) before emptying into the Gulf of Tunis. There are many lake basins, however, both artificial and natural; the latter (*chott* and *sebkha*) occur throughout the Sahel and Jifarah plains as well as the pre-Saharan depression, but their water regimen is temporary, fed as they are by the rains in the winter and dried up by the intense evaporation in the summer, when they are covered by a salt crust.

Climate. Because of the country's geographic position and latitude, the Tunisian climate appears to be greatly affected by both the presence of the Mediterranean and the closeness of the Sahara, from which the *chelili*, a dry and very hot sirocco, blows during the spring and summer months. As in all Mediterranean regions, precipitation occurs primarily in the winter, with an annual intensity which varies from a little more than 40 in. [1000 mm] of rainfall in the northern regions that are also reached by humid breezes from the Atlantic, to less than 8 in. [200 mm] in the Gulf of Gabès. In winter (January), the temperature hovers around 50–52°F [10–11°C] on the coast, decreasing slightly inland because of the altitude, while in the summer (July) it reaches 79–81°F [26–27°C] on the coast and even 90°F [32°C] in the interior where, however, the range is greater due to the continental conditions. There are at least three types of climate in Tunisia: Mediterranean or humid subtropical climate in the north, steppe-like or dry subtropical climate in the central areas, and dry tropical desert climate in the southernmost parts of the country.

Flora and fauna. As a result of the climatic conditions, the vegetation reflects the character of the three types of environment described, although human intervention has often made profound changes in the original landscape by reducing the extension of the wooded areas and seeking to make the arid ones (Sahel, Jifarah) verdant again by introducing irrigation. Grand Mediterranean forests with Aleppo pine, cork, and holm oak still cover large tracts of the Tell, alternating with cultivated areas and meadows. The latter are more common in the central part of the country and the sub-Sahara depressions where the steppe predominates and esparto grass grows in abundance. In the desert, there are palm trees in the oases.

The number of wild animals is now limited: antelopes and gazelles still live in the sub-Saharan regions and a wealth of birds populate the lagoons and coastal ponds.

Protected areas include the Djebel bou-Hedma National Park (established in 1936) west of Gabès, where the rare gum tree (*Acacia tortissima*) grows, and the Garaet Ichkeul Natural Reserve (established in 1978), slightly west of Bizerte.

Population. After the Europeans departed, the Tunisian population regained some of its ethnic homogeneity, with a clear-cut prevalence of the Arabs over the Berbers, whose language is still widely spoken in the Krumiria and in Djerba. Ottoman domination had some slight influence, but the spread of Sunni Islam has greatly cemented the country's national unity, to which government policy and the permanence in power of the Néo-Destour party has undoubtedly also contributed.

Demographic structure and dynamics. The 1984 census recorded the Tunisian population at 7 million, almost double what

it was thirty years earlier on the eve of independence. At that time there was a very strong European presence, with over 250,000 people, the majority French (180,000) and Italian (67,000), which today is insignificant, largely as a result of a xenophobic policy instituted during the 1960s, when foreign property was confiscated, including that of 45,000 Italians (1964).

In the 1980s annual population growth was steady at 2.5%, as a result of which the population reached 8 million, with an average density of about 127 per mi^2 [49 per km^2], which increases to about 220 per mi^2 [85 per km^2] if the inland waters and desert areas of Tatahouine and Kebili are excluded. The most densely populated areas (aside from the metropolitan capital district with its more than 800,000 inhabitants and a density close to 2600 per mi^2 [1000 per km^2]) are the urban centers of Soussa, Sfax, and Bizerte, and the most intensely farmed areas (northern part of the Gulf of Hammamet, island of Djerba, and the well-irrigated coastal plains of the Sahel and the Jifarah). The desert regions are practically uninhabited.

However, the urban population is not a substantial part of the total population. After Tunis, Sfax is the most populated city, followed by Bizerte, Gabès, and Soussa—all on the sea. The capital rises on the inner bank of a lagoon around which a canal connects it to the outer harbor of La Goulette. Béja in the Medjerda valley, Gafsa near the rich phosphate deposits at the threshold of the chott region, and the important Islamic center of Kairouan should also be mentioned among the other cities of Tunisia. In this country the old historic city centers, while still preserving the *casbah*, the *suk*, and the *medina*, are often surrounded by the European quarters built during the colonial period; but even more recent constructions echo European architectural patterns, and around the periphery of the large cities it is not rare to see the temporary structures of shantytowns, attesting to the extensive urban immigration.

Elsewhere, the rural dwellings in scattered villages reflect the typical designs of the Maghreb: stone or clay structures with terraced or wagon roofs. In the southern steppe regions and the Sahara oases, the use of tents and simple huts is widespread among the nomadic and seminomadic peoples.

Social conditions. Because of sharp demographic pressure, living standards are not yet satisfactory, as the high illiteracy rate (34.7%) and low individual income rate (US$1560 in 1990) attest. Educational facilities are improving and, in addition to more than 200 Koranic schools, there are the universities of Tunis, Sfax and Soussa.

Economic summary. At the time of independence (1956), the Tunisian economy appeared to be heavily affected by the long period of French occupation which had, in practice, made the country a colony for settlement and for agricultural and mineral exploitation. The most productive lands were in European hands, and the foreign investments themselves contributed to the development of an agricultural economy producing essentially for export. The rapid abandonment of the country by the European residents, who also held leading positions in the public administration, had very negative consequences that were reflected by a slowdown in economic activity, an increase in unemployment, and a considerable flow of emigration abroad.

The first economic directions were characterized by the adoption of a self-sufficiency development model based on principles

Climate data				
Location	**Altitude (ft asl)**	**Average temp. (°F) January**	**Average temp. (°F) July**	**Average annual precip. (in.)**
Tunis	216	48	78	16.0
Bizerte	7	52	77	24.8
Gafsa	984	48	85	6.0
Sfax	13	52	78	7.8
Conversion factors: 1 ft = 0.3 m; 1 in. = 25 mm; °C = (°F − 32) × 5/9				

Administrative structure

Administrative unit (governorates)	Area (mi²)	Population (1984 census)
Aryanah	601	374,192
Béja	1,373	274,706
Bin Arus	294	246,193
Bizerte	1,422	394,670
Gabès	2,770	240,016
Gafsa	3,470	235,723
Jendouba	1,197	359,429
Kairouan	2,591	421,607
Kasserine	3,113	297,959
Kebili	8,524	95,371
El Kef	1,916	247,672
Mahdia	1,145	270,435
Médenine	3,315	295,889
Monastir	393	278,478
Nabeul	1,076	461,405
Sfax	2,912	577,992
Sidi Bou Saïd	2,700	288,528
Silyanah	1,788	222,038
Sousse	1,012	322,491
Tatahouine	15,011	100,329
Tozeur	1,822	67,943
Tunis	134	774,364
Zaghwan	1,068	118,743
TUNISIA	59,803(*)	6,966,173

(*) Excluding 3,505 mi² of inland waters.
Capital: Tunis, pop. 596,654 (1984 estimate)
Major cities (1984 estimate): Sfax, pop. 231,911; Aryanah, pop. 98,655; Bizerte, pop. 94,509; Djerba, pop. 92,269; Gabès, pop. 92,258; Sousse, pop. 83,509; Kairouan, pop. 72,254.

Conversion factor: 1 mi² = 2.59 km²

of a planned economy supported by cooperative structures. In the 1970s, when this model failed to yield the expected results, it was dropped in favor of a market economy with room for private initiative and foreign investments.

Demographic pressures and the growth of domestic consumption, however, were factors that brought on a negative trade balance in the agricultural and food sectors, so that more recent development plans have sought to stimulate production and exports as well as promote tourism. A notable contribution to the national economy has been the remittances sent home by emigrants working abroad.

In terms of the structure of the economy, the agricultural sector, while monopolizing about a third of the economically active population, accounts for less than 20% of the national income. The industrial and mining sector, on the other hand, contributes 32% to the GNP with the same number employed.

Agriculture and livestock. Strongly affected by the climate (years of drought are not uncommon), agriculture can nevertheless avail itself of large irrigated areas, both in the Medjerda valley and in the Sahel and Jifarah coastal plains. Cultivated areas, including land suitable for sowing and tree-growing, cover 28.7% of the country, meadows and pastures occupy about 18%,

and approximately 4% consists of wooded areas, especially in the northern elevations. In practice, at least half of Tunisia is nonproductive land. The Medjerda valley, the terraced land of the Sahel, and the interior plateaus accommodate cereal crops (wheat and barley), whose production, maintained at a combined level of some 1.8 million t, was severely reduced by drought in 1988. Ligneous cultures are widespread in the northern valleys (vineyards), the Sahel and the Jifarah (olive trees), as well as on the coasts of Sfax (citrus groves) and the Gulf of Hammamet. The production of vegetables and pome fruits is varied, while in the southern oases dates are cultivated for large-scale export. Other crops of some importance are sugar beets, tobacco, and almonds. The Mediterranean forests also supply cork and wood and the inland steppes produce esparto fibers.

Livestock resources include 7 million head of sheep and goats, generally herded in accordance with transhumance practices, whereas cattle (some 600,000 head) are bred on stock farms in the wet northern regions. The extension of the offshore underwater shelf along the Tunisian coasts (Sicilian channel and Gulf of Gabès) is a rich fishing ground with an annual catch of about 110,000 t (tuna, sardines, anchovies, etc.). Sponges and coral are harvested from the warm waters of the Gulf of Gabès. The fishing fleet is based in the ports of Soussa, Mahdia, and Gabès.

Energy resources and industry. All told, Tunisia's mineral resources are modest, except for the rich phosphate deposits in the south-central region of the country (Gafsa), where 6.6 million t of phosphate rock is mined annually. In the Sahara (El Borma), moderate quantities (31.5 million barrels) of petroleum are extracted and carried by pipeline to the port of Sekhira. The subsoil also supplies metal ore (iron, lead, zinc, and silver) as well as methane and halite.

Electric energy is produced by both hydroelectric and thermal power stations and the annual output is about 5 billion kWh.

Tunisia's industrial sector is still fairly modest. Particularly developed are the industries connected with mining, such as those producing superphosphates, phosphoric acid, and fertilizers (Gabès, Gafsa, and Sfax), as well as the petrochemical, iron and steel (Tunis, Bizerte), and lead metallurgy industries. Of some importance are the textile (cotton fabrics and yarn) and food industries (oil extraction mills, canning, etc.). Other industrial facilities include cement works, automotive assembly plants, and factories which produce consumer goods (household appliances and the like).

Commerce and communications. The commercial balance of trade with other countries reflects a prevalence of imports,

Socioeconomic data

Income (per capita, US$)	1,560 (1990)
Population growth rate (% per year)	2.5 (1980-89)
Birth rate (annual, per 1,000 pop.)	30 (1989)
Mortality rate (annual, per 1,000 pop.)	7 (1989)
Life expectancy at birth (years)	66 (1989)
Urban population (% of total)	54 (1989)
Economically active population (% of total)	32.6 (1988)
Illiteracy (% of total pop.)	34.7 (1991)
Available nutrition (daily calories per capita)	2,964 (1989)
Energy consumption (10⁶ tons coal equivalent)	5 (1987)

predominantly with some European countries (France, Italy, and Germany). The communications network includes over 16,000 mi [27,000 km] of roads (1988) and 1400 mi [2200 km] of rail. Thanks to the shape of its coasts, Tunisia also has an efficient system of ports, most of them naturally well protected. Together the ports of Sfax and La Goulette (Tunis) move approximately 8.8 million t of shipping a year. The chief airports are located in Tunis, Monastir, and on the island of Djerba. The latter is also one of the major tourist centers in the country, which attracted more than 3.2 million visitors in 1989.

Historical and cultural profile. *From the Carthaginians to the Arabs.* In antiquity, its central and dominant position in the Mediterranean basin made the land of Tunisia a disputed region. The first inhabitants were Libyan or Berber, and the presence of Egyptians and Phoenicians has been documented. In the 9th century B.C., the latter founded the colony of Carthage, which was destined to become the center of a far-flung commercial empire on the coasts of North Africa, Spain, Sicily, and Sardinia. The clash with Rome (264–146 B.C.) led to the destruction of the city, which was rebuilt as a Roman colony at the time of Caesar, flourished during the Empire, and with the triumph of Christianity asserted itself as the center of the African church. Although scant finds remain from the Punic era, there is considerable evidence of Roman urbanization, not only at Carthage (the amphitheater and baths of Antoninus Pius), but also at Thugga (modern Dougga), Thuburbo Majus, Mactaris (now Maktar), and Sufetula (Sbeitla).

Vestiges also remain of the Byzantine domination (such as the Basilica of St. Cyprian in Carthage and the monastery at Monastir), which in the 5th century had replaced the Vandals when the Roman Empire collapsed. Then came the Arab conquest of northern Africa which, despite the fierce Berber resistance, included the region corresponding to modern Tunisia.

In 670 the Arabs founded the holy city of Kairouan, which down to this day has preserved major examples of Islamic architecture: the Great Mosque (9th century), the mosque of the "Three Doors," and the walled enclosure surrounding the old city (medina). The building of the city marked the birth of the Maghreb's Muslim history. It occurred in an inhospitable wooded and marshy area, as is here recalled:

Serpents and ferocious beasts! We are the Companions of the Prophet; make way because we shall set our home in these parts, and from now on we shall kill all of you whom we find here! Witness now the marvelous spectacle of the marching past of the lions, wolves, and serpents going away and carrying their young with them; respecting the orders given, the animals are permitted an honorable retreat. When the exodus was completed, Uqba stopped at this place and had the trees cleared.... He then drew up the plan for the Government Palace (dar-al imara) and the Great Mosque....

The capital was later moved from Kairouan to Tunes, today called Tunis, a city which reaped the legacy of Carthage. There, various dynasties succeeded each other, and it was under the Hafsids (13th to 16th centuries) that Tunisia emerged from the rest of the Maghreb as an autonomous state entity. In 1574 the country was conquered by the Turks and became a province of the Ottoman Empire, but the ties to Constantinople weakened

with the passage of time, so that the power of the local administration under the "bey" actually assumed the form of a hereditary monarchy. In fact, the Hussenite dynasty reigned under rising and falling fortunes from 1710 to 1957.

French protectorate. Ever heavier interference was exercised on Tunisia, not so much by the Ottoman Empire as by European imperialism, specifically that of France, which made the country so extremely financially dependent as to succeed in imposing its own protectorate on Tunisia in 1883: the bey's position was not touched, at least formally, but real power was in the hands of the French. Unlike Morocco and Algeria, Tunisia was not the theater of bloody rebellions against a tribal background. However, ahead of the rest of the Maghreb, a nationalist movement developed, headed by exponents of the urban and intellectual strata of society, which united autonomist claims with the requirements of modernization. This movement gave birth in 1920 to a political party, the Destour, but its program was soon supplanted by events, and in 1934 the left wing established a new party, the Néo-Destour, under the leadership of Habib Bourguiba. The new party substituted the rallying cry of total independence for demands for simple reforms under the protectorate. France tried to crush the movement, but the latter was indirectly favored by the events of World War II, not least because of the French defeat in 1940 and the subsequent German occupation (from November 1942 to May 1943).

The Allied reoccupation included France and led by way of repercussion to repression of the nationalist movement, forcing its leaders into exile or underground. But in the years that followed, the process of decolonization extended to the entire Afro-Asian world, compelling the French government to negotiate. Tunisian independence was declared in 1956 and in the following year Bey Lamine was deposed and a republic proclaimed with Bourguiba as president.

Tunisian independence. The orientation of the new state was avowedly socialist, and in 1964 the party in power adopted the name Destour Socialist Party (PSD). From the very outset the new regime sought to avoid both liberal deviations and hasty collectivization plans, and has shown substantial evidence of stability, even if recent Tunisian history has witnessed difficult moments, as during the student oppositions of 1968 and 1971–72. In 1978, the General Union of Tunisian Workers (UGTT) organized the first general strike since the proclamation of independence, to which the government responded by declaring a state of emergency and using the army. The internal situation became tense in 1981 with the insurrection of the city of Gafsa maneuvered by Libya, and with the popular revolt in 1984 when the price of bread doubled. Also in 1981, however, the first multiparty elections reconfirmed the guiding role of the PSD. Six years later, Bourguiba, who had been named president for life in 1974, was removed and his position assumed by General Ben Ali, who instituted a policy of democratization and compromise with the Islamic fundamentalists who favor a unitary policy.

MEDITERRANEAN AFRICA

Images

1. Remains of the baths of Antonius Pio (2nd century B.C.) in Carthage, Tunisia. Standing guard by the Mediterranean, these impressive ruins show a building of octagonal halls laid out around a large square area supported by nine pillars and represent one of the last important tokens of Roman Carthage—metropolis of the empire, capital of proconsular Africa, and important economic and cultural center.

2. Dunes of the Erg d'Admer in the southwestern Algerian Sahara. These large expanses of sand (ergs) often cover as much as a thousand square miles. Unlike the other natural environments of the Sahara like the pebble-strewn or rock-strewn desert (serir or hammada), ergs abound in plant species: the grasses that grow there provide the best grazing of the desert after the rains.

3. Mt. Toubkal (13,661 ft [4165 m]), the highest peak in Morocco and all of western Africa, lies in the Grand Atlas range; it rises in the form of a truncated pyramid with a tabular summit of superimposed dark Precambrian lava. The Toubkal National Park, one of Morocco's two natural parks, is located here.

4. The Algerian tassili, tabular elevations in the Sahara sur-rounding the Hoggar massif, are deeply furrowed by the erosion of Quaternary river-beds which have actually excavated deep gorges. Dwelling here are the Tuaregs. The many prehistoric carvings and rock-paintings make it the major center of rupestrine art in northern Africa.

5. Desert region of the western Sinai. A closed triangular peninsula delimited by the Mediterranean and the Gulfs of Suez and Aqaba, the Sinai is characterized in its southern part by calcareous rock elevations, which gradually decline toward the sandy and dune-hemmed coasts of the Mediterranean. The generally arid nature of the territory of Egypt permits intensive cultivation only in the Nile valley, where the majority of its population is also concentrated.

6. View of the Nile near the first cataracts above Aswan in Egypt and the typical sailing vessels known as feluccas. The Nile is the longest river in the world (4136 mi [6671 km]); its source is the Kagera, the main tributary of Lake Victoria. Emerging from the equatorial lake region as the Victoria Nile, it flows northwestward to the Ripon Falls, then veers north through Kyoga Lake and, after passing the Karuma and Murchison Falls, empties into Lake Albert. From here it runs through a mountainous region as it enters the Sudan, where it is called Bahr al-Jebel; after crossing steppes and swamps, it turns north-ward, arriving at the confluence of the Blue Nile, its major tributary, near Khartoum. It then cuts through the Nubian desert and is joined by its last tributary, the Atbara, after which it crosses into Egypt. Characteristic of the Nile are the seasonal floods caused by the enormous quantities of water, rich in fertilizing silt, which the Blue Nile and Atbara heap onto its bed between June and September.

7. Aerial view of a dried-out wadi in the Algerian Sahara. These channels of inconstant watercourses are characteris-tic of the desert, where rainfall is rare but violent. During flooding, the water has strong erosive and transport power, dragging along its course large quantities of detritus. Some wadis reach the sea; they have a scant yet perma-nent outflow, but most of them empty into brackish salt flats (chotts) or disappear under-ground, feeding the aquifers which contribute to the forma-tion of oases.

8. The palm groves of the fer-tile oases of Maradah, which is also a center rich in petro-leum deposits in the southwest-ern region of the Cyrenaican Plateau, represent one of the rare interruptions in the arid Libyan desert. Inside these oases the population density is high, but outside the rocky and sandy areas are only tem-porarily inhabited by nomadic and seminomadic groups.

9. The opening of the Suez Canal, started in 1859 and inaugurated ten years later, gave Egypt a geopolitical sta-tus of great international importance and, as a result, the country is on the major commercial shipping lanes of all other countries in the world. The course of the canal, 105 mi [169 km] long between Port Said in the north and Suez in the south, was designed to take simultaneous advantage of the low altitude of the isthmus and the pres-ence of the Timsah and Amari lakes. The first attempt to open an artificial navigable channel between the Mediterranean and the Red Sea goes back to ancient times, specifically to 600 B. C. during the reign of Pharaoh Nechao II.

10. A casbah in the Dra valley of Morocco. Casbahs are for-tified villages located along wadis, which were built by local chiefs long ago. They usually contain the fortress building (shown in the photo-graph) which housed the local administration and whose function it was to dominate the

surrounding territory, a mosque, and a group of dwellings, the outermost of which constituted the village bastions. In the rural areas, where concentrated settlements prevailed, the typical home is a very simple masonry structure of quadrangular design and a flat or wagon roof, in the Mediterranean style, or a slanted roof (in the mountainous areas), but also with more modest roofing of branches and straw.

11. An image suggestive of the famous "fantasia" of Berber horsemen, the cavalry charge that the Moroccan nomads execute with extraordinary bravura. These historical evocations, which recall in costume and gestures the times of guerrilla warfare against the invaders, animate inter-community dances, seasonal festivities and sometimes religious celebrations as well.

12. Unlike other ethnic groups of northern Africa, the Tuareg nomad society is traditionally matriarchal, in the sense that women enjoy respect, independence, and full social equality. One of the most conspicuous signs of their prestige is the practice of only the men covering their faces with a veil and not the women, as Islamic custom prescribes.

13. Night scene of a square in Algiers, where modern, European-style quarters appear side by side with the ancient casbah, that picturesque Arab citadel of narrow and tortuous streets that characterizes the historic sections of northern Africa's cities. In the alleys of this hill, rising sharply from the sea, practically impregnable through the houses arranged in a funnel-like pattern and built so close as almost to touch, the anticolonial revolt of the Maghreb broke out, which in 1962 led to independence. In the background of the picture rises the monument to the Martyrs of the Revolution, built in 1982.

14. The Palace of the Orient rises in the ancient medina of Tunis, the Muslim quarter which has maintained the traditional urban pattern of narrow and tortuous streets on which the whitewashed terraces of the houses front. It is an example of a typical patrician interior, with the walls covered with decorative multicolored ceramics of geometric and floral designs. In the distance we can see the cupola of the Great Mosque, the oldest Islamic temple in the west, built in the 9th century.

15. A poster in the streets of Tripoli displays an image and political statement of Muammar al-Qaddafi, the controversial leader elected president of Libya in 1977. Qaddafi came into the international limelight in 1969, when he led the military coup that overthrew the monarchy of King Idris Senussi and proclaimed a republic, but became even more well-known at the end of the 1970s and during the early 1980s, when he headed the radical front of Arab countries that opposed the United States and the West in general.

16. The Center of Cairo as seen from the citadel, with the Hassane and Rifai mosques. The first, erected between 1356 and 1363, is more properly a madrasa, or school for higher education in the theological and juridical sciences; in fact, the cross-shaped layout reflects the special requirements of the parallel teaching of the four recognized rites of orthodox Islamic doctrine. The second was built between 1896 and 1912 to counterbalance the more important first mosque and as a burial place for famous personalities.

17. Small plots cultivated in the High Atlas of Morocco. The agricultural panorama, with its intertwining of olive groves, vineyards, and citrus orchards, and its variegated mosaic of mixed crops and land parcels is in the best Mediterranean style. Gradual agrarian reform in the countryside is asserting individual property as an important economic instrument, replacing the traditional forms of land collectivization which characterized the economy of agricultural villages.

18. A seller of spices in the village oasis of Tinerhir in the Middle Atlas of Morocco. The sale of spices, a true vocation in the Arab world, is widespread in the suks of all Moroccan cities. In the shops, which are open all hours and always stocked with goods, one is served with eagerness and competence.

19. A pottery craftsman fashioning a water jug. Along the alleys and arched galleries of the Muslim quarter of Cairo, tiny shops no bigger than an alcove produce manufactured articles made of wood, straw, wool, and precious metals. Equally characteristic are the bars, which are reserved for men in keeping with the traditional separation of the sexes that forbids women to frequent mixed public places.

20. A foundry in Algiers. In 1962, the year of national independence, the Algerian economy was in great difficulty because of the heavy exodus of Europeans, which suddenly deprived the country of a large part of its management executives. The subsequent program of economic planning, with its investment of capital derived from petroleum sales and exploitation of the country's rich mineral resources, gave special impetus to the iron and steel industries.

21. The consistent finds of graffiti and rock-paintings attest to the presence in Libya of prehistoric dwellers as early as Paleolithic and Neolithic times. Depicted here is a tropical-type fauna almost extinct today, documenting the presence of a people of hunters and herders.

22. A nocturnal panoramic view of the Great Pyramids at Giza in Egypt, a grandiose complex of three tombs built in 2600 B.C. of calcareous stone blocks to exalt the divine nature attributed to the pharaohs Cheops, Khafre, and Mycerinus. They are lined up from northeast to southwest in decreasing order of age and size, reflecting an architectural model based on an improvement of the old "step" pyramids.

23. Colonnade of the Luxor temple and a statue of Ramses II. Erected on the site of the ancient city of Thebes, the temple was built by Amenhotep III and Ramses II; because of its clear plane structure and unit design, it is considered to be the greatest monument of the New Kingdom. The temple contains chambers which become increasingly narrow as one proceeds inward, almost as if the designers wanted to protect the mystery of the sacred triad of Ammon, Mut, and Khonsu to which it was dedicated.

24. Broad view of Leptis Magna, showy testimony of the Roman presence in ancient Libya. The urban complex, in the style of the Emperor Augustus, was started in the 1st century A.D. by the Emperor Septimius Severus as part of an ambitious reorganization of the empire and its colonies, and completed by Caracalla. The forum, considered the largest in North Africa, is bounded on three sides by an arched portico containing a temple in Italian Romanesque style. Striking features are the high podium with a grandiose entrance staircase and eight red granite columns standing on white marble dadoes with sculptured scenes of battling giants.

25. The minaret of the Ibn Tulun Mosque, one of the best examples of Islamic architecture in Cairo, was built between 876 and 879. From the spiral-shaped top over a square design one looks down on the inner courtyard, where brick columns support horseshoe arches decorated with stucco carvings according to a special technique. In the Muslim countries, the minaret from which the muezzin calls the faithful to prayer is an important religious symbol as well as a typical feature of the urban landscape.

3

4

اللجان في كل مكان

• لا نيابة عن الشعب والتمثيل تضليل.
• المجلس النيابي حكم غيابي.
• المجالس النيابية تزييف للديمقراطية.
• الحزبية إجهاض للديمقراطية.
• من تحزب خان.
• لا ديمقراطية بدون مؤتمرات شعبية.
• اللجان في كل مكان.
• الديمقراطية رقابة الشعب على نفسه.
• الديمقراطية هي الحكم الشعبي
  وليست التعبير الشعبي.

من الكتاب الأخضر

21

22

SAHEL

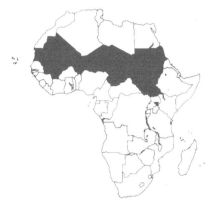

Africa offers more striking contrasts, in terms of its natural environment as well as its people, than perhaps any other part of the world. However, the transition between diametrically opposed conditions occurs gradually, across climatic and environmental zones.

The prevailing horizontality of the relief promotes a variability that is perfectly mirrored on either side of the Equator, producing a symmetry between the two opposite halves of the continent: North and South. This means that every environment tends to represent a transition to the ones next to it: such is the case with the savanna and the steppe, which frame the rain forest and the desert, respectively. In the latter case, since the Sahara extends from the Atlantic to the Red Sea, the steppe also stretches between these two opposite and distant shores (some 3700 mi [6,000 km]) like a rippling ribbon, shifting each year in response to changes in precipitation. Every year the rains fall, very briefly, in a long and torrid summer; sometimes they fail, and the environmental and economic repercussions of the consequent drought can be serious.

This broad and changeable strip of land (from 110–280 mi wide [180 to 450 km]), which receives an average of 10–24 in. [250–600 mm] of rain a year, was aptly referred to by the Arabs as the *Sahel,* or the "shore" of the Sahara. Geographically, it covers large portions of the nations that extend from the Atlantic to the Red Sea: northern Senegal, southern Mauritania, Mali, Niger, and Chad, the central part of the Sudan, and small portions of Burkina Faso and northern Ethiopia.

The Sahel is inhabited by a population of some 16 million, equivalent to about 15% of the total in the eight countries partially encompassed by this climatic zone. The countries most affected in demographic terms by the problems created by the Sahel are: Mauritania (with 40% of its population), Mali and the Sudan (with a quarter each), Chad (with a third), and Niger (no less than three quarters). Also worth emphasizing is the low proportion of urban population (only 15%) as compared with "rural" inhabitants—who include both sedentary farmers and nomadic herders (slightly more than a third of the population), practicing specific forms of transhumance. There is a certain interchange between the two ways of life, however, influenced both by climatic variability and by the breakdown, seen more and more frequently in the last few decades, in the balance between the natural resources of the Sahel (the soil covering of which is rather fragile) and the intensity of both grazing and cultivation.

The delicate nature of the environmental balances that characterize the Sahel is also demonstrated by the recurrent droughts which scourge this region, and which are by no means a peculiarity of our own time. According to several travelers' accounts, in the last century the Sahel was already being afflicted by prolonged droughts, lasting for up to 10 or 12 years. These were separated by periods of often heavy rains that caused flooding of major waterways such as the Senegal and Niger rivers, and even of Lake Chad itself, already characterized by major oscillations in its level and thus by a considerable variability in surface area (at present the area of the lake has been reduced to slightly more than a tenth of normal). Our own century has also not been without long phases of drought—with consequent famine, epidemics, loss of livestock, and migrations of people towards the wetter southern regions—that have occurred several times, but especially during the periods 1910–1916 and 1940–1949. The last drought period, which affected the Sahel during the ten years between 1965 and 1974, was particularly severe in 1972–73, intensifying to the point of a true catastrophe that again had grave consequences: in addition to drying up numerous wells and lake basins, it caused the degradation and disappearance of grazing pastures and the scrub vegetation that is also used to feed livestock. This produced a substantial drop in both animal herds and agricultural yields. The repercussions on society, health, and the economy were so severe as to stimulate direct intervention by international organizations, not to mention certain individual wealthy countries. These sources supplied emergency aid, and established projects designed to restructure traditional economic activities and make sensible use of the potential of natural resources available to the countries of the Sahel.

This section will discuss Mauritania, Mali, Niger, Chad, and the Sudan; Senegal and Burkina Faso will be considered together with the nations around the Gulf of Guinea (western Africa), and Ethiopia with those of eastern Africa.

CHAD

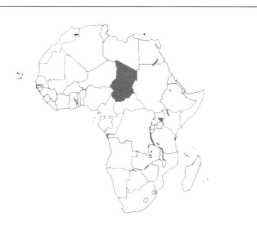

Geopolitical summary

Official name	République du Tchad/Jumhuryat Tashad
Area	495,624 mi² [1,284,000 km²]
Population	4,030,000 (1975 census); 5,428,000 (1988 estimate)
Form of government	Presidential republic with National Consultative Assembly consisting of a single party
Administrative structure	14 prefectures
Capital	Ndjamena (pop. 594,700, 1988 estimate)
International relations	Member of UN and OAU; associate member of EC
Official language	French and Arabic; also widespread Sudanese languages
Religion	Muslim 50%, animist 44%, Catholic and Protestant 6%
Currency	CFA franc

Natural environment. The country, which extends from the Tropic of Cancer to 7°30' N latitude, almost at the center of northern Africa, is bordered on the north by Libya, on the east by the Sudan, on the south by the Central African Republic, and on the west by Cameroon, Nigeria, and Niger.

Geological structure and relief. From a geomorphological standpoint, Chad is quite unusual. The northwestern sector contains the mighty Tibesti massif, dominated by the great volcanic heights of Emi Koussi, which culminate at an altitude of 11,205 ft [3415 m]. The site of intensive volcanic activity during the Pliocene and Quaternary, with outflows of rhyolitic and basaltic lava, the entire plateau—which sits on a basement of Paleozoic sandstone that in turn rests on a Precambrian crystalline shelf—is studded with volcanic outcrops that extend to considerable heights, like Pic Touside (10,712 ft [3265 m]) and Tarso Emisou (11,075 ft [3376 m]), creating in the middle of the Sahara desert

a highly evocative environment rich in geomorphological contrasts. The eastern section of the country, up to the border with the Sudan, consists of vast high tablelands cut into sandstones of the Paleozoic age which rise to impressive heights—such as Mt. Ennedi (4300 ft [1310 m])—and are dissected by deep valleys. Farther south the Precambrian crystalline basement rock is directly exposed; it too rises to considerable elevations, for example on the Guéra massif (5900 ft [1800 m]). Between this and the Tibesti massif in the center of the country, the topography drops considerably in the Bodélé, a wide basin (originally a lake), that was once connected via the Bahr el Ghazal to Lake Chad, of which it constituted an extension.

Hydrography. The southwestern end of the country is occupied by the Lake Chad basin, fed predominantly by the Chari river (1000 mi [1600 km]) which rises in the central spine of Africa, and by its left-bank tributary, the Logone, whose spring-fed branches flow off the high plateau of Adamaoua. The Chari also receives water on its left bank from the Ouham, while on the right it is fed by the Bahr Kéita and the Bahr Salamat, which flow from the Darfur massif (Sudan). Also descending from these mountains is the Batha, which flows into the bed of Lake Fitri. The average depth of Lake Chad is only 10–12 ft, and its area varies from 3900–6200 mi² [10,000 to 16,000 km²] due to irregularities in inflow, evaporation, and the quantity of water withdrawn for agricultural purposes.

Climate. Climatic conditions in Chad vary considerably from region to region. The humid tropical conditions of the southern regions, characterized by two seasons (a dry winter and wet summer), with rainfall that can reach 47 in. [1200 mm] per year at Moundou and Sarh, merge into the tropical desert climate of the northern regions, with a wide daily temperature range and little precipitation. Proceeding from south to north, the wet season becomes increasingly shorter (only 2.5 in. [60 mm] of rain at Faya-Largeau) and finally disappears. The dry season, dominated by the typical Sahara wind called the *harmattan*, becomes increasingly hot: at Abéché (altitude 2000 ft [600 m] asl), maximum temperatures in April can reach 113°F [45°C].

Flora and fauna. The differences in climate are matched by similar changes in natural vegetation, the limits of which reveal a shift to the south as the desert gradually advances. In southern Chad, at about 10°N latitude, the vegetation is of the wooded savanna type, with baobab, karité, and acacia trees, while the shores of Lake Chad are covered with papyrus and aquatic plants. The central sector of Chad corresponds to the Sahel, with steppe vegetation consisting primarily of grasses and thorny scrub. The fauna is abundant and varied, especially in the savanna, which is home to elephant, giraffe, lion, and leopard; antelope and gazelle live in the desert and the Sahel. Two large nature preserves currently exist, Zakouma and Sinianka-Minia, in which the large African animals are well protected.

Climate data

Location	Altitude (ft asl)	Average temp. (°F) January	Average temp. (°F) July	Average annual precip. (in.)
Ndjamena	965	75	82	29.8
Abéché	1788	80	84	20.2

Conversion factors: 1 ft = 0.3 m; 1 in. = 25 mm; °C = (°F − 32) × 5/9

Administrative structure

Administrative unit (prefectures)	Area (mi^2)	Population (1988 census)
Batha	34,277	431,000
Biltine	18,084	216,000
Borkou-Ennedi-Tibesti	231,735	109,000
Chari-Baguirmi	32,003	844,000
Guéra	22,755	254,000
Kanem	44,205	245,000
Lac	8,616	165,000
Logone Occidental	3,356	365,000
Logone Oriental	10,822	377,000
Mayo-Kébbi	11,621	852,000
Moyen-Chari	17,439	646,000
Ouaddaï	29,429	422,000
Salamat	24,318	131,000
Tandjilé	6,965	371,000
CHAD	497,624	5,428,000

Capital: Ndjamena, pop. 594,700 (1988 estimate)

Major cities (1988 estimate): Sarh, pop. 113,400; Moundou, pop. 102,800; Abéché, pop. 83,000.

Conversion factor: 1 mi^2 = 2.59 km^2

Population. Chad has a very diverse ethnic makeup; north of the 15° parallel, in the regions of Tibesti, Borkou, and Ennedi, live the Tebu, descendants of "black" populations and Arab strains, strongly influenced by Islam who once lived by caravan activities and now are simply nomadic herders. The largest ethnic group, the Sara, live in the southern regions between the Logone and the Chari; Sudanese in origin, they live by farming and fishing and are of either animist or Christian faith. Other dark-skinned Sudanese groups are the Massa, who live along the Logone, and the Mundang in the Bongor region. The Arabs and the Fulbe, who represent the "white" population, occupy the Sahel zone and are nomads who keep cattle and camels.

In the 1950s, Chad had slightly more than 2 million inhabitants; in thirty years this figure has risen to more than 5 million, at an annual rate of increase of 2.5–3%. The average population density is very low (11 per mi^2 [4 per km^2]), and is distributed unevenly over the various regions of the country. The most densely populated areas are in the south; they occupy only one quarter of the territory, at densities which in certain regions (such as the Chari and Logone valleys) exceed 78 inhabitants per mi^2 [30 per km^2]. The least populated region is the northern Sahara area, where the population is concentrated in and around the oases, especially in the northeastern region. Urban centers consist of small cities which contain only about 31% of the total population; the majority of people lives in villages or in circular mud huts. Only the major centers of Ndjamena, Moundou, Sarh, and Abéché currently have more than 50,000 inhabitants. The capital, Ndjamena (known as Fort-Lamy in French colonial times), extends along the right bank of the Chari, and Chad's few industrial activities are concentrated there; the city is an important commercial center, with a river port and an international airport. Moundou, located on the Logone river, is the capital of the southern agricultural district, and an important

crossroads. Sarh (colonial Fort-Archambault), located on the upper Chari, was well known as a jumping-off point for big-game hunts. Abéché, located in the eastern part of the country, is the former capital of the Muslim sultanate of Ouaddaï, and still houses a very old Islamic school. The city is a commercial center and livestock market, particularly for sheep and camels.

Chad's population is growing at an average annual rate of about 2.5%, but the high mortality rate (2.2%) compensates for a birth rate of 4.2%. Chad is one of the world's poorest countries: its unfortunate geographical location, arid climate, and prolonged droughts are aggravated by an unstable political situation resulting from the long civil war. Illiteracy is still very widespread, afflicting 82.2% of the population. The school system did not exist prior to independence; it comprises six years of compulsory primary education, and higher education at the University of Ndjamena and other technical colleges.

Economic summary. Chad's economy is extremely impoverished, suffering not only from the prolonged drought, but also from the consequences of the civil war, which has seen enormous financial resources diverted to military ends (approximately 5% of gross national product). The ethnic conflicts between northern populations—seminomadic, Arab, and Muslim—and the Christian or animist farmers who live in the southern regions, date back to the colonial period. The French deported laborers, particularly from the southern populations, to work on plantations in equatorial Africa, and this exploitation continued in other ways even after independence.

There is symbolic value in the fate of this Sahelian country, of which Eugenio Turri has written:

...plans were made for the future of the country, to which France had promised technical, financial, and military assistance. Wells were being dug, especially in the Sahel, to help develop stock raising, one of the country's fundamental activities. On this wave of enthusiasm, there was talk of the future Chad as a new Arizona, capable of providing meat to Africa and Europe. The development models followed were those of the West, where a number of 'miracles' were in progress, based on an economic policy that, according to prevailing ideologies, could be transferred unchanged to the Third World. The problems of development took precedence over everything else, giving rise to a philosophy which in the long run proved damaging, and for which the nations of the Sahel are still paying the price today. Chad's economic policy in the Sahel focused on 'developing' pastoralism in the commercial sense of the word, while in the more wealthy south, where paysannats had been created as a form of village-level cooperation, the idea was to expand the growing of cotton. Fundamentally, the intent was to supply goods for export and thereby replenish the government's coffers, which more than ever needed money to launch all the development programs. Then a lot of other agricultural and mining development projects were set up. The protagonists of this movement to construct a modern state were young Africans with European training, educated in Paris or at the missionary schools, impelled by a great deal of faith but tied to political and tribal factions that threatened the existence of the new state. What then happened in this unfortunate country, torn to pieces by twenty years of conflict and civil war, is a sad chapter in the unfinished history of independent Africa.... The independence of Chad became a shattered dream....

Socioeconomic data

Income (per capita, US$)	207 (1990)
Population growth rate (% per year)	2.4 (1980–89)
Birth rate (annual, per 1,000 pop.)	44 (1989)
Mortality rate (annual, per 1,000 pop.)	19 (1989)
Life expectancy at birth (years)	47 (1989)
Urban population (% of total)	29 (1989)
Economically active population (% of total)	35.1 (1988)
Illiteracy (% of total pop.)	82.2 (1991)
Available nutrition (daily calories per capita)	1852 (1989)
Energy consumption (10^6 tons coal equivalent)	0.1 (1987)

Agriculture and livestock. The predominant activities in Chad are agriculture and stock raising, which engage 60% of the economically active population and provide half the country's income but are still pursued with archaic methods that do not meet nutritional needs. International financial assistance, provided mostly by France, has helped rebuild domestic animal stocks that were destroyed by the drought. Of primary economic importance is cotton as a single-crop culture, which was introduced by the French to the southern regions along with rice, and is the biggest export.

Land suitable for cultivation and plantations accounts for only 2.5% of the land area, while meadows and pastures, covering 35%, are poor and arid. The principal agricultural region is the southern section of the country, defined by the Chari river and the borders with Cameroon and the Central African Republic. The main products available to feed the population are millet, corn, wheat, rice, sesame, sweet potatoes, manioc, dates, and beans. Cotton seed and cotton fiber, sugar, and peanuts are grown for export. Stock raising is practiced in the central region of the country located between Lake Chad and the Ouaddaï, and is based on modest stocks of cattle, horses, donkeys, sheep, and camels.

Energy resources and industry. Mineral resources are practically nonexistent. The only mineral produced in modest quantities is salt, which is extracted from the area north of Lake Chad. Deposits of tin have been located at Mayo Kébbi, with uranium, petroleum, lead, iron, and bauxite in the southern region. Electrical energy is produced exclusively from imported oil: installed power is 31,000 kW (1989), and production reached 52 million kWh (1989) at the Sarh, Moundou, and Ndjamena power stations. Industrial production contributed 14% to the gross national product (US$64 million in 1986), and employs only 10% of the working population (35%). The most important sector involves the processing of agricultural and animal products, such as sugar, vegetable oils, beer, cotton textiles, and meat. The two cities with some industrial activity are Ndjamena and Moundou. Fishing takes place in Lake Chad and along the two major rivers. Some 30% of the working population is employed in this sector, and it produces 32% of the gross national product.

Commerce and communications. Along with a meager internal commercial network, there is also an international trade structure. The balance of payments is overwhelmingly negative: the value of exports is not even half the value of goods imported. France is the principal supplier of food and transport equipment, Nigeria supplies petroleum, and Cameroon provides some minerals. Exports include cotton (8.7%) and meat (6.4%), and go to Nigeria and France. The transportation network is poor: there are no railroads, and the total length of all-weather roads is less than 1860 mi [3000 km]. Air traffic is based at the international airport of Ndjamena and 12 regional airports.

Historical and cultural profile. *Ancient Chad.* During the last Ice Age, Lake Chad occupied a much larger area than it does now; the surrounding region was covered with vegetation, and the climate favored human settlement. North of the lake, rock engravings from the 9th millennium B.C. attest to the presence of populations similar to modern Bushmen. Gradually increasing aridity was accompanied by migratory movements, leading to fusion between the original populations and the "newcomers" of Negroid stock. Nonetheless, it has proved difficult to follow the development, diffusion, and migrations of the various peoples. To find a unified political structure in the regions around the lake, we must move forward to the time of the kingdom of Sao, which flourished between the 8th and 12th centuries A.D. Thereafter, these areas were part of a number of kingdoms of the central Sudan: Kanem-Bornu, Ouaddaï, Baguirmi, and Darfur. These were caravan states, Islamized to various degrees, which grew rich by trading in slaves and through that activity became conduits between Arab Africa and black Africa. They often fought among themselves over territory, and it was only in the second half of the 19th century that the regions corresponding to modern Chad, Nigeria, and Cameroon were unified under the rule of the Egyptian slave-trader Rabah. In *L'enigma Ciad* [The Enigma of Chad], Gian Carlo Costadoni has written this about the kingdom or sultanate of Kanem-Bornu:

> *We do not know which populations founded the kingdom. One hypothesis is that they were Tubu (a seminomadic people who now live in Tibesti), together with other small northern Sudanese ethnic groups; or perhaps they were Berbers driven south by the Muslim invasion. Undoubtedly they were a northern, non-Negro people. They were nomads and herders, traders, marauders, and animists: they conquered the pre-existing black population and created a centralized society. In the first years of the 10th century, Muslim proselytization began to penetrate into Kanem, and with it came Arab influence. Starting in the 11th century the first known dynasty, called the Saif, which emerged from an original council of the heads of powerful families, imposed unity and peace on this mosaic of peoples, cultures, religions, and social systems that formed today's Chad; it reigned without interruption until 1846.*
>
> *Theirs was a primitive form of central government: a feudal monarchy culminating in a sultan who was revered as a god. The council of state consisted of 12 princes, each with territorial and administrative authority. The royal finances were fed by taxes in cash and in kind, and by the slave trade. Islamic justice was confined to the major cities, and was in no position to replace customary law.*

In the meantime the Europeans had been advancing, first as explorers fascinated by the mystery of the "inland sea of Chad," but soon as leaders of occupation forces.

Modern Chad. From 1895 to 1913, France controlled the territory of Chad. In 1920 it became part of French Equatorial Africa, and ten years later the northern Tibesti region was annexed. Another decade later, Chad became the first colony to join the cause of De Gaulle's Free France, for which it then served as a base for military actions on Libyan soil. The framework law of 1956 began the process of decolonization, which ended in 1960 with the proclamation of independence. From the start, the new nation suffered from severe instability, due to poor integration between the Muslim northern regions and the black south: this instability forced François Tombalbaye, president of the republic, into an authoritarian, pro-French policy.

It was French support which allowed the government to stand up against the FROLINAT (Front de Libération Nationale) guerrillas in the northern regions, who were backed by the Arab countries and especially by Libya. In 1979 FROLINAT gained power by naming as head of state Goukouni Oueddei, the major points of whose political program are reproduced below:

- *To fight by every available means against the neocolonialist regime imposed by France, which attempts to perpetuate domination, oppression, and massive exploitation.*
- *To defend territorial unity.*
- *To pursue neutralism as a policy, eliminating any foreign military presence from the territory.*
- *To install a democratic, popular, and progressive government which guarantees the basic freedoms of the individual (expression, assembly, press, labor organization, and religion).*
- *To institute radical agrarian reform by organizing the peasants into agricultural cooperatives.*
- *To nationalize key sectors of the national economy in order to build an independent economy.*
- *To create a democratic and progressive culture and educational system with a national character.*
- *To establish diplomatic relations with all countries, with the exception of Rhodesia, Israel, and South Africa, pursuing an independent national policy based on the 10 principles of the Bandung conference.*

But not even this could restore peace: in 1982 Oueddei was succeeded by Hissène Habré, and a year later civil war broke out again; Libya intervened directly, invading part of Chadian territory. After alternating military successes, the two nations reestablished relations in 1989.

Culture. In the colonial period, a number of archeological sites of the "Chad civilization" were discovered; this extremely vague term covers everything from prehistoric finds to the Sao period. Remains of this culture include various kinds of pottery with relief decoration, and figurines of ancestors, masked dancers, and animals. More interesting, however, are decorative objects made of bronze or copper, cast using the lost-wax method. The old craft traditions have survived down to modern times at a high artistic level, as shown by the jewelry made by the Kotoko tribe. As far as literature is concerned, there exists in Chad an ancient wealth of orally transmitted texts, including songs of love and war, incantations, stories and legends, proverbs and riddles. There are also traveling poets, called *dugu* in Kanem and *anbanin* in Ouaddaï, as well as storytellers associated with village chieftains, similar to the Senegalese *griots*. A first attempt at a national literature in French was made in 1962 with *Au Tchad sous les étoiles* [Chad under the stars], by Joseph Brahim Said.

MALI

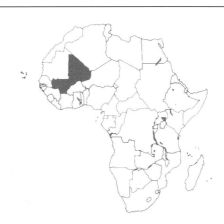

Geopolitical summary

Official name	République du Mali
Area	478,714 mi^2 [1,240,192 km^2]
Population	7,620,225 (1987 census); 8,156,000 (1990 estimate)
Form of government	Presidential republic with a one-party National Assembly
Administrative structure	Seven regions, plus the capital district
Capital	Bamako (pop. 646,163, 1987 census)
International relations	Member of UN and OAU; associate member of EC
Official language	French; Arabic is also widely spoken, along with Hamitic languages (Berber) and Sudanese languages (Mande)
Religion	Muslim 90%, animist 9%; approx. 100,000 Catholics and Protestants
Currency	CFA franc

Natural environment. Mali is located in the interior of northwest Africa between the Tropic of Cancer and the Equatorial belt; its northern region embraces a huge stretch of the Sahara, while the south-central part of the country contains the wide valley of the Niger river and its tributaries. It is bordered on the north by Algeria, the east by Niger, the southeast by Burkina Faso, the south by Ivory Coast and Guinea, and the west by Senegal and Mauritania.

Geological structure and relief. The topographical framework consists of fairly low upland plains formed by recent sedimentary rocks, overlying an older crystalline basement. The northern region is covered by large plateaus about a thousand feet high, culminating in the Adrar des Iforas mountain range; to the west, the massif is bounded by the deep furrow of the Tilemsi valley, beyond which the crystalline Timetrine plateau fades into the broad depression of the Niger valley. Protruding into the southwestern region are the northern edges of the Guinea massif, with Mt. Mina (2690 ft [820 m]), on whose eastern slope the Black Volta river rises, the Kéniéba Mountains, and the Bambouk plain, none of which exceeds 1970 ft [600 m] in elevation.

Hydrography. The hydrography of Mali is well developed only in the southern region, which contains 1115 mi [1800 km] of the course of the river Niger, and the upper reaches of the Senegal river. The Niger, which like the Senegal rises in the Fouta Djallon massif in Guinea, enters Mali with a southwest-northeast orientation, then makes a wide marshy loop before changing direction to the southeast. Three segments can be distinguished, each characterized by different flooding behavior: an upper reach down to Ségou, the great inland delta of the Macina and Mopti regions, and the stretch downstream from the delta.

The huge area watered by the inland delta represents a fortunate exception in the arid Sahel region, and allows considerable agricultural and stockraising activity. Only 280 mi [450 km] of the course of the Senegal river lies within Mali; its principal tributary is the Falémé, at the border with the nation of Senegal.

Climate. From a climatic point of view, the territory of Mali is characterized by three distinct situations: in the northern region beyond the Sahel fringe, a desert climate prevails: temperatures are very high in summer and during the day (from 77 to 126°F [25–52°C]), and relatively low in winter and at night (from 32 to 41°F [0–5°C]), and there is generally no precipitation. The central Sahel region receives more than 9 in. [250 mm] of rain each year, while in the southern and southwestern areas, annual average rainfall exceeds 24 in. [600 mm], and the rainy season starts at the end of April and lasts through October. Temperatures are generally high throughout the country. Temperature ranges become more extreme as one moves toward the more continental areas: average values for minimum temperature in January are 63°F [17°C] at Bamako and 57°F [14°C] at Gao, while maximum May temperatures are 102°F [39°C] and 109°F [43°C], respectively. Throughout the year, the northeast trade wind dominates the northern region; the *harmattan* blows consistently from November to April (dry season).

Flora and fauna. The vegetation in the Sudanese area is dominated by wooded savanna with plentiful tall grasses, majestic baobabs and fruit trees; this alternates with residual patches of tropical forest and fringing forests along the rivers. Steppe vegetation is extremely common in the Sahel zone: here the tall grasses are replaced by thorny *Graminaceae* grasses, and by giant euphorbias and acacias.

Fauna is abundant and varied: the savanna and the edges of the tropical forests are inhabited by monkeys, ruminants, felines, and large reptiles. Animals similar to those of the savanna can also be found in the Sahel region. The waterways are home to crocodiles, hippopotamuses, and many species of fish. The Baoulé National Park, southeast of Bamako, provides protection for an enormous wealth of animals.

Population. The principal ethnic groups which make up the population of Mali today belong predominantly to Sudanese races. These include the Bambara, settled in the southern regions, who represent the most numerous and most developed group and live almost exclusively by herding; the Songhai, Muslim farmers dwelling on the banks of the river Niger, who created around the city of Gao the first nucleus of an empire that bore their name; and the Fulbe, a pastoral people inhabiting the inland Niger delta region. Minority groups include the Dogon of the Bandiagara plateau, the Volta branches of the Bobo, Senufo, and Minianka tribes, and the Mauri and Tuareg in the Sahara. Other natives of Mali include the Mandingo tribes, living in the area of the southwest upland plains; with their glorious past, they were much in demand in the slave-trading era for their physical and intellectual capabilities.

Mali was one of the earliest areas of West Africa settled by humans. With some eight million inhabitants and an average population density of about 18 per mi^2 [7 per km^2], the country may be considered sparsely populated. Population distribution is highly differentiated: the densest areas are those in which precipitation is highest (more than 90% of the population lives south of the 10-in. [250-mm] isohyet). The south-central region through which the Niger river flows has the highest concentration, with areas of 260 per mi^2 [100 per km^2], while the Sahara region in the north is practically uninhabited (3 per mi^2 [1 per km^2]). The cities and major centers, located along the principal rivers, contain 19% of the population. Bamako, which became the capital upon Mali's independence in 1960, is located on the banks of the Niger and is an important communication nexus; it is the terminus of the railroad line linking it to the port of Dakar (800 mi [1300 km] away), and has an international airport. The other large cities are located downriver: Ségou, administrative capital of the delta and headquarters of the Niger Authority; farther downstream the town of Mopti is the most important commercial center in the Macina region; and Timbuktu, north of the Niger, rich in architectural history, was one of the most important centers of Islamic life and a focus for the caravan trade between black Africa and the Mediterranean.

In *Io Africa* [I Africa], Folco Quilici has written about the past greatness of Timbuktu:

> *In the dilapidated and almost deserted mosques, it is difficult to imagine the glory of Timbuktu seven centuries ago. Difficult but not impossible: everyone in the city lives with a memory of its past greatness, even though now the place is poor, isolated, and in ruins.*
>
> *"These mosques of ours," one man told me, "are famous in our Muslim history; the one that you are photographing was the University of Samkorè, a place of learning as well as the house of God.... Famous teachers taught here; they came from Spain, from Syria, and from Morocco ... the libraries in our city were so rich in books that Timbuktu was called the capital of Islamic thought and culture. But today, wherever we look, we see nothing of that greatness...."*

Approximately 250 mi [400 km] from Timbuktu is Gao, ter-

Climate data				
Location	Altitude (ft asl)	Average temp. (°F) January	Average temp. (°F) July	Average annual precip. (in.)
Bamako	1600	76	80	43.7
Kayes	157	76	93	28.3
Timbuktu	987	71	90	9.0
Conversion factors: 1 ft = 0.3 m; 1 in. = 25 mm; °C = (°F − 32) × 5/9				

Administrative structure

Administrative unit (regions)	Area (mi²)	Population (1987 census)
Gao	124,290	383,734
Kayes	76,335	1,058,575
Koulikoro	34,676	1,180,260
Mopti	34,258	1,261,383
Ségou	21,650	1,328,250
Sikasso	29,521	1,308,828
Timbuktu	157,865	453,032
Bamako (district)	103	646,153
MALI	478,839	7,620,225

Capital: Bamako, pop. 646,163 (1987 census)
Major cities (1984 estimate): Ségou, pop. 99,000; Mopti, pop. 78,000; Gao, pop. 55,000; Timbuktu, pop. 20,483 (1976 census).

Conversion factor: 1 mi² = 2.59 km²

minus for water traffic on the inland delta, and former capital of the Songhai kingdom.

Mali's population has more than doubled since independence, achieving annual population increase rates of 2–2.5%; nonetheless the mortality rate has remained high due to the persistence of infectious diseases and malaria (the latter especially in the Niger delta area). The persistent drought of the last few years, particularly in the Sahel belt, with its consequent adverse effect on nutrition, has placed a limit on population growth and is impelling younger Malians to emigrate to Ivory Coast, Niger, and Senegal. Mali is one of the poorest countries in the world, with enormous foreign debt and an annual per-capita income of US$302 (1989); the very high illiteracy rate (89.9%) also contributes to extremely slow cultural progress among the population.

Economic summary. Mali's economic structure is extremely weak, hampered by an unfavorable physical environment and a dearth of natural resources. The absence of communication routes constitutes another serious obstacle to development of the Malian economy. Almost two thirds of the nation's territory is occupied by desert, with recurring periods of intensified drought in the Sahel. Only the great Niger valley, with its alluvial plains regularly covered by floods, could sustain intensive agriculture. In the Macina region (the Niger's inland delta), large-scale cotton and peanut cultivation was developed in the colonial period, and the Niger Authority has been created to implement a major agro-industrial project.

Agriculture and livestock. Mali's production structure is based on agriculture, which employs more than 73% of the working population on cultivated land that represents less than 2% of the country's area. Millet, rice, and sorghum are the main crops, followed by peanuts and cotton (the latter intended for limited export and often raised to the detriment of subsistence crops). Cereals such as corn and wheat are grown to a lesser extent, along with manioc, sugar cane, vegetables, and fruit. Livestock represent the country's second most important economic resource, with an impressive number of head (almost 5 million cattle and approximately 12 million sheep and goats).

Fish in the inland waterways also constitute a considerable resource, while forests occupy only 5.6% of the land area.

Energy resources and industry. Mining is not significant, representing a small part of gross national product. Some newly discovered resources are still difficult to exploit because there is no infrastructure. Bauxite is present in the southeastern region near Bamako, and on the Mandingo plateau, and there is iron near Kita and Bafoulabé; limited quantities of other minerals also exist. Gold and diamonds support a vigorous trade, much of it illicit. The only mineral wealth is salt extracted in the Taoudenni area (Sahara), which moves to the Gulf of Guinea in camel caravans. Development of mining activities is impossible without hydroelectric plants, but although plans for such installations exist, there are no resources to build them. Major dams along the Niger are already operating at Selingue and Sotuba, and dam projects along the Senegal river are planned.

Industrial activity is still modest, employing 7% of the total working population and accounting for 13% of gross national product. The major activities involve processing local agricultural products (especially cotton), animal products (hides and wool), and fish (canning fish caught in inland waters). Also present are facilities producing consumer goods, with textile and food plants in the Bamako region and in the Niger Authority district, sugar factories at Dougabougou and Seribala, cement plants operating at Diamou, and shoe-making plants and tobacco factories in Bamako.

Commerce and communications. Communication is difficult due to the lack of an adequate transportation system capable of linking Mali to major foreign commercial centers. The only railroad, which connects the country to the port of Dakar, is 400 mi [646 km] long (1987). Total length of the all-weather road system is approximately 8700 mi [14,000 km] (1989). The principal communications axis, therefore, is represented by the Niger, Senegal, and Bani rivers, which are navigable for several months of the year. The trade balance is heavily negative, since export volume is completely unable to cover the massive importation of products for domestic needs. Tourism is being developed in connection with the history of the region, and the ancient cities of Timbuktu and Djenné.

Historical and cultural profile. *The Kingdom of Mali.* When it achieved independence, the former colony of French Sudan decided to take the name Mali, signifying an idealized continuity between the new nation and the powerful African empire of that name, whose first nucleus was located within its

Socioeconomic data

Income (per capita, US$)	302 (1990)
Population growth rate (% per year)	3.5 (1980-89)
Birth rate (annual, per 1,000 pop.)	50 (1989)
Mortality rate (annual, per 1,000 pop.)	19 (1989)
Life expectancy at birth (years)	48 (1989)
Urban population (% of total)	19 (1989)
Economically active population (% of total)	31.8 (1988)
Illiteracy (% of total pop.)	89.9 (1991)
Available nutrition (daily calories per capita)	2181 (1989)
Energy consumption (10^6 tons coal equivalent)	0.2 (1987)

borders (at the village of Kangaba, between the "black river" and the "yellow river"). This was the home of the Keita clan, who had converted to Islam before the year 1000, and, with the support of the Almoravids, had created a small Mandingo kingdom. The first great king of this dynasty was Sundiata (1218–1255), a semi-legendary figure who conquered the earlier empire of Ghana. After him, the little kingdom was transformed in slightly more than fifty years into a huge empire: Kankan Mussa (1307–1332) ruled from the middle Senegal to the middle Niger to the Gambia basin; he moved the capital to Timbuktu and encouraged commerce, the arts, and architecture. But soon the downward slide began for the Mali empire, ending in the 17th century when the Keita were left with nothing but their original village of Kangaba, where their ascendancy had begun. The decline of the Mali empire did not, however, mean the end of Sudanese civilization. For centuries, the Songhai kingdom had been flourishing around the city of Gao; in the 1500s it stretched from the northern borders of modern Nigeria to the southern frontier of Morocco. The Songhai conquered Timbuktu in 1546, and made it into a center of Islamic culture that was famous as far away as Egypt for its erudition in theology, law, and grammar.

The historian Robert E. Lovejoy has written:

> *From the end of the 15th century to the collapse in 1592, the Songhai succeeded in annexing a larger geographical region than any previous state. Their success would not be repeated until the 19th century. From a political standpoint, and perhaps also in terms of economic prosperity, the western Sahel reached its zenith under the reign of the Songhai.*

The Songhai empire was followed, in the 18th and 19th centuries, by kingdoms established by the Bambara at Ségou and Kaarta, the Fulbe in the Macina region, and the Mossi in Yatenga, all of them demonstrating the vitality of the Sudanese region. The real decline began later, in the second half of the last century, under the combined pressure of invasions by the Tacruri from Senegal, and French colonial expansion. Again starting off from Senegal, the French, led by Faidherbe, launched their expeditions towards the interior of the continent, and by 1895 had extended their hegemony to the river Niger.

Independence. The colonies that were gradually established consisted of untidy ethnic mosaics referred to at various times as Senegambia-Niger (1902), Upper Senegal-Niger (1904), Upper Volta (1919), and finally French Sudan (1920). It was indeed the Sudan that played a leading role in the fight for independence by France's possessions in western Africa: in 1946 the Rassemblement Démocratique Africain (RDA) was established in Bamako, led by the Union Démocratique Soudanaise (UDS) under Mobido Keita. Independence was gained in 1960, after which the country renamed itself the Republic of Mali, with Keita as president. In a speech given in London on June 7, 1961, Keita explained his nation's foreign policy as follows:

> *I must make it clear that our position in international affairs is one of positive neutrality, which absolutely does not mean a balancing act.... A tightrope-walker policy makes a country lose its personality, since it can be blackmailed by both blocs and becomes simply an instrument to be utilized. In our opinion this is dangerous both for the country itself and for the Great Powers.*

An initially socialist orientation changed over the years to increasingly authoritarian positions, due to the difficulty of implementing a large-scale socioeconomic reform policy throughout the country. A military coup in 1968 brought to power a Military Committee of National Liberation which nevertheless followed the same road as the previous regime.

Culture. Like that of black Africa in general, Sudanese art goes back to the period before European colonization, an experience which sapped the life from earlier social and religious systems and thereby threatened the very foundations of artistic inspiration. The major form of expression today is wooden sculpture prepared for ritual or funerary purposes; recurring subjects include human figures performing propitiatory acts, ancestor statues, and symbolic animals.

Some groups, like the Dogon and Bambara, achieved in these figures levels of stylization and sculptural vigor whose influence on the artistic avant-garde of Europe's 19th century is easily understood. Equally successful are the dance masks and a series of household objects reserved for a chief's personal and household use (staffs, stools, food containers, etc.), in which realistic decoration approaches the abstract. The oral literature of the Sudanese region, transmitted by the *griots*, includes legends, fantastical tales, and historical stories which, in the case of the Bambara and the Malinke, have been reported in the research of the German Africanist Leo Frobenius (1873–1938). Here is one of the many stories transmitted by the *griots* (literally "rememberers") that speaks of the ancient deeds of the Mandingo:

> *I am a griot. I am Dieli Mamadu Kuiaté, son of Bintu Kuiaté and Dieli Kedian Kuiaté, master of the art of speaking. From time immemorial the Kuiaté have served the Keita princes of the Mandingo: we are the scribes of the spoken word, we are the scribes who keep secrets of many centuries. The art of speaking holds no mysteries for us. Without us the names of the kings would fall into oblivion, we are the memory of the people; with our words we give life to the deeds and doings of the kings before the younger generations.*
>
> *I learned everything I know from my father Dieli Kedian, who in turn learned it from his father. History keeps no secrets from us. We teach to the profane only that which we wish to teach them; it is we who hold the keys to the twelve gates of the Mandingo.*
>
> *I know the list of all the rulers who have acceded to the throne of the Mandingo. I know how the black men are divided into tribes, because my father passed on to me all his knowledge.*
>
> *I know why one man is called Kamara, another Keita, yet another Sibidé or Traoré; every name has a meaning, a secret significance. I have taught kings the history of their ancestors so that the life of the Ancients might serve as an example to them, because the world is old, but the future comes from the past.*
>
> *My spoken words are pure and free of any lies; they are the words of my father, and the words of my father's father. I shall speak to you the words of my father exactly as I received them; the griots of the king know no lies. When a dispute arises among the tribes, it is we who put an end to all quarrels, because we are the guardians of the oaths that our ancestors spoke.*

The nucleus of an indigenous theater, of obscure origins, is still perceptible today in many ritual and humorous performances.

MAURITANIA

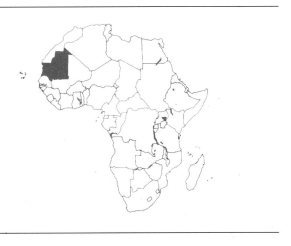

Geopolitical summary

Official name	Al-Jumhuriya al-Islamiya al-Muritaniya
Area	397,850 mi² [1,030,700 km²]
Population	1,803,193 (1988 census)
Form of government	Presidential republic under military control, ruled by a Military Committee for National Salvation
Administrative structure	12 regions plus the capital district
Capital	Nouakchott (pop. 393,325, 1988 census)
International relations	Member of UN, OAU, Arab League; associate member of EC
Official language	French and Arabic
Religion	Islam; several thousand Catholics
Currency	Ouguiya

Natural environment. Mauritania's geomorphology is predominantly governed by the presence of a large area of the Sahara desert, which occupies most of the country and extends to the Atlantic coast, and by the fairly homogeneous topography (maximum elevations rarely exceed 1640 ft [500 m]) characterized by extensive areas of plateau or gently rolling hills. Only in the southernmost sector does the presence of the final stretch of the Senegal river, which allows a certain amount of vegetation to develop, give the landscape a less desolate appearance. Traversed by the Tropic of Cancer, Mauritania faces the Atlantic Ocean on the west and is bordered on the southwest by Senegal, on the south and east by Mali, on the northeast by Algeria, and on the northwest by Western Sahara, in almost every instance along purely arbitrary lines; only the border with Senegal follows the natural line of the river of the same name.

Geological structure, relief, and hydrography. From a geological point of view, the country is characterized, in the farthest inland areas, by outcroppings of a Precambrian base consisting of highly metamorphic rocks (gneiss, amphibolites, and mica schists) with massive granite intrusions, covered by sedimentary strata of Paleozoic age—such as the sandstone plateau of Adrar (average elevation 1640 ft [500 m]) and the lower Tagant massif—which border the numerous Sahara depressions. The Mauritanian desert is primarily sandy; its landscape is dominated by *ergs*, enormous sand dunes oriented northeast and southwest that are continually being shifted by the *harmattan* winds of the Sahara. The only relief of any significance is Kedia Idjil (3007 ft [917 m]), located in an iron-rich region. Towards the south the desert becomes less arid, merging into the steppes of the Sahel, where the vegetation becomes more lush near the Senegal river and in the Chemama region. The coastline is 465 mi [750 km] long, barren in the northern portion, and entirely straight with no harbors; there is a single broad bay between Cape Blanc and Cape Timiris.

The Saharan environment is typified by the presence of widely scattered *uidian* (wadis), which fill with water during the rainy period and dry out in a few hours, and temporary ponds, called *grara* if they quickly vanish and *guelta* if they persist for several months. Water from the Senegal is used for irrigation, and its flow is kept steady by two lake basins: Lake Rkiz in the north, and Lake Guier in Senegal itself.

Climate. The climate of the Sahara region is obviously characterized by extremely arid, desert conditions, with annual precipitation of less than 4 in. [100 mm] and relative humidity of 40–50%, with minimums of 10% in summer. At Nouadhibou only 0.2 in. [5 mm] of rain falls all year, concentrated in four days; in the capital, total rainfall amounts to 1.9 in. [48 mm] over 12 days of unsettled weather (at Atar, farther inland, annual precipitation is less than 1.8 in. [46 mm] over 10 days of rain). Temperatures are not as high on the coast, where the climate is sub-Saharan (average temperatures at Nouadhibou are 67.6°F [19.8°C] in January and 73.5°F [23.1°C] in July, and 72.8°F [22.7°C] and 79.3°F [26.3°C], respectively, at Nouakchott). In the interior, with its more markedly continental conditions, the temperature gets considerably higher (at Atar, 220 mi [350 km] from the ocean, temperatures reach 71.1°F [21.7°C] in January and 94.1°F [34.5°C] in July). In the Sahel zone on the southern fringe of the desert, the climate becomes steppe-like with a brief rainy season, which becomes longer closer to the Senegal river.

Flora and fauna. Except for the oases which adorn the desert depressions, vegetation coverage is continuous only to the south of the Sahara. The pre-desert steppe of the Sahel is dominated by spiny acacia with their umbrella-like crowns, dum palms, arborescent euphorbias, thorny mimosas, tamarinds, and (as far as herbaceous plants are concerned) hardy grasses. Vegetation becomes more abundant near the Senegal river and on the coast, facilitating farming activities. The pre-desert steppe is also characterized by a fauna consisting of herbivorous mammals such as

Climate data

Location	Altitude (ft asl)	Average temp. (°F) January	Average temp. (°F) July	Average annual precip. (in.)
Nouakchott	7	73	79	6.4
Fdérik	974	69	96	0.2
Conversion factors: 1 ft = 0.3 m; 1 in. = 25 mm; °C = (°F − 32) × 5/9				

antelope, camel, and wild ass, as well as insects and reptiles, which escape the heat of the day by adopting nocturnal habits. A national park has been established along the northern coast (Arguin Bay).

Population. Mauritania is a crossroads of three different cultures: Arabs and Berbers, who once dominated trade and now constitute the ruling class; nomads (Tuaregs); and settled Sudanese-Guinean farmers of the southern region, members of black Sudanese tribes including the Wolof, Sarakollé, Toucouleur, Tekrur, and Fulbe.

Mauritania's independence created a nation of great size but low population—792,000 native inhabitants plus 55,000 European residents in 1963, two thirds of them nomads and the rest scattered in small villages and oases, and concentrated into only two cities with barely 10,000 inhabitants (Port Étienne, now Nouadhibou, and Kaédi). With an annual growth rate of 2.4%, the population has now almost tripled (to nearly 2,000,000), and over the last 20 years many people have been driven from the countryside by drought; as they have poured into the cities in search of work and food, they have greatly increased the proportion of urban population (now 45%).

The country's average population density is extremely low (less than 5 per mi^2 [2 per km^2]), since most of Mauritania is covered by desert (non-productive uncultivated land constitutes about 50% of total area) and by the meager seasonal pastures of the Sahel. The southern sector extending along the right bank of the Senegal river is the only part of the country with a density greater than the national average, and even there only the administrative regions of Gorgol (with its capital Kaédi) and Guidimaka (with its seat Selibabi) exceed 28 per mi^2 [11 per km^2].

Nomads at present make up no more than 30% of the population, and are now slowly abandoning the oases—some of them formerly important caravan centers, like Aoudaghost (capital of

the tribal kingdom of Lemtuna) and Ouadane near Chinguetti (now a religious center)—in order to seek employment in the cities. Hence the cities are expanding, some at an impressive rate—the capital Nouakchott, which was founded in 1957, had only 40,000 inhabitants in 1970, and in 1988 had about 400,000 —and others more slowly, like Nouadhibou (pop. 59,158 in 1988), Atar (21,366), and Zouérate (25,892). The southern provinces contain numerous villages scattered along the right bank of the Senegal.

The population is growing rapidly, especially the black Sudanese group, which is increasing in importance in the country. The annual population growth rate of 2.4% is among the highest in Africa, the result (as in all developing countries) of a high birth rate (4.8%) and a relatively low mortality rate (1.9%). Living conditions for Mauritanians are influenced by traditional social relationships, in which the Arab/Berber group, herders and traders, have always dominated the sedentary black Sudanese farmers. Educational availability is still very low, with 72% illiteracy; there is no national university, and information is disseminated by one daily newspaper and two radio transmitters. Caloric intake barely reaches subsistence levels, and per capita income (US$500) is again among the lowest on the continent.

Economic summary. As a result of the socialist orientation of its first president, Mauritania broke off diplomatic, economic, and monetary relations with France, withdrew from the OCAM (Common African and Malagasy Organization) and thus broke with the EC, seeking assistance from the socialist countries of that era. Partition of the former Spanish Sahara with Morocco caused severe economic difficulties, due to the enormous costs incurred in fighting against the Polisario Front and frequent sabotage of the railroad between Zouérate and Point Central, which transports phosphates from the mine to the port. Two coups in 1980 and 1984—in which the neighboring powers, Algeria and Libya, were not without influence—and the consequences of a long period of drought, made Mauritania's food shortages even more dramatic.

The economy is characterized by the existence of two incompatible productive and social models. On the one hand traditional activities—based on nomadic herding, hoe cultivation, and leather crafts and carpet-making—have generally survived. In the meantime, however, a mining and industrial economy has developed, with focal points in the mining towns of Zouérate and Akjoujt and the phosphate port of Point Central, and plans to develop a domestic steel industry.

Agriculture and livestock. Agriculture, which is concentrated in the southern Chemama district on 0.2% of the land area (and once supplied 60% of the gross national product), has lost almost half of its economic importance (now 38% of the GNP). The sector still employs 67% of the working population, but produces only a small portion of food requirements (millet, rice, potatoes, sweet potatoes, dates, and corn). Livestock represents a substantial resource, with 1.263 million head of cattle (1990), 7.52 million sheep and goats, and 820,000 camels. The fishing industry has access to excellent waters nourished by the cold current of the Canary Islands, but must contend with the technical superiority of the Japanese, Russian, and Spanish fleets. Gum arabic, once a profitable product of the thorny acacia, is now of almost no economic significance.

Administrative structure

Administrative unit (regions)	Area (mi^2)	Population (1988 census)
Adrar	83,106	61,043
Assaba	13,896	167,123
Brakna	14,321	192,157
Dakhlet Nouadhibou	11,580	63,030
Gorgol	5,404	184,359
Guidimaka	3,860	116,436
Hodh ech Chargui	64,076	212,203
Hodh el Gharbi	22,002	159,296
Inchiri	18,914	14,613
Tagant	35,898	64,908
Tiris Zemmour	98,546	33,147
Trarza	25,862	202,596
Nouakchott (district)	386	393,325
MAURITANIA	397,850	1,803,193

Capital: Nouakchott, pop. 393,325 (1988 census)
Major cities (1988 census): Nouadhibou, pop. 59,158; Kaédi, pop. 30,515; Zouérate, pop. 25,892; Atar, pop. 21,366.

Conversion factor: 1 mi^2 = 2.59 km^2

Socioeconomic data

Income (per capita, US$)	500 (1990)
Population growth rate (% per year)	2.4 (1980–89)
Birth rate (annual, per 1,000 pop.)	48 (1989)
Mortality rate (annual, per 1,000 pop.)	19 (1989)
Life expectancy at birth (years)	46 (1989)
Urban population (% of total)	45 (1989)
Economically active population (% of total)	32.1 (1988)
Illiteracy (% of total pop.)	72 (1991)
Available nutrition (daily calories per capita)	2528 (1989)
Energy consumption (10^6 tons coal equivalent)	1.44 (1987)

Energy resources and industry. Mineral resources are outstanding, based on the presence of huge quantities of an extremely pure iron ore at the two deposits of Zouérate and Fdérik. The copper ores of Akjoujt, however, are difficult to exploit for lack of a rail connection to the coast. Rock salt, a traditional resource, is still important even though international demand has declined, simply because it is located close to Mauritania's iron ore deposits and can therefore be rapidly transported to the coast by the mine railroad. Enormous gypsum deposits have been discovered in the salt marshes of Ndaghamcha. There are also promising signs of uranium near the Algerian border, petroleum offshore on the continental shelf, phosphates in the Gorgol region, and ilmenite in the Trarza region.

Manufacturing activity is weak and contributes little to the economy, employing only 10% of the economically active population (55,000 workers), and constituting 19% of GNP. Focal points include the major industrial center of Nouadhibou, a transshipment port for iron ore and rock salt from the Idjil mountains, with petroleum refineries, fish-meal production facilities, and packing and canning plants. Consumer goods production industries, and a large sugar factory reactivated in 1982 with Algerian aid, are located in the capital.

Commerce and communications. This service sector employs 23% of the working population and contributes 43% of GNP; its importance derives from exports of iron (85%), fish (9%), and copper (5%), and on the ability, derived from its state-controlled economic policy, to restrict imports and keep the balance of trade approximately neutral. Imports consist of food products (25%), transportation resources (14%), and petroleum products (9%).

Communications are poor, consisting of a single rail line 428 mi [690 km] long (1988), a negligible road system (5053 mi [8150 km]), two important harbors at Nouadhibou and Nouakchott along with the specialized ports of Point Central (iron) and Casnado (ore shipping), and three airports, two in the principal cities and one at Néma.

Historical and cultural profile. *Mauritania in antiquity.*

Until the 6th millennium B.C., the Maghreb and Sudan regions were linked together by the area corresponding to present-day Mauritania. At that remote period the Sahara was not an inhospitable desert, but an expanse of grasslands inhabited by a black African people known as the BaFuru. But gradual desertification severed these connections, and the Mauritanians in fact originated from a fusion between the autochthonous black populations, consisting of nomadic hunters or herders, and agricultural

peoples, particularly the Zenata and Zanaga groups. This fusion occurred relatively recently, in the first centuries of the Christian era, after the Roman armies had moved through the region. The Mauritanians or Mauri, known in antiquity as the Moors, made their living as caravan traders on the salt and gold routes. They were converted to Islam by the Zanaga Berbers, and later gave rise to the Muslim sect known as the Almoravids, dedicated to an intransigent defense of orthodoxy, which in the 11th and 12th centuries united under its control a patchwork of nations extending from Spain to Morocco, and from Algeria to the Senegal river, obviously including Mauritania. The western coast of Africa was also touched by the first European explorations, and it was in fact along the Mauritanian coast that the Vivaldi brothers apparently disappeared in the 13th century. Exploration was followed by the first commercial exchanges: slaves and gold in return for textiles, trinkets, and grain.

The colonial period. In the 14th century the Moors were subjugated by the Bedouin Ma'qil, who swept down from Morocco and spared only the emirate of Trarza, which retained control of the coastal strip. This became the base from which European colonizers, first the Portuguese and then the French, penetrated to the interior. France prevailed; its sovereignty over the region was recognized in 1783 and later in 1814. But formal recognition was one thing and the actual process of conquest was quite another; it was extremely slow and bitterly contested. In 1858 General Faidherbe imposed a French protectorate on the emir of Trarza, and in succeeding decades this was extended to all of Mauritania, whose borders were not defined until 1900 under agreements between France, Spain, and Portugal.

Independence. The Mauritanians revolted several times: after their territory was annexed to French West Africa (1920), after entry into the French Union (1946), and after relative autonomy was achieved with approval of the framework law of 1956. Their claims were often supported by the expansionist aims of Morocco, which did not abate even after independence was proclaimed in 1960. The new republic, led by the Mauritanian People's Party and its president Mokhtar Ould Daddah, thus found itself weakened by this dispute right from the start, and also had to engage in a war to divide up the former Spanish possessions in the Sahara, which in 1976 were shared with Morocco. Although this settled the dispute with Rabat, it also marked the beginning of a conflict with the Polisario Front, the Western Sahara liberation movement, which continued for three years until a peace agreement was signed. In 1978 Daddah was overthrown by Colonel Moustapha Ould Mohamed Salek; he was removed from office in 1980 by the prime minister Mohamed Khouna Ould Haidalla, who revamped the entire institutional apparatus to conform to Islamic law. Battered by a severe economic crisis and a ten-year drought, in 1984 Mauritania suffered another coup d'état under Maawiya Ould Sid' Ahmed Taya, an ardent opponent of the peace treaty with the Polisario Front.

Culture. The country has no particular forms of artistic expression, except for rock-cut graffiti of prehistoric date and coins from the Hellenistic epoch. A certain artistic sensibility is expressed today in crafts such as leather goods, cushions, baskets, carpets, and especially the characteristic painted bags. Goldsmithing has also developed, and gold is sometimes even used to decorate everyday wooden objects.

NIGER

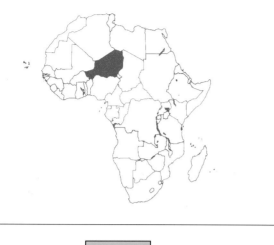

Geopolitical summary

Official name	République du Niger
Area	457,953 mi^2 [1,186,408 km^2]
Population	7,249,596 (1988 census); 7,450,000 (1990 estimate)
Form of government	Presidential republic under military control, subject to a Supreme Military Council
Administrative structure	8 departments
Capital	Niamey (pop. 399,000, 1983 estimate)
International relations	Member of UN, OAU, *Conseil de l'entente*; associate member of EC
Official language	French; Berber and Sudanese languages also widely used
Religion	Predominantly Muslim (about 85%); also animists (15%) and approx. 30,000 Christians
Currency	CFA franc

Natural environment. Niger is bordered on the north by Algeria, on the northeast by Libya, on the east by Chad, on the south by Nigeria, on the southwest by Benin and Burkina Faso, and on the west by Mali.

It occupies the south-central portion of the Sahara, between the Ahaggar and Tibesti massifs, containing not only the Djado plateau but also the high crystalline and volcanic massif of the Aïr, with elevations exceeding 6500 ft [3000 m]. The Aïr separates two vast desert depressions: the Ténéré on the east and Talak on the west, which fall away southward towards the Lake Chad basin and the Niger valley, respectively. Lastly, the south-central region of the country consists of alternating small table-lands and depressions, carved out of Mesozoic and Tertiary sedimentary rock.

Except for the small portion of Lake Chad that extends into Niger's territory (in the southeast corner), the only permanent surface waterway is the river Niger, which flows within the country's borders in the southwest for about 310 mi [500 km]. It is only sporadically navigable, due to the rapids scattered along its course as far as Say. Unlike the Niger's right-bank tributaries, those on its left bank are seasonal in nature, and are almost always dry (wadis). The inland drainage basin feeding Lake Chad is also crisscrossed by numerous wadis, the only exception being the Komadugu, which flows into the lake and contains water for about 10 months of the year. Lake Chad, located at the borders of Niger, Chad, Cameroon, and Nigeria at an average elevation of 925 ft [282 m], has a surface area that varies over the course of the year from less than 3860 mi^2 [10,000 km^2] to more than 9650 mi^2 (25,000 km^2), because of intense evaporation. Its principal tributaries are the Chari in the south and the Komadugu Yobe in the west. It is extremely shallow, varying from 3 to 20 ft [1–6 m] in depth depending on the season.

Located just south of the Tropic of Cancer, Niger's northern sectors have a predominantly desert tropical climate, characterized by substantial daily temperature swings (temperature can vary from 100.4°F [38°C] during the day to 50–53.6°F [10–12°C] at dawn), and by very sparse precipitation; in some areas of the Sahara, there may be no rainfall for years. The southwestern and southern regions, however, are affected by trade winds from the Gulf of Guinea, resulting in annual precipitation of about 24–31 in. [600–800 mm] (29 in. [750 mm] at Niamey), concentrated during the summer, and by more moderate daily temperature swings. The *harmattan*, a dry, cold wind from the Sahara, is especially common during the winter.

In the Saharan region in the north, vegetation is confined to the oases, where carefully cultivated date palms flourish. As the desert gives way to the Sahel steppe, the predominant flora consists of grasses, euphorbias, and thorny scrub. Along the southwestern strip, dominated by the presence of the river Niger, the vegetation consists of wooded savanna, characterized by acacia, baobab, karité, and mogano. The fauna of the desert areas partly reflects that of the steppes, but includes only drought-tolerant species such as antelope and camel. A number of reptile species live in the desert, spending long periods of semi-lethargy buried in the sand. Lions, leopards, jackals, gazelles, giraffes, numerous bird species, and monkeys are found in the southern regions. The banks of the rivers and of Lake Chad are inhabited by warthogs and hippopotamuses. The Nigerian section of the W national park on the right bank of the Niger river protects the richest faunal inventory and natural environment in West Africa.

Population. The principal ethnic groups are the Hausa, Djerma, Fulbe, and Tuareg, along with lesser groups consisting of various Arab tribes. The Hausa (52% of the population) represent the dominant ethnic group; they live throughout the Sahel, subsisting on agriculture and herding, and in the urban centers,

Climate data

Location	Altitude (ft asl)	Average temp. (°F) January	Average temp. (°F) July	Average annual precip. (in.)
Niamey	640	77	84	30.0
Agadez	1627	67	91	3.0
Conversion factors: 1 ft = 0.3 m; 1 in. = 25 mm; °C = (°F − 32) × 5/9				

Administrative structure

Administrative unit (departments)	Area (mi^2)	Population (1988 census)
Agadez	244,805	203,959
Diffa	54,123	189,316
Dosso	11,967	1,019,997
Maradi	14,892	1,388,999
Niamey	259	398,265
Tahoua	41,177	1,306,652
Tillabéry	34,594	1,331,611
Zinder	56,136	1,410,797
NIGER	457,953	7,249,596

Capital: Niamey, pop. 399,000 (1983 estimate)
Major cities (1983 estimate): Zinder, pop. 82,800; Maradi, pop. 65,100.

Conversion factor: 1 mi^2 = 2.59 km^2

where they are traders and craftsmen and retain Islamic traditions. The next most populous group are the Djerma-Songhai (approx. 15%), settled in the capital area, who represent one of the most developed groups. Other populations are the Fulbe (10%), nomadic herdsmen in the Sahel and tradesmen in the cities, and the Tuareg, seminomadic herders who live in the Saharan oases.

Approximately 80% of the population is concentrated in the southwestern corner of the country through which the middle of the Niger flows, where population density reaches 130 per mi^2 [50 per km^2], and in the southeastern region around Lake Chad; in the northern desert areas, average population density is less than 0.5 per mi^2 [0.2 per km^2]. The severe droughts that periodically afflict Niger have resulted in a continuous migration of nomadic populations from the north-central regions towards the south. Settlements are rural (for about 80% of the population), and there are no inhabited centers with urban traditions; the cities developed during the colonial period from the amalgamation of primitive rural villages, which have maintained their traditional characteristics. The capital, Niamey, located on the left bank of the Niger, has experienced substantial population growth in the last few years, primarily because of migration from those areas of the country suffering from famine. A colonial city, it consists of a nucleus of European origin, with residential neighborhoods and administrative and commercial buildings, and a series of villages along the riverbank, where the indigenous population lives. An important market for farm products and animals, an active commercial center, and the site of Niger's few industries, the city has a river port and a busy international airport. Other towns of some importance include Zinder, Maradi, Tahoua, and Agadez.

The periods of drought that have repeatedly struck Niger, and the increased rate of population growth in the last few decades, have compromised the country's self-sufficiency in food production, making it one of the Sahel countries that receives international aid for its famine-stricken population. With a high level of foreign debt and per capita income of US$300 (1990), Niger is one of the world's poorest countries: the infant mortality rate is 14.57% (1986), and average life expectancy is only 45 years. Illiteracy is high (86.1% of the population), even though there are two universities (at Niamey and Say), and some impetus has

been given to technical instruction with the creation of vocational training courses.

Economic summary. Niger's economic backwardness is primarily the result of particularly adverse environmental conditions. In the first decade following independence, the country seemed to be on the way to self-sufficiency in food production with the development of agriculture and herding, but the terrible drought between 1968 and 1974, and the even worse one in the next decade, hit livestock and crops particularly hard, impoverishing even the savanna and reducing Niger to a very low economic level. Recently ore deposits have been discovered, especially of uranium; extraction and export of these ores has begun, and along with better efficiency in agricultural production, this should bring a certain degree of economic autonomy.

Agriculture and herding employ approximately 70% of the working population, in a country that is 88.3% uncultivated and nonproductive. The main production category is traditional subsistence crops such as millet (2,038,300 t in 1991) and sorghum (519,200 t), followed by rice (a crop that was introduced along the banks of the river Niger but which requires suitable irrigation systems), manioc, a few vegetables, and dates. The principal plantation crop is peanuts, almost all grown for export; cotton, sugar cane, and tobacco are of lesser importance.

Animal herding is the typical economic activity of Niger's Sahel belt. The livestock inventory includes a considerable number of goats (7.617 million head in 1990), cattle (3.609 million) and sheep (3.539 million), followed by donkeys, camels, horses, and pigs. Freshwater fish are caught in the Niger and its tributaries and in Lake Chad.

Mineral exploitation, based primarily on cassiterite deposits, was expanded with the discovery of uranium at Arlit (on the fringes of the Aïr) in 1971, and at Akouta in 1978, now being extracted by multinational companies (largely French). Uranium, produced initially at a rate of 4400 t per year, is Niger's leading export; tonnage dropped after 1980 due to a fall in demand from the industrialized nations (production in 1990 was 3114 t). There are also deposits of iron at Say and Kantché, and of phosphates at Anou Araren. Near Bilma and Agadez are salt mines (3300 t in 1989), which supply the entire country. Efforts are also under way to extract coal (in the Agadez region) and petroleum (around Lake Chad).

Industrial activity is still minimal and is concentrated in the food-processing sector (peanut oil production plants), and in light manufacturing (cotton processing factories). Textile plants, a

Socioeconomic data

Income (per capita, US$)	300 (1990)
Population growth rate (% per year)	3.4 (1980–89)
Birth rate (annual, per 1,000 pop.)	51 (1989)
Mortality rate (annual, per 1,000 pop.)	20 (1989)
Life expectancy at birth (years)	45 (1989)
Urban population (% of total)	19 (1989)
Economically active population (% of total)	51.5 (1988)
Illiteracy (% of total pop.)	86.1 (1991)
Available nutrition (daily calories per capita)	2340 (1989)
Energy consumption (10^6 tons coal equivalent)	0.34 (1987)

cement plant, and factories to process phosphates and refine uranium are also in operation. Niger's balance of trade is consistently negative: exports are confined to uranium (60–70%), peanuts, and livestock, while consumer goods and appliances, machinery, and fuels are all imported. Trading partners are primarily France, other European countries, and Nigeria. Communication routes include a road network of approximately 6200 mi [10,000 km] (1986), but there are no railroads. The Niger river is navigable, on a seasonal basis, from Gaya and Malanville to Niamey.

Historical and cultural profile. Inhabited since remote antiquity by black populations of various origins, the region corresponding to modern Niger did not develop, prior to the colonial era, any governmental entities with typical unifying features, but rather was partially incorporated into kingdoms which occasionally arose in surrounding areas. For example, the southeastern region was pulled into the orbit of the powerful caravan realm of Kanem-Bornu, which extended around Lake Chad between the 13th and 19th centuries. The southwestern end, on the other hand, became part of the Mali empire (13th–14th centuries) and Songhai empire (15th–16th centuries), due to the ease of river communication along the Niger to Gao and Niamey. But the most original civilization that flourished in this region, although it was somewhat scattered, was that of the Hausa, who came to European attention through the accounts of Arab travelers and through local chronicles. Located in the border area between modern Niger and Nigeria, the Hausa city-states declined between the end of the 17th and beginning of the 18th centuries as they were invaded by the Fulbe. During that same era the Europeans began exploring the area, beginning with the Scottish physician Mungo Park, the first to find the source of the Niger. Germany, France, and Great Britain became interested in the river basin, and during the era when Africa was being partitioned, the last two powers defined their respective spheres of influence with agreements in 1890 and 1898. Two years later the French defeated the Arab adventurer Rabah, who had pieced together a huge empire in Chad and the lands around it, and continued their military expeditions for two decades until they had completely subjugated the Niger region. In 1922 Niger became a colony within French West Africa, with its capital at Zinder (transferred to Niamey in 1926); thereafter it was bound up with the fate of France's colonial empire, first as an Overseas Territory of the French Republic, then as a member of the French Overseas Community, until finally gaining independence in 1960. The new republic, led by the Parti Progressiste Nigérien, repressed the leftist movement called Sawaba ("liberty") which favored a clean break with the former mother country, and instead pursued a policy of alignment with France. In 1974 president Hamani Diori was overthrown by a coup that installed Colonel Seyni Kountche as president; he was succeeded 13 years later by Ali Saibou. In its attempts to lift the country out of its dramatic economic crisis, the military regime found itself confronted by both political and natural obstacles, such as the recurring waves of drought.

A variety of 16th-century texts dating to the Songhai imperial period, written in the Songhai language and Arabic script, are important evidence of the history of this era. A rich oral tradition handed down within the various ethnic groups, transmitted by itinerant storytellers called "griots" (and in some tribes also in a particular kind of puppet theater), includes legendary stories of giants, animals, heroes, and pre-Islamic divinities.

SUDAN

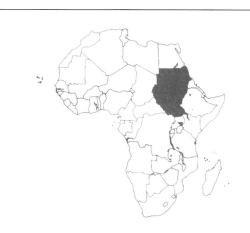

Geopolitical summary

Official name	Al-Jumhuriyat as-Sudan
Area	967,243 mi² [2,505,813 km²]
Population	20,564,364 (1983 census); 24,485,000 (1989 estimate)
Form of government	Military government controlled by a Revolutionary Council for National Salvation
Administrative structure	6 regions divided into 18 provinces, plus the capital province
Capital	Khartoum (pop. 557,000, 1983 census)
International relations	Member of UN, OAU, Arab League; associate member of EC
Official language	Arabic; Hamitic, Sudanese and Nilotic dialects are widely spoken
Religion	Muslim 73%, animist 18%, Christian 9%
Currency	Sudanese pound

Natural environment. Sudan, the largest nation in Africa, is bordered by the Red Sea on the northeast, Egypt on the north, Libya on the northwest, Chad on the west, the Central African Republic on the southwest, Zaire, Uganda, and Kenya on the south, and Ethiopia and Eritrea on the east.

Geological structure and relief. The territory of Sudan presents a fairly simple geomorphological structure, consisting of a strip of Precambrian crystalline basement rocks which form the framework of the African continental plate, slightly undulating and traversed by several systems of fractures. In addition to allowing granitic magmas to rise, these fractures have also led to the formation of unmistakable volcanic structures, some still active at the beginning of the Quaternary era, and others now partially covered by more recent marine sediments (Mesozoic and Tertiary limestones and sandstones). Sudan's most obvious geomorphological feature is the wide, complex valley depression carved out by the Nile, which passes through the entire country from south to north, describing several extensive loops. One of these occurs in the southern part of the country in the wide, swampy lowlands of the Sudd, where the White Nile receives a number of tributaries from both the right bank (Sobat) and the left (Bahr el Ghazal); this depression is bordered on the southwest by the highlands that constitute the watershed with the Congo basin, and on the south and east by the high plains containing the great equatorial lakes which reach their highest point at 10,453 ft [3187 m] on Mt. Kinyeti. The central region of Sudan, on the other hand, is predominantly mountainous, with the great volcanic massif of Jebel Marra (10,128 ft [3088 m]) to the west and the high plateaus of Kordofan at the center, extending to moderate elevations of about 2600–3300 ft [800–1000 m], occasionally reaching to approximately 4500 ft [1300–1400 m] (Jebel ed Dair, 4759 ft [1451 m]).

C. Cesari described the Kordofan's physical aspects as follows:

In the Kordofan the landscape consists of a continuous undulation, perhaps furthered by the geological constitution of the soil, which is a sandstone lightly tinged with iron peroxide.

For the traveler, water is a matter literally of life and death; in lands where water is scarce, the traveler fears neither savages nor wild beasts; what he dreads is a dry well, at which he would most certainly meet a terrible end.

In the Kordofan, however, this fear need not be all-consuming: in those months of greatest drought water can be found every two days, provided the natives do not always display the same hostility that they exhibited towards us, for then one might indeed die of thirst....

The Kordofan stands some 600 meters [2000 ft] above sea level, and about 280 m [920 ft] above the Nile. Not a single river, or torrent, or brook waters this immense area, which is some 800 kilometers [500 mi] long, and somewhat less in width. The average temperature over the year is no less than 34 degrees Centigrade [93°F], because it is one of the hottest areas of Egyptian Africa. No mountains, no hills: plains and valleys succeed each other in a constant rhythm. At its surface the land is so sandy that animals walking on it sink some 30 centimeters [12 in.] into it. The wet season begins in June and ends in September. The rains are irregular and never abundant—a sad exception for this part of Africa.

East of Kordofan, the Nile again follows a more rectilinear course; east of the river rise the foothills of the Ethiopian highlands, from which flow the waters of the Blue Nile that joins the White Nile at Khartoum. Lastly, in the northern part of Sudan, the Nile forms another great bend interrupted by frequent cataracts (six in all), flowing into the great artificial basin of Lake Nasser on the Egyptian border. In this part of the country, in a fully Saharan environment, the only topographical relief of any significance is the hills which frame the Nubian desert, rising to elevations of more than 6500 ft [2000 m] at the edge of the tectonic basin occupied by the Red Sea, whose extremely tortuous coastline is peppered with numerous islets and coral reefs.

Hydrography. Sudan's hydrography centers on the Nile, which covers 1804 mi [2910 km] of its total length of 4136 mi [6671 km] within the country's borders. When it enters Sudan, the river is well fed with equatorial precipitation; its flow is fairly strong and generally constant, usually peaking in the summer. Despite contributions from numerous tributaries, the permeability of the land and intense evaporation in the rain-poor Sudd depression cause a considerable reduction in its rate of flow. Fortunately the contribution of one of the right-bank tributaries, the Sobat, raises the White Nile above the Malakal threshold, while farther north the Atbara (between the Fifth and Sixth Cataracts), provides an additional and critical volume of water. Before the construction of the Aswan dam, the Nile's behavior was highly predictable: every June the river slowly filled with clear water from the equatorial precipitation that flowed into the White Nile. From late July to October the river flowed strongly, fed by winter snows from the Ethiopian highlands, and carried fertile volcanic soil that the Blue Nile had eroded away (as it still does today) after emerging from Lake Tana. Average flow at Wadi Halfa is about 95,000 ft^3/sec [2700 m^3/sec], ranging between extremes of 25,000 ft^3/sec [700 m^3/sec] at low water, and more than 318,000 ft^3/sec [9000 m^3/sec] in full flood.

Climate. The great north-south length of the country, stretching almost from the Tropic of Cancer to the equator, explains its variety of climatic conditions. The north is dominated by the Sahara and ruled by the dry *harmattan* wind; here the extremely meager precipitation (a few thousandths of an inch per year) is concentrated in only two or three days in July. Average monthly temperatures range between 61°F [16°C] in January and 90°F [32°C] from July to September (with a maximum of 122°F [50°C]). Daily temperature swings are considerable. On the coast the climate is influenced by the Red Sea highlands; these produce a slight increase in precipitation (4.4 in. [111 mm] at Port Sudan), which is distributed over one major rainy season from October to December, and a shorter one in July and August. Average monthly temperatures here are 75°F [24°C] in January and 91°F [33°C] in July, with considerable relative humidity. The central region, however, has two distinct seasons, one wet and one dry. Average annual rainfall ranges from 6.3 in. [161 mm] at Khartoum to 9.1 in. [230 mm] at Kassala on the northern edge of the Ethiopian highlands, to exceptional amounts of 24 in. [600 mm] on Jebel Marra, the wettest part of the country. Lastly, the climate in the south is heavily influenced by equatorial air masses, which increase precipitation to 60 in. [1500 mm] between May and August (rainfall at Wau is 44 in. [1122 mm]).

Flora and fauna. Vegetation faithfully reflects changes in climate and latitude. The desert, never completely devoid of vegetation, is interrupted by the long ribbon-like oases fringing the two banks of the Nile, while the band of Sahel steppe extending to the Bahr el Ghazal depression is the realm of the grasses, low acacia trees, and thorny shrubs. Farther south and on higher ground is the savanna, forested to various degrees, with several kinds of acacias and typical species such as the tamarind and baobab. In the higher areas of Darfur and Jebel Marra, the savanna is enriched with euphorbias and a few wild olive trees.

Climate data				
Location	Altitude (ft asl)	Average temp. (°F) January	Average temp. (°F) July	Average annual precip. (in.)
Khartoum	1240	74	89	6.4
El Fasher	2394	66	81	12.2
El Obeid	1919	68	79	14.8
Juba	1509	83	78	14.8
Port Sudan	16	74	94	4.4
Conversion factors: 1 ft = 0.3 m; 1 in. = 25 mm; °C = (°F – 32) × 5/9				

Along the rivers the vegetation takes the form of fringing forests, while the swampy Sudd region is dominated by "elephant grass," reaching heights of up to 13 ft [4 m]. Papyrus and aquatic lettuce are also common, often so dense as to interfere with river traffic.

A traveler's account dating from 1930 contains this description of the vegetation of the Sudd:

January 4. – I do not wish to exaggerate, nor to lose my good temper, but remembering what was told to me by the white men at Malakal (the last outpost we visited) concerning the delights of crossing the Sudd (a process which takes more than three days), there is indeed little to cheer about. All that can be done is to ignore the opinion of others, and speak from one's own experience.

... the Sudd, 500 or more kilometers long [300+ mi], and the same distance across at its widest point... is an interminable marsh covered by a prodigious growth of papyrus, growing five or six meters [16–20 ft] above the water, and sometimes even ten [33 ft]. These papyrus transform the marsh, to the eyes of the aviator flying above it, into an intensely green and flat field furrowed by a network of channels, one of which is the White Nile. But the aviator prays fervently to his God that he will not have to land on the Sudd, since although the papyrus stalks can support the birds which enliven this green sea, his airplane would sink into the water and the papyrus would close in above it. I said "green sea," and the Sudd is indeed a sea, or rather a great inland lake covered with grass, foreshadowing the true lakes that lie not too far beyond: Albert, Rudolf (Turkana), Victoria, Nyanza, etc. In this sea the White Nile acts like a current that has succeeded, to some extent, in making a way for itself, digging a channel, but one never wider than about a hundred meters [some 300 ft], and often less. In the past this channel often closed up and became blocked, and the tangle of dead papyrus created obstacles to navigation. So ships would proceed with difficulty in a kind of semi-solid medium, often finding the way before them closed off, especially during floods, when great numbers of papyrus stalks would be uprooted and then form floating islands.

Taking advantage of a low human population density, the fauna includes a wealth of wild species, including gazelles at the edges of the desert, lions in the western savanna, monkeys in the equatorial regions, and ungulates such as antelope, giraffe, elephant, hippopotamus, rhinoceros, and zebra. Bird life is extremely varied, but crocodile populations have been decimated.

Population. An independent country since 1956, following more than half a century as an Anglo-Egyptian "condominium," Sudan is torn by an age-old confrontation between the Islamized, Arab, mercantile north (40% of the population), and the animist and Catholic south, inhabited by Sudanese blacks and Nilotic populations. In the northeastern region live the Beja, a Hamitic people descended from the kingdoms of Cush and similar to the Ethiopian Danakil; the Arabs, divided into two main groups— Jaali (Nubia) and Juhayna (central and western)—live in the northern and central regions. The area to the east of the White Nile is inhabited by groups of ancient origin (Ingassana, Bertat, Burun), and in the Darfur are several non-Arabized groups, such as the Zaghawa, Masalit, and Fior (13% of the population, with Sudanese Paleonegritic characteristics). South of the 10th parallel begins black Africa (30% of the population), with Nilotic groups living in the Sudd (Dinka, Nuer, Shilluk, Anuak), and

Administrative structure

Administrative unit (regions)	Area (mi²)	Population (1983 census)
Bahr el Ghazal (*)	82,508	2,265,510
Central	54,865	4,012,543
Darfur	191,599	3,093,699
Eastern	131,493	2,208,259
Equatoria (*)	76,475	1,406,131
Khartoum (province)	8,095	1,802,299
Kordofan	146,891	3,093,294
Northern	184,151	1,083,024
Upper Nile (*)	91,166	1,599,605
SUDAN	967,243	20,564,364

(*) Combined to form the Southern Region in 1985.
Capital: Khartoum, pop. 557,000 (1983 census)
Major cities (1983 census): Omdurman, pop. 526,287; Khartoum North, pop. 341,146; Port Sudan, pop. 206,700; Wad Medani, pop. 141,065; El Obeid, pop. 140,024; Kassala, pop. 98,751; Atbara, pop. 73,000.

Conversion factor: 1 mi² = 2.59 km²

Sudanese (Sandè) and Paleosudanese groups (Moru, Mittu, Bongo) west of the Upper Nile and also south of the Sudd. Since time immemorial, the policy of the northern Arab populations, who emigrated to the south to engage in commerce and the slave trade, has been to exterminate the Nilotic populations, whose numbers dropped from 8 million to 4.5 million during the wars of the Mahdi, a historical Muslim chieftain (1880–1900). In recent years there have also been migratory movements from neighboring countries affected by civil wars (such as Chad, Uganda, and Ethiopia), along with pilgrims returning from Mecca, who have settled permanently.

Demographic structure and dynamics. The mean population density of Sudan is about 26 per mi² [10 per km²], but the population is very unevenly distributed. The north-central regions are the most densely populated, occasionally exceeding 260 per mi² [100 per km²]; the eastern administrative regions, with their center at Wad Medani, have 5 inhabitants per mi² [2 per km²]. The Gezira region located between the two branches of the Nile, with twice the country's average density, is the real center of attraction for migrants. The low density in the south is due to past persecutions and slave trading, and to the presence of swampy soil that makes cultivation difficult. The construction of the Jonglei canal, which is designed to cross the entire Sudd region, may change that situation in the future. Eighty percent of Sudan's population lives in rural environments, characterized by Arab-influenced villages and mud houses located in twisting alleys along the Nile as far south as Khartoum. The nomadic peoples (14% of the total population) concentrated in the territory around Kassala and in Darfur and Kordofan provinces, and the semi-nomads live in mud houses surrounded by meager cultivated plots, gathered together in oases, or in tentlike huts with a stick framework covered by reed matting made from palm leaves. South of the capital appear the first conical, straw-roofed huts, clustered together in villages enclosed by circular palisades. Some 22% of the population is urbanized, living in cities with a center that is typically European or Muslim, containing shops, the market, government buildings, and the palaces of former sultanates and missions. The periphery consists of native quarters similar to savanna villages, often located near water sources. Two cities have more than a half million inhabitants: they are Khartoum and Omdurman, which together with Khartoum North constitute a single conurbation. The other cities are of more limited but still substantial size (Port Sudan, Wad Medani, El Obeid). Sudan's population is predominantly young: 51.2% is less than 20 years old, while only 4.5% is more than 59 years old. This is the result of a population growth rate that in the 1960s reached 3.5% (now only 2.8%), which increased the number of inhabitants from 10,260,000 in 1956 to 26,263,000 in 1988 (estimated). The birth rate is still high (4.4%), while the mortality rate has been reduced by general improvements in health care and sanitation, and by the fight against endemic diseases.

Social conditions. With a per capita gross national product of US$150, a life expectancy of 50 years, and available nutrition of only 1996 calories per day, life for the Sudanese is still very precarious, often skirting the limits of survival. Only 21.6% of the population has received any education, despite the fact that primary schooling (from age 7 to age 12) is mandatory and free. There are two universities in Sudan, both in Khartoum: one is government-run, while the other is a branch of Cairo University. Social conditions are aggravated by the presence of 2 million refugees from adjacent countries torn by civil war (Chad, Uganda, Ethiopia), and by the emigration of skilled workers to the Persian Gulf states in search of higher wages.

Economic summary. The huge size of the country, the diversity of its climate, the uncomplicated geomorphology, and rich hydrography all represent great inherent economic potential for the development of Sudan. In colonial times, resource exploitation was limited to the expansion of cotton growing in the Gezira; since independence, attention has been concentrated on large-scale irrigation projects and electrical energy installations (dams at Er Roseires on the Blue Nile and Khashm el Girba on the Atbara). A grand scheme to build the Jonglei canal along the Upper Nile valley, designed to transform the Sudd depression into the breadbasket of Africa, has run into many obstacles: lack of capital, limitations imposed on loans from international agencies, and the presence of anti-government guerrillas (symptomatic of the lack of national unity).

At present the economy is based on archaic agricultural practices (although attempts are being made to update them), on very large livestock holdings, on an industrial framework involved almost exclusively with agriculture and food production, and on a lack of exploitable raw materials, even though geological prospecting indicates the presence of petroleum, silver, copper, and zinc. But economic development has been slowed down by divisions within the country that hinder the pursuit of shared objectives, and thus ultimately by the lack of democracy. In addition, there is little sympathy with the current policy for national recovery (increases in the price of food have caused general strikes and demonstrations).

Agriculture and livestock. Agriculture, which is the principal productive activity, contributes 37% of the gross national product and employs 72% of the economically active population. It has extremely high potential; cultivated land area is continually increasing, although at present it covers only 5% of the country's land area, with 32.1% meadows and pastures. Uncultivated and non-productive land constitutes 37.9%. Production for internal consumption consists of poor cereals—such as durra (a variety of sorghum), sorghum itself (3,235,100 t in 1991), corn, millet, manioc, and sweet potatoes—on which the population subsists. Commercial agricultural production includes date palms (with 286 million lb [130 million kg] of dates produced in 1990) in the north; cotton fiber and seed in the middle Nile valley, the Gezira, and Kordofan; sugar cane in the Gezira; and peanuts, sesame, and tobacco.

Livestock represents an important resource for the country, and is concentrated in the central region, with cattle (21.028 million head in 1991), sheep and goats (35,977,000), and camels (2,757,000). The principal forest resources (on 18% of the land) are gum arabic (44,000 t in 1987) and timber, produced in the large stretches of land occupied by forest and wooded savanna. Fishing is a traditional activity along the Nile and other inland waterways, but is negligible along the Red Sea coast.

Energy resources and industry. Sudan's energy resources are basically linked to its water resources and to the production of electrical energy (1061 million kWh in 1989). Half of this is hydroelectric, provided by the dams at Er Roseires and Sennar; the rest is thermal, generated with imported petroleum products. Nevertheless, energy production is still very limited, a factor which hinders more rapid development. Mineral resources are meager, consisting of chromite (5500 t in 1988), salt (55,000 t) obtained from the salt pans at Port Sudan, copper from several deposits at Hofrat en Nahas and one at Bishara, iron at Fodikwan, magnesite, and petroleum.

Industrial activity plays a very limited role in the national economy, contributing only 15% to the GNP and employing 10% of the economically active population. Those industries that do exist are predominantly related to agriculture and food, and are involved in processing local products. Sudan's sugar factories are capable of meeting domestic needs, producing 463,100 t of sugar in 1990; they are located in the Gezira region at Sennar, at Guneid along the Blue Nile, and at Khashm el Girba on the Atbara river, near the Ethiopian border. Vegetable oil plants and a modern textile industry are located in Khartoum, while the cement factory at Atbara alone produced more than 121,000 t (1988) of construction material. Distilleries, breweries, and tobacco factories round out local production, while a modern refinery at Port Sudan processes imported oil, and three thermoelectric power plants (at Port Sudan, Atbara, and Malakal) complete the country's energy-generating potential.

Socioeconomic data	
Income (per capita, US$)	150 (1990)
Population growth rate (% per year)	2.8 (1980–89)
Birth rate (annual, per 1,000 pop.)	44 (1989)
Mortality rate (annual, per 1,000 pop.)	15 (1989)
Life expectancy at birth (years)	50 (1989)
Urban population (% of total)	22 (1989)
Economically active population (% of total)	32.4 (1988)
Illiteracy (% of total pop.)	78.4 (1991)
Available nutrition (daily calories per capita)	1996 (1989)
Energy consumption (10^6 tons coal equivalent)	1.5 (1987)

Commerce and communications. The service sector employs 18% of the economically active population, and contributes less than half of the GNP. The balance of trade is negative; exports consist of cotton, peanuts, sorghum, sesame, livestock, and gum arabic, while machinery and vehicles, fuels, food products, and manufactured goods are imported. Sudan's trading partners are the USA, Italy, Japan, and China on the export side, with Great Britain, Germany, the Netherlands, and Japan for imports. Communication routes are meager; most of the road system (45,618 mi [73,577 km]) (1985) is impassable during the rainy season, and in any event represents very poor coverage given the nation's enormous size. The railroad system, with 3412 mi [5503 km] (1983) of track, is truly vast compared with those of other African countries, but still insufficient for internal needs. Maritime navigation can use Port Sudan, and soon a new harbor under construction at Suakin. Air traffic is based at the airports of Khartoum and Port Sudan. Tourism is very minor, although there is considerable potential based on natural environments and the cultural remains of the Nubian civilization.

Historical and cultural profile. *Ancient Nubia.* The history of Sudan, like that of Egypt with which it is inextricably linked, unfolded along the fertile Nile valley. Hunting scenes cut into rock faces 9000 years ago attest to the presence, and the artistic sensibilities, of Sudan's Negroid populations in the Paleolithic period. Ceramic household articles are later in date (probably the 5th millennium B.C.), and it is interesting to note the gradual emergence of Egyptian influence in later ceramic styles. The Egyptians were indeed irresistibly drawn to the upper course of the Nile. Their expeditions during the 2nd millennium found a governmental entity already in place: the kingdom of Cush, with its capital at Kerma, inhabited by farmers and herders. This kingdom, which extended over the Nubian region between the First and Third Cataracts of the Nile, became a stable part of Egypt's sphere of influence; the same happened with the kingdom of Napata, which developed in the 9th century B.C. around the Fourth Cataract. But these kingdoms, referred to as "Ethiopian," were never entirely dependent on Egypt, and at a certain point the roles were reversed: the 25th Dynasty (715–663 B.C.) was called the "Cushitic" or "Ethiopian" dynasty. More years of prosperity followed as the capital was transferred even farther south, to Meroë, in about 590 B.C.; but the next few centuries saw a slow decline in the Nubian region, one which also produced a withdrawal from the civilization of the pharaohs and was expressed, in the artistic sphere, by the appearance of new influences: from Rome, Byzantium, and even India.

Christian and Arab empires. The kingdom of Meroë was affected only marginally by Roman expansion in Egypt, and remained independent until the capital was destroyed in the 4th century A.D. by the Ethiopian king Ezana. Migrations and influxes of new peoples, such as the Blemmyes and Nuba, changed the face of the kingdom, as did the later penetration of Monophysite Christianity (6th century) and Islam (7th century). The Arabs demanded periodic payment of tribute, but were not able to Islamize Nubia with the same success with which they had converted Mediterranean Africa. As a result, the territory that is now Sudan remained divided between two Christian kingdoms that, although isolated, survived for some time: the kingdoms of Dongola (until 1315) and Aloa (until 1504).

With their fall, the country's epicenter shifted farther south, corresponding to the central regions of modern Sudan. A new administrative entity was brought to life by the Fung, a people that emerged from intermingling between the northern Arab element and the southern black populations. Their kingdom, with its capital at Sennar, stretched from ancient Nubia to the regions of Darfur and Kordofan, even threatening neighboring Ethiopia. But at the end of the 18th century their dominion also was swept away, overthrown by a period of anarchy in which little Muslim sultanates quarreled with one another as they sought a monopoly on the slave trade. After achieving independence in 1811, the Egyptians began to push into the Sudanese region, where they dreamed of finding both veins of gold and black soldiers to stiffen the backbone of their troops. The conquest began under the Khedive Mehmet Ali, who founded Khartoum in 1823; he was succeeded by his son Ismail, who ran up against the slave-trader Rabah, the man who had claimed the Darfur region as a personal possession from which he controlled the slave trade. It was this problem that ignited a conflict between the Egyptian government and the local tribes, who in 1881 revolted under the leadership of Muhammad Ahmed, called the Mahdi. This revolt took on the appearance of a national uprising precisely because of the Mahdi's accomplishments, including the defeat of General Gordon and his British soldiers in 1885.

Colonization and independence. Sudan had now risen to prominence on the international scene, and the "Fashoda Incident" a few years later also demonstrated that it stood at the intersection of the routes followed by the European powers in their drive to lay claim to the African continent. At the turn of the century Sudan was declared an Anglo-Egyptian "condominium," and it remained so until World War II; but in practice the British governed it as nothing more than a colony, although some attention was paid to the country's economic development. Since Britain's primary aim was to isolate Sudan from the rest of the Muslim world, nationalists reacted by taking refuge behind a defense of Islam's traditional values. A Sudanese literature was created in the classical Arabic style, and poetry was often modeled on Arabic authors of the Umayyad and Abbasid periods. Literary journals and publications began appearing in the 1930s, and made a decisive contribution to the cause of national independence.

In truth, British policy had produced within Sudan a rift between the supporters of complete independence and those who favored unification with Egypt; the latter idea appeared to be close to realization several times, first under Fuad I, then under Faruk, and lastly under Fuad II. But Nasser's revolution in Egypt upset Sudan's political balance in favor of autonomy, and the country's independence was recognized in 1956. Difficult times awaited the new nation, threatened in its southern provinces by a conflict between the black populations (Catholics and animists) and an Arab elite determined to impose forced Islamization. After a series of disturbances, the situation appeared to return to normal when power was assumed by a military group with "Egyptian" leanings. Colonel Nimeiry managed to settle the more serious disputes, but stability turned out to be a still distant goal, especially after Sharia (the Koranic law) was reintroduced in 1983. The military regime of Omar Hassan el-Bashir, which took power in yet another coup d'état in 1989, also appeared to take a position favorable to the Islamic fundamentalists.

SAHEL

Images

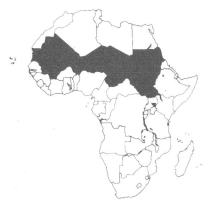

1. *The Tuareg, an ancient people who arose from intermarriage between proto-Berber stock and Negroid elements, are nomads who raise camels and transport salt, living in the seemingly endless territory between the west central Sahara and the great bend of the Volta. In their typical dress, the men cover their faces by repeatedly winding a long, thin cloth around their heads to protect themselves from the sun and blowing sand, leaving a narrow slit for the eyes. The continual wanderings of the Tuareg through a land apparently so hostile to any form of life are a significant instance of the constant struggle for survival facing Africans every day.*

2. *The arid Sahara desert occupies five sixths of the entire territory of Mauritania: it is an enormous sandy plain, some 1000–1300 ft [300–400 m] above sea level, interrupted by moving dunes of various sizes, continually sculpted by the northerly winds that blow through the country for almost the entire year. The only opportunities for survival therefore depend on the rare oases, watered by seasonal wadis and springs of brackish water.*

3. *Panoramic view of Lake Chad, near the rock formation called the Karal. Last remain-* der of a larger inland sea that extended to the outermost edges of the basin, the lake today consists of a succession of bodies of water, semi-submerged islets, swamps, and characteristic "Inselbergs," the tips of ancient eroded plateaus. Native peoples fish in the lake on boats similar in shape to pirogues, made of papyrus stalks tied together and with a tapering prow, like those depicted on Egyptian reliefs. Alternatively, exploiting the fact that the water level is usually low, they advance in long, silent rows and catch fish by hand in special baskets also made of papyrus.*

4. *The Chari river, some 750 mi (1200 km) long, forms at the confluence of two spring-fed tributaries, and for a brief stretch constitutes the border between Chad and Cameroon, after which it empties in a wide delta in the southern portion of Lake Chad, of which it is the major tributary. Although it is subject to seasonal floods, its year-round level is extremely low, and the river is navigable only in small wooden boats used for fishing and to transport agricultural products.*

5. *The great sandy region of the Ténéré, in the heart of Niger, is perhaps the most arid and inhospitable part of the entire Sahara, interrupted only* by a few seasonal wadis that drain toward Lake Chad, and by water-bearing strata that form the rare oases. Blown by the wind, the sand accumulates to form dunes in depressions or near rocks and hillocks, thus producing the vast dune fields called* ergs.

6. *A group of Tuaregs crosses the Arakao dunes, deep in the desert of Niger. The perennial aridity of the Sahara, an Arabic word meaning "nothingness," is due essentially to the lack of rain. This is because it is located in that band of the earth's surface located between 15 and 35° N latitude, which receives almost no moisture as a result of atmospheric dynamics and its considerable distance from the ocean. The rare precipitation that does fall is extraordinarily violent, making the climate even more intolerable; temperature swings are extreme, with very hot temperatures averaging 122°F (50°C).*

7. *Traditional Tuareg dress consists of a loose tunic or jacket of European style, worn over baggy Saharan trousers extending from the crotch to the ankles, so the wearer can sit down comfortably. Over these the Tuareg nomads then wear another colored garment. On their head coverings, the men often wear the same metal roundels and sheets that usu-* ally decorate the characteristic wooden camel saddle with its cruciform pommel.*

8. *In the western and southern regions of Mali, where agriculture and fishing are widely practiced, the most common dwellings are huts with thatched roofs and walls made of reeds and mud, sometimes constructed on stilts to protect the inhabitants from wild animals and floods. The nomadic and seminomadic groups occupying the semi-desert areas of the North, however, prefer tents, which they feel are better suited to their environment.*

9. *The mounted bodyguards of the Sultan of Agadez (Niger), who retains his traditional authority only in religious matters, are distinguished by their sumptuous bright red robes, draped like those of the Tuareg. The palace that they guard is the most important meeting-place and festival site in the city; these events bring together many different ethnic groups with their varied customs and brightly colored dress.*

10. *Evening panorama of the river port of Mopti on the River Niger in Mali, an important fishing center as well as a major marketplace for African handicrafts. Except in the rainy season, the river's level*

is extremely low, but it nonetheless abounds in large fish, which are caught with rudimentary unbaited lines. Both river traffic and fishing activities rely on canoes.

11. In a village of the Dogon, an ancient population now concentrated in the southern region of Timbuktu (Mali), a woman prepares millet for drying on the terrace of a granary, the building with the characteristic pointed thatched roof. Dogon villages are dense agglomerations of square and circular buildings made of stone and clay, often extending vertically along rock walls to which they adapt almost seamlessly, matching every undulation in the uneven terrain. The Dogon belong to one of Africa's earliest civilizations; they contributed to the formation of the empire of Ghana, and then formed part of the empire of Mali (10th–16th centuries).

12. A street in Bamako, capital of Mali, on the left bank of the Niger. Like all the major cities of formerly French Africa, Bamako began as a military base; its geographical position and its railroad link with Dakar, inaugurated in 1904, then allowed it to develop rapidly as a commercial market, river port, road junction, and educational center.

13. The great steel bridge at the confluence of the two Niles, near the Sudanese capital of Khartoum. The Nile basin is one of the pivotal elements in African geography, and the focus for Sudan's greatest economic vitality, since it is the only fertile region in which stockraising and intensive cultivation are possible. The Nile consists of two great branches: the White Nile, which rises in East Africa and flows through a marshy region with no tributaries; and the Blue Nile, which rises at Lake Tana in Ethiopia and carries the greater quantity of water, representing almost two thirds of the total during flood periods.

14. The oasis of Djado (Niger), at the edge of a sandy plateau interrupted at its foot by the massif of the same name, offers the surreal image of an old fort now almost completely abandoned. In reality the village, in which several rock incisions of considerable archaeological interest are preserved not far from the palm trees, is inhabited by people from nearby Chirfa for at least part of the year (from August to October), when they harvest dates from the palm trees.

15. The economy of Mauritania, still in the developing phase, is based not only on mineral resources but also on nomadic herding and on agriculture; the latter, using technology from various sources including China, is practiced predominantly in the country's southern regions, and produces millet, sorghum, and corn for local consumption. The only agricultural cash crop is dates. The extreme aridity of the soil, and frequent periods of drought, are preventing the country from developing economically beyond the subsistence level.

16. One of the most important economic resources of Chad is its flocks of sheep and cattle, scattered over the wide meadows of the Sahel and in the savanna belt. The principal obstacle confronting the herders, commonly called baggara, is the scarcity of water sources during the dry season, which forces them to move long distances and exposes them to possible loss of livestock. In addition, deforestation of certain areas of the Sahel, induced deliberately to increase the production of gum arabic, has often allowed the desert to advance, destroying grazing land.

17. The iron mine at Zouerate. Mauritania contains numerous mineral resources, especially in the inland areas of Idjil and Akjoujt, where iron, rock salt, phosphates, copper, and ilmenite are extracted. Minerals create an opportunity for exports, facilitated in part by the railroad which connects Fort-Gouraud to the Atlantic harbor of Port-Étienne. This transportation link is the foundation of Mauritania's iron and steel production capability (its only truly viable industry), and thus constitutes one of the country's principal sources of income.

18. The Chinguetti oasis in Mauritania, considered the seventh holy place of Islam, enjoys considerable spiritual prestige on the basis of its religious schools and its extraordinary libraries, which house ancient volumes of the Koran, engraved and decorated by hand. Among the most artistically important of these is the "Bu 'Ain Gafra," which in Arabic means "he who has the yellow eye," so called because the text is ornamented with a golden circle on which witnesses swore their oaths before Muslim judges. The region that is now Mauritania first appears in Islamic history when it was conquered by the Zenhaga Berbers in the 11th century A.D.

19. An example of rock art from the prehistoric Sahara. With astonishing faithfulness, the artist has depicted herds of animals with small and large horns, demonstrating that in the Neolithic this area was much wetter and was covered with rich forests suitable for hunting and stockraising. The most radical invention of this period was unquestionably pottery: it allowed prehistoric peoples to cook and to store food and water, freeing them from dependence on game and on climatic conditions. The demographic consequences were just as significant: settlements became larger and more stable, and human beings began to domesticate animals and cultivate the earth.

20. These spear-points and arrowheads made of chipped stone, found in the northern regions of Mauritania, prove that as early as the lower Paleolithic (5000–6000 B.C.), when the Sahara desert was definitely less arid and inhospitable than it is now, this region was already inhabited by human beings and by large animals. Hunters fashioned these weapons by hand, using flint, quartz, and other hard stones, and used them to hunt elephant, wild sheep, hippopotamus, antelope, and ostrich.

21. Panoramic view of the ancient royal necropolis of Merowe (Sudan), a Nubian metropolis and capital of classical Ethiopia from the 4th century B.C. to the Abyssinian conquest in 350 A.D. Some of the superstructures are shaped like mastabas, while others are pyramidal in the Egyptian style, with burial chambers as large as the bases. Construction began in the 2nd century B.C. when the kingdom of Merowe reached its greatest political and economic splendor, fueled by large-scale production of iron and by the skills of its metalworkers.

22. Nighttime view of Nouakchott, capital of Mauritania since 1957. Although it faces the Atlantic, it is a typical "desert city," where modern buildings in the French colonial style are gradually replacing the traditional one-story brick and mud dwellings. Note the mosque dome in the background.

23. The great mosque at Djenné (Mali) was constructed in 1907 by the French colonial government on the model of the previous sanctuary, built in the 13th century and destroyed in 1830 by Sultan Cheikou Amadou as punishment for what he considered the depravity of the city's inhabitants. This unique structure of stone and clay, its towers bristling with wooden stakes and its minarets covered with ostrich-egg shells, is an example of the "Sudanese" architectural style. On the right side, in profile, note the two projections standing above the entrance to the mosque, which serve as amulets to repel djinns (evil spirits).

7

8

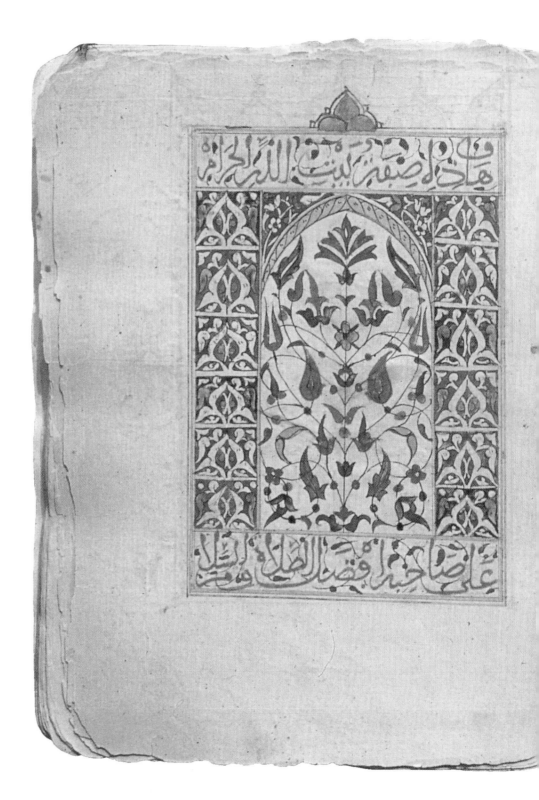

غزوات النبي صلى الله عليه و سلم
احدى و عشرون غزوة وهي

EASTERN AFRICA

The term "Eastern Africa" denotes a large geomorphological region characterized by dominant mountainous areas (the Ethiopian plateau and the lake highlands) and by a system of tectonic trenches among the world's largest. These deep fissures in the rigid continental crust, known as the Great Rift Valley, originate in the Syro-Palestinian depression and the Dead Sea and then run almost due south. The western edge then extends along the Sudanese coast of the Red Sea, which itself is a large submerged section of the trench separating two geologically related regions: Africa and the Arabian peninsula. Farther south, the trench penetrates into the heart of the Ethiopian plateau, skirts the highlands of the great African lakes to the east, and ends at Lake Nyasa. Another fracture runs east of the lake highlands, encompassing the characteristically elongated lakes Albert, Edward, and Tanganyika. These gigantic cracks in the Earth were caused by buckling in the continental crust, which then fractured and gave way along the anticlinal axis. The phenomenon, datable to a time between the Mesozoic and Cenozoic, is part of that system of movements of crustal plates known as "continental drift."

Considerable volcanic activity probably occurred along the rift, affecting the entire Ethiopian highlands and covering the more ancient basement rocks with lava flows that later, extensively shaped by erosion, resulted in dome-shaped table-mountains called *ambas*. To the east and south, the Ethiopian highlands slope down towards the Somali peninsula, known as the Horn of Africa. At Ras Asir (formerly Cape Guardafui) on its tip, the mountainous northern coast facing the Gulf of Aden meets the flat east coast bathed by the Indian Ocean.

This great orographic complex comprises large closed basins often occupied by lakes and depressions—like the bone-dry Danakil depression—and is an important hydrographic nexus. It contains the sources of the Nile, with its major right-bank tributaries (Atbara, Blue Nile), and several spring-fed branches of the Congo. These flow into the Mediterranean and the Atlantic, respectively; a far smaller volume of water runs down the eastern slope to the Red Sea. Only two major Somali rivers flow into the Indian Ocean: the Webi Shebelle, and the Juba.

Although it lies near the equator, Eastern Africa experiences a wide variety of climatic conditions, governed largely by altitude. A hot, wet, tropical climate exists along the coast and on the only real plain, the Benadir in southern Somalia. Temperatures are especially torrid in the Danakil depression. Altitude moderates the heat considerably, and above 6500 ft [2000 m] the temperature is largely uniform throughout the year. Higher elevations also experience abundant precipitation—annual totals drop below 40 in. [1000 mm] only at the bottom of the tectonic trenches and on the Somali high plateau. The arid coastal zones, however, rarely receive over 8–12 in. [200–300 mm] of rainfall annually. Rainfall is governed by the moist southeast winds that blow from April to September. The vegetation reflects these climatic conditions; arid savannas cover the highlands, with plant communities that change with altitude. Xerophytic forms are prevalent at intermediate altitudes, and evergreens predominate in more rainy areas, but they generally do not coalesce into the true forests that one would expect at these subequatorial latitudes.

Eastern Africa has always been a crossroads of peoples and cultures. Primitive Negroid populations have been joined by immigrant Hamitic elements, then Semitic peoples from the Arabian peninsula, as well as Nilotic, Nilo-Hamitic, and even Jewish groups (the Falasha). Relations were especially close with Mediterranean Africa, via the valleys of the upper reaches of the Nile. Stimulated by contacts with Ptolemaic Egypt, the Amharic people established the ancient kingdom of Aksum, which traded with both the Mediterranean and Persia, and converted to Christianity in the 4th and 5th centuries. The impregnable highlands and the prestige of the Coptic Christian kingdom that inherited the mantle of Aksum protected Ethiopia from outside intervention, and guaranteed a long period of independence, interrupted only by the Italian occupation (1936–1940). The Horn of Africa, on the other hand, lay open to the Red Sea and the Indian Ocean and was greatly influenced by Islam beginning in the 7th century. This area was also more vulnerable, starting in the late 19th century, to European colonizing assaults on Eritrea (Italy) and Somalia (Italy, France, and Great Britain), all of them motivated—especially after the opening of the Suez Canal—more by strategic considerations than by local economic resources. The end of colonialism left enormous economic and political problems unsolved, triggering violent ethnic conflicts in the region.

ERITREA

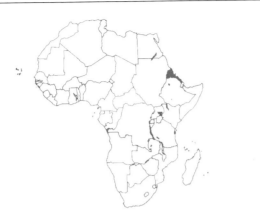

Geopolitical summary

Official name	—
Area	46,761 mi² [121,143 km²]
Population	3,039,465 (1988 estimate)
Form of government	Parliamentary republic
Administrative structure	—
Capital	Asmara (pop. 296,000 in 1991)
International relations	Independent since May 25, 1993; recognized by Italy; member of UN
Official language	Tigrean; Arabic and Italian also widely spoken
Religion	Muslim (50%); Monophysite Christian (35%); animist (15%)
Currency	The Ethiopian birr is currently being used (1993)

Natural environment. The present borders are roughly those of the former colony: to the west Eritrea is bordered by Sudan along the mountains that slope down to the Nile plains, to the south by Ethiopia, partly along watercourses and partly along arbitrary lines, and to the east by Somalia at the Bab al Mandeb.

Eritrea includes the northernmost part of the Ethiopian plateau, which drops off abruptly along a long coastal plain facing the Red Sea. This body of water gives Eritrea its name (from the Greek *erythros*, "red"), bestowed by the Italians in 1890.

Geological structure and relief. The highest area is the central region, where the Ethiopian plateau extends its northern outliers, exceeding 6500 ft [2000 m] and facing the Red Sea along a steep escarpment incised by deep valleys. The western part consists of highlands that descend to the west from the plateau, furrowed by numerous tributaries of the Atbara, itself one of the major tributaries of the Nile. A narrow coastal plain that extends to the border with Somalia includes a large portion of the Danakil depression (380 ft [116 m] below sea level). The high-

Climate data

Location	Altitude (ft asl)	Average temp. (°F) January	Average temp. (°F) July	Average annual precip. (in.)
Asmara	7,613	57	66	21.8
Aseb	23	77	88	1.2
Massawa	10	78	94	7.6

Conversion factors: 1 ft = 0.3 m; 1 in. = 25 mm; °C = (°F − 32) × 5/9

lands consist for the most part of a basement of crystalline rocks, surmounted by layers of sandstone over which in turn extend formations of volcanic rocks such as tufa, trachyte, and basalt; the result is an alternation of tablelands and mountains of harder rock, isolated on all sides by erosion to form steep slopes. Erosion has also produced the deep incisions that converge to the west in the valleys of the Tekeze and Mereb, tributaries of the Atbara, and to the north in the courses of the Baraka and the Anseba, which merge into a river that dries up before reaching the sea. On the coastal plain there are outcroppings of marl, gypsum, and conglomerate, while in the arid Danakil, intense evaporation has produced extensive salt formations.

Climate and vegetation. The climate reflects the variety of Eritrea's geomorphology. Extremely high temperatures, high humidity, and sparse rainfall (almost totally absent in the Danakil depression) characterize the entire coastal strip.

At Massawa the average temperature varies between 77 and 95°F [25–35°C], and precipitation is less than 8 in. [200 mm]. The plateau, on the other hand, enjoys generally mild or even cool temperatures, with small annual variations but pronounced diurnal swings; rainfall, which is quite extensive on the eastern slopes exposed to winds from the Indian Ocean (30–35 in. [800–900 mm] per year), diminishes gradually toward the west, where it occurs mostly in summer, preceded by more moderate spring rains ("little rains"). Asmara, at an elevation of 7613 ft [2321 m], has average temperatures of 59–63°F [15–17°C], with precipitation exceeding 20 in. [500 mm]. On the arid coastal steppes, which become desert-like in the Danakil, woodlands and pastures give way, on the eastern edge of the plateau, to land suitable for grain cultivation. Farther west the plateau is again characterized by a steppe environment with xerophytic woodlands (predominantly acacias) the most common plant community; in the interior valleys, however, the natural vegetation is interrupted by wide stretches of crops, in some cases irrigated.

Population. Although the western plateau has been subjected since antiquity to the influence of Nilotic populations, and today is home to a Sudanese ethnic substrate, the rest of Eritrea is inhabited by Ethiopid peoples, identified primarily by language. Hamitic dialects are spoken by many tribes scattered in peripheral areas: to the northwest in the Anseba basin, to the southeast along the Red Sea coast, and in the central region between Massawa and the Ethiopian border in the Danakil. The rest of the population speaks Semitic languages, especially Tigrean (Tigre is the name of a large historical and linguistic area in both Eritrea and Ethiopia, and of the adjacent Ethiopian province south of the Mareb river). The Hamitic peoples tend to be Islamic herders, while the Semites are Monophysite (Coptic) Christians.

Population density is extremely low (65 per mi^2 [25 per km^2]); the greatest concentrations occur around the capital, Asmara (pop. 296,000) and in the valleys that cut into the plateau. The hot, dry coastal strip is very sparsely populated; this is the location of the port city of Massawa (pop. 15,441 in 1984), which is linked to Asmara by a road and railway constructed during the Italian occupation. Other towns include Keren, Akordat (pop. 26,000 in 1980), and Adi Ugri, all on the plateau, and Aseb (pop. 27,985 in 1982), the second largest maritime port, strategically located near the strait of Bab el Mandeb. Most Eritreans live in farming villages scattered throughout the more fertile regions. Especially typical of the plateau, the village is associated with the ancient tradition of a tribally based social organization. A substantial percentage of people is nomadic, especially in steppe regions. In terms of population dynamics, there has been a considerable increase from 1951 (approximately a million inhabitants) to 1988 (more than 3 million); this increase is greater than in neighboring Ethiopia, to which Eritrea was politically united for the last forty years. The population has grown despite a bloody civil war that caused great loss of life among young people and led to a certain amount of emigration, especially to Italy, to which Eritrea is linked by decades of history. An Italian minority still lives in the capital, having settled in the former colony before World War II.

Economic summary. Eritrea is a country with few resources, in which the Italian colonial interlude resulted not so much in economic exploitation as in a policy of infrastructure development and technological innovation that might have set it on the road to modern life. Reunification with Ethiopia called a halt to this trend, and constituted an underlying cause of the conflict that led to separation and independence. For Italy, Eritrea was a bridgehead for penetration into eastern Africa; for Ethiopia, it was a precious outlet to the sea. Today, Eritrea's recovery is linked to its function as a way station between Ethiopia and international markets.

Agriculture and livestock. The hot, dry coastal plain offers few opportunities for crops (cereals, vegetables, and cotton), and only in those rare areas where irrigation is possible. The eastern slopes of the plateau, blessed with summer rains, are favorable for grains as well as coffee beans. The most common crops in the higher regions of the plateau are cereals (wheat, barley, durra, and corn) and garden vegetables; cattle raising is also widespread. Similar conditions are present on the western slopes, where tobacco and agave flourish in the valley bottoms along with dum palm, the main plant product of the colonial era: buttons were produced from the nuts, and fiber from the leaves.

There is little usable forest, since woodlands dominated by acacias are so widespread. The most valuable species are several aromatic plants, such as gum arabic and incense.

On the plateau, cattle are raised on a semi-sedentary basis; as early as the colonial period, the milk supported a cheese-making industry around Asmara. The extensive steppes are suitable for a much less profitable nomadic type of herding (cattle and goats), practiced purely for subsistence. Fishing is of some importance along the Red Sea, especially for pearl-bearing shellfish and *trochus* shells; Massawa is a traditional market for pearls and mother-of-pearl. In 1988 a total of 4686 t of fish was landed.

Energy resources and industry. The only economically sig-nificant metal ore is gold; iron ore extraction, started in the colonial period, never produced any important results. The salt pans of Massawa and Aseb are a much more substantial asset. In the absence of any fossil fuel reserves, energy is produced for the most part from imported crude at thermal power plants in the two principal ports (there is a refinery at Aseb). Hydroelectric plants in the highlands make only a modest contribution. Asmara and Massawa are the two principal industrial centers. The capital, in particular, has developed not only food-processing industries but also leather and shoe-making factories based on local products, a textile industry (using cotton) and a paper industry (11,000 t in 1989). Asmara is also the center for processing mother-of-pearl collected along the coast. There is a significant fish-canning industry in Massawa. Other smaller industrial centers are Keren and Akordat.

Commerce and communications. For trade and other aspects of Eritrea's economy, there are no data distinct from the overall information for Ethiopia. After the long political crisis, the nation undoubtedly faces a difficult period of restructuring and reestablishing economic and commercial relations, both within eastern Africa (especially with Ethiopia), and with other international trading partners. It is fundamentally an extremely poor country with few exportable products, in great need of outside aid for its very survival.

Transportation facilities have essentially remained as they were at the end of the colonial era, centering on the vital 190-mi [306-km] rail artery that links Massawa to Asmara, climbing to an elevation of 7500 ft [2300 m] and continuing across the highlands to Akordat. There are just over 600 mi [1000 km] of roads, again built by the Italians. Asmara has the only international airport, and Massawa is the principal commercial port, followed by Aseb. In 1991 the merchant fleet consisted of some 30 ships with a total gross register tonnage of 75,000 t.

Historical and cultural profile. Since prehistory Eritrea has shared the destiny of Ethiopia; it was an integral part of the kingdom of Aksum (the region's greatest historic and religious center that was established in the highlands near the Eritrean border), and the successor kingdom of Ethiopia. Only in the 7th century, with the rise of Islam, did the coasts become dominated by Arabs, who established a principality on the Dahlak islands off Massawa. In the 16th century the Turks captured Massawa, and from there attempted (unsuccessfully) to occupy Ethiopia. Islamic colonists thus became established in the coastal region, and Arab domination lasted until the second half of the 19th century, when Egypt acquired the coastal regions (1865) and in turn attempted to conquer the highlands but were repulsed by Ethiopian troops. In 1882, Italy established its first African colony at Aseb, supplanting the Egyptians and later occupying Massawa, Asmara, and Keren. The colony of Eritrea was founded in 1890 following treaties with Ethiopia, and became the base for the first, unsuccessful, Italian attempt to conquer Ethiopia (1896, battle of Adwa), and for the second one in 1936.

After World War II Eritrea was reannexed by Ethiopia, but a separatist movement soon came to life, focusing on guerrilla action by the Eritrean People's Liberation Front, which allied itself with rebels in neighboring Tigre. Independence was obtained on May 25, 1993, following a free referendum held a month earlier.

ETHIOPIA

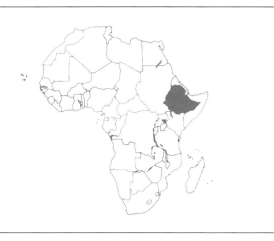

Geopolitical summary

Official name	Hebresabawit Yatyopya
Area	436,234 mi^2 [1,130,139 km^2]
Population	42,019,418 (1984 census); 44,265,529 (1988 estimate)
Form of government	Democratic popular republic under military rule. The head of state is also secretary of a single political party to which all members of the Parliament (*Shengo*) belong.
Administrative structure	13 regions
Capital	Addis Ababa (pop. 1,424,575, 1984 census)
International relations	Member of UN and OAU; associate member of EC
Official language	Amharic; other Hamitic and Semitic languages are also spoken along with English and Italian
Religion	Christian (55%, predominantly Ethiopian Orthodox); Muslim (35%); animist (10%)
Currency	Birr

Natural environment. Ethiopia occupies the highest part of eastern Africa: a complex of high plateaus stretching between the upper Nile valley and the Red Sea, divided lengthwise by the Danakil depression and the Galla trench. The territory is bordered on the northeast by Eritrea, on the east by Djibouti and Somalia, on the south by Kenya, and on the west by Sudan.

Geological structure and relief. The Ethiopian plateau consists geologically of a basement of very old crystalline rocks on which repeated marine transgressions during the Mesozoic and Cenozoic deposited massive blankets of sediment. Large-scale tectonic movements (the same thought to be responsible for the separation between the African and Arabian plates) caused the entire region to rise and bend into the shape of a gigantic anticline. Along its axis, fracturing and downfaulting occurred, cre-

ating the great tectonic trench of eastern Africa and generating extensive outflows of volcanic lava, which still cover large areas of the Ethiopian highlands. Erosion has cut into these structures in various ways, isolating their highest portions to form crests or individual peaks (*ambe*), defining plateaus, and excavating steep-sided valleys. The highlands average over 6500 ft [2000 m] in elevation, with some peaks exceeding 13,000 ft [4000 m] (Rasdajan, 15,154 ft [4620 m]). These elevations are more pronounced along the edges of the tectonic trench, which divides the Ethiopian region into two distinct parts: the west is the Ethiopian plateau in the strict sense, which features the highest elevations and slopes toward the Nile basin to the west and the Eritrean coast to the north; to the east of the trench, the mountains form a sort of inclined plane (Ogaden plateau) which drops toward the Somali lowlands. The northern section of the trench widens out into the arid Danakil plain, where active volcanism is evident, and the adjacent depression of the Salt plain at 380 ft [116 m] below sea level, an endorheic basin separated from the Red Sea coast by a few modest volcanic hills. The southern part of the trench narrows to form the Awash river valley, and contains several lake basins (including Lake Galla) with no outlet to the sea. Eritrea's recent independence has deprived Ethiopia of its only outlet to the Red Sea.

Hydrography. Ethiopia's hydrography is conditioned by its relief. Most of the plateau to the west of the Danakil trench flows into the Nile, one of whose major spring-fed tributaries, the Blue Nile, rises in Lake Tana (the largest lake in Ethiopia), forming a deep and picturesque canyon valley. Another major Nile tributary is the Atbara, which rises a little north of Lake Tana and flows north between the Amara and Tigre plateaus. The Awash river feeds Lake Abbe, an inland basin at the Djibouti border. The lands east of the Galla trench are drained by two large rivers that flow into the Indian Ocean: the Juba and the Webi Shebelle. Lesser rivers fade out into the arid central Somali steppes. A substantial portion of Ethiopia is endorheic, including the Danakil and the basins of Lake Galla and Lake Turkana, the northern tip of which, at the mouth of the Omo river, belongs to Ethiopia.

Climate. The climate is highly dependent on the presence and magnitude of mountains, which mitigate temperatures and deflect moist air masses. Extremely high temperatures occur only in the Danakil depression and in the lowest parts of the southern plateau at the Somali border, with maximums exceeding 122°F [50°C] in the Salt plain. The rest of the country can be divided into well-defined climatic bands governed by elevation, as in other mountainous tropical countries; the Ethiopians call these the *quollà* (2000–6000 ft [600–1800 m]), *voina degà* (6000–8200 ft [1800–2500 m]), and *degà* (above 8200 ft [2500 m]). The Ethiopian plateau, most of which lies above 6500 ft [2000 m], is characterized by uniform temperatures, with minimal annual swings. An example is the case of Addis Ababa, the capital, located 8020 ft [2445 m] above sea level, where January and July temperatures differ by no more than a degree or two from the annual mean of 63°F [17°C]; because of the elevation, however, diurnal swings are much greater. Harar, located near the western rim of the Galla-Somali plateau, is said to enjoy the country's most pleasant climate, with an annual temperature range of 54–77°F [12–25°C], averaging about 68°F [20°C]. Ethiopia's morphological conditions also affect winds and atmospheric humidity. The general wind circulation pattern is of the monsoon type: in the summer, moist air from the Indian Ocean blows over the highlands from the east, while during the winter, the plateau is affected by dry northwest continental winds. The moist summer air masses pass over the Somali lowlands and condense over the mountains as rain, which tends to become more plentiful toward the interior, reaching 62 in. [1600 mm] on the highest mountains. The eastern part of the Ogaden plateau and the Danakil trench are arid (less than 8 in. [200 mm] of precipitation per year).

In 1883 the explorer Gustavo Bianchi, searching for a route between northern Ethiopia and Aseb, crossed the Danakil, which he described as follows:

How is it possible, the mind asks, that not even a rivulet, or the water of some spring, flows among so many mountains? That not even a droplet of God's goodness emerges from them? But it does not.... The plain is arid, always arid, like the one we are on. No helpful detritus covers it: like this one, it is barren, naked. Although it is old, quite old, as a formation, for us it is Nature in its infancy. It does not yet know how to do anything, it does not know how to work; it merely claims, asks. But what can be done for it? ... Nothing; absolutely nothing. Only time, perhaps, only austere and rigid age might nurse and educate it. ... The eastern cones are distant, veiled by the atmosphere: what beautiful waves! And the plain extends, immense, to the east. And through every opening in that horizon, every break between shadow and shadow, between profile and profile, extends another plain.

Maximum precipitation occurs in July and August, but it is preceded in April and May by a period of less intense rainfall called the "little rains." From October to March the entire Ethiopian region is dominated by high pressure and drought; only where summer rainfall exceeds 23 in. [600 mm] is it possible to practice agriculture alongside stock raising, the dominant activity in arid regions where irrigation is impossible.

Flora and fauna. Vegetation varies depending on elevation and rainfall. Steppe and xerophytic shrub communities predominate in lower-lying areas: these are the environments suited for nomadic pastoralism, extending from the semidesert Danakil and the arid Ogaden into the *quollà* zones, where the steppe gives way to savanna, with gallery forests along watercourses and woodlands of acacias, euphorbias, palms (including dum palm), and tamarinds. These features continue into the upper levels of the *voina degà*, with denser forests containing species typical of the Ethiopian plateau, such as juniper and *Podocarpus*; the *degà* is dominated by upland meadows and pastures. Actual tropical forests occur only in the wettest areas, such as the Semien, and along tributaries of the White Nile.

Climate data				
Location	Altitude (ft asl)	Average temp. (°F) January	Average temp. (°F) July	Average annual precip. (in.)
Harar	–	65	64	34.9
Jima	5,707	62	67	62.5

Conversion factors: 1 ft = 0.3 m; 1 in. = 25 mm; °C = (°F – 32) × 5/9

Population. The present-day population comprises ethnic groups resulting from long-ago intermingling of the original Negroid stock and Europoid (Hamitic) populations coming from the north along the Nile tributaries. These were joined later, in historic times, by Semitic populations from the nearby Asian continent. This has resulted in a recognized "Ethiopian" ethnic type, with typically Caucasian features, but differentiated by variations in skin pigmentation and by the distribution of Negroid physical characteristics, such as curly hair, thicker lips, and so on. This ethnic type is common over much of the plateau and internally differentiated mostly by cultural distinctions. The Semitic contribution was the force behind the development of an advanced culture, which founded the ancient kingdom of Aksum that stretched over much of the Abyssinian plateau between Eritrea and Shoa. The Coptic Christian religion took root here, consolidating a cultural identity that managed to resist outside influences, especially the penetration of Islam. The Amhara and the Tigreans are the major Semitic groups; they constitute almost half of Ethiopia's population and are unquestionably dominant in politics and culture. The nation's official language is Amharic. The population of the Hararge region is also of Semitic stock, but has been exposed to Islamic influence. The second major ethnic group consists of non-Semiticized peoples, speaking a Cushitic language, which have remained associated with Hamitic culture. They occupy the southern part of the plateau on either side of the tectonic trench, and constitute various ethnic groups within the Galla (Borana, Arussi, and others), making up about a third of the population. The Sidamo and Danakil also speak a Cushitic language. The rest of the population consists of Somalis, Paleonegroid minorities (the Shangalla), and Nilotic and Nilo-Hamitic groups scattered along the southwestern edges of the country. One significant ethnic and cultural minority was the Falasha, Ethiopian Jews who once lived north of Lake Tana and for the most part have now emigrated to Israel. The European population includes several thousand Italians who stayed behind after the colonial period ended.

Demographic structure and dynamics. Including Eritrea, which is now independent, the population of 20 million in 1950 increased to 42 million in 1984 (date of the last census) and an estimated 47 million in 1988, with an annual growth rate of 3%. Distribution is uneven, reflecting the different opportunities offered by the physical environment. The average density of 98 per mi^2 [38 per km^2] (101 per mi^2 [39 per km^2] for present-day Ethiopia alone) is greatly exceeded in the highlands (326 per mi^2 [126 per km^2] in Shoa, the capital region, 205 per mi^2 [79 per km^2] in the Arussi region, and 153 per mi^2 [59 per km^2] in Gojam). Concentrations are also fairly high in Kefa, Wollo, Tigre, and Gonder, while the other regions have densities lower than the national average (91 per mi^2 [35 per km^2] for Sidamo, 44 per mi^2 [17 per km^2] for Hararge). The lowest figures are found in the Galla trench, especially in the arid Danakil region (23 per mi^2 [9 per km^2] in Bale). Often the boundaries between areas of greater and lesser population density correspond to specific geographic features, such as the edges of the escarpments that separate the plateaus (which are favorable for agriculture) from the Danakil trench inhabited by nomadic herders.

Rapidly increasing population has not led to massive growth of cities, as has occurred in so many other African countries.

Only Addis Ababa, the capital, has more than a million inhabitants. All the other cities have fewer than 100,000, and only six (Dire Dawa, Jima, Nazret, Gonder, Harar, and Dessie) have more than 50,000. Ethiopia has retained its village-centered civilization, linked to agriculture and to exchanges of primary products in the context of a largely autarkic economy. Markets, characteristic of larger centers, perform important social as well as economic functions. Every Coptic town has a church, another element of cultural cohesion. The settlement pattern also reflects the ancient feudal organization of Ethiopian communities, but the hierarchy of inhabited centers has recently been redefined by the development of communication routes, initiated by the Italian colonization.

The capital, Addis Ababa, at the geographical center of the country, is not only the terminus of the railroad to Djibouti, but also the focal point of the main roads that run through the highland regions and descend to the Eritrean and Somali coasts. The secession of Eritrea deprived Ethiopia of the city of Asmara, and of Massawa and Aseb, two harbor towns on the Red Sea. However, communications with Aseb, which is the most heavily used outlet to the sea for central Ethiopia, are provided by the new main road that extends to Addis Ababa through Dessie, and have diminished the importance of the harbor at Massawa, served by the old railroad from Asmara. Harar, formerly important as a focal point of the eastern plateau, has given way to Dire Dawa, favorably located on the way to Djibouti, while Jima has continued to expand its role as a major center of the western highlands. Other cities are of local importance as administrative and commercial centers for their respective provinces.

Economic summary. The economic backwardness of this eastern African country is the result of many factors. Most important is geographical isolation and the lack of a colonizing experience, which (except for the brief Italian interlude) slowed

contacts with the outside world and delayed intensive exploitation of resources. Second is population growth, among the highest in the world, which has found no outlet either internally or outside the country. Lastly, the bloody civil wars, which began in 1962 with Eritrean demands for independence which later spread to Tigre, depleted many resources and created severe internal disorder. This situation was aggravated in 1974 by the abrupt transition from a conservative government, linked to the monarchy and to cooperation with western countries, to a Marxist regime that in reality masked a military dictatorship, incapable of transforming a society still based on feudal privileges into a more open and progressive system. Ethiopia has thus not only remained bound to an economy based on meager subsistence agriculture and herding, but has also squandered the proceeds of its swelling foreign debt on unproductive expenditures (such as military equipment), and on often futile reform efforts. The problem is vividly illustrated by the huge proportion (77%) of the economically active population still engaged in agriculture and livestock, the very low per capita GNP (US$120 in 1990), and an enormous foreign trade deficit.

Agriculture and livestock. Agriculture, which has by no means exploited all available land, finds especially favorable conditions in the mild climate of the plateau, particularly in the vegetation zone called the *voina degà*, between 2600 and 8200 ft [800–2500 m]. Traditional cereals such as durra (sorghum) flourish here; along with wheat, barley, sweet potatoes, and legumes, these are the basis of the population's diet. The only industrial crop in this zone is coffee, a plant that appears to have originated on the Ethiopian plateau and is cultivated in the Jima and Hararge regions. In the lower zone, the *quollà* (between 2000 and 6000 ft [600–1800 m]), the general aridity makes agriculture less profitable but creates better conditions for herding. Irrigation can nevertheless be used to grow high-value cash crops such as cotton, citrus fruit, tobacco, and oilseeds. Cotton is an important crop in the Awash valley, where irrigation water is provided by the dam at Koka (Nazret).

Livestock raising is associated with sedentary agriculture on the plateaus, while it is practiced in the form of pastoralism in the savannas of southern Ethiopia, where it is the predominant activity of the Hamitic populations. The inhabitants of the Danakil are also associated with nomadic herding. Meadows and pasturelands cover more than half of Ethiopia's territory, supporting 30 million head of cattle, together with more than 41 million sheep and goats. Exploitation of the forests, although confined to the wetter southern regions of the plateau where there are patches of tropical forest, does make a certain contribution to the economy. Lastly, fishing is practiced along the Red Sea coasts.

Energy resources and industry. There is no lack of mineral resources in Ethiopia, although they have not yet been completely explored. Gold and platinum are the only metals exploited, in Wollega and Gojam respectively. No petroleum has been recovered yet, however, although its presence has been detected in the Danakil. The salt pans of the Danakil have some economic significance. In terms of energy, Ethiopia possesses great hydroelectric potential; recent projects on the Awash and Blue Nile have focused on this. Energy production has therefore risen from 455 million kWh in 1969 to 819 million kWh in 1989; at least 655 million kWh of this was produced by hydroelectric plants, and the remainder by thermal plants using im-

Administrative structure

Administrative unit (regions)	Area (mi²)	Population (1988 estimate)
Arussi	9,139	1,860,606
Bale	49,042	1,126,697
Gemu Gofa	15,574	1,395,331
Gojam	23,632	3,632,276
Gonder	30,717	3,270,440
Hararge	105,238	4,657,859
Illubabor	17,898	1,078,308
Kefa	21,861	2,740,773
Shoa	32,932	10,714,244
Sidamo	46,227	4,241,527
Tigre	25,060	2,700,921
Wollega	27,206	2,770,598
Wollo	31,708	4,075,959
ETHIOPIA	436,234	44,265,539

Capital: Addis Ababa, pop. 1,424,575 (1984 census)
Major cities (1984 census): Dire Dawa, pop. 98,104; Nazret, pop. 76,284; Gonder, pop. 68,958; Dessie, pop. 68,848

Conversion factor: 1 mi² = 2.59 km²

Socioeconomic data

Income (per capita, US$)	120 (1990)
Population growth rate (% per year)	3 (1980–89)
Birth rate (annual, per 1,000 pop.)	52 (1989)
Mortality rate (annual, per 1,000 pop.)	18 (1989)
Life expectancy at birth (years)	48 (1989)
Urban population (% of total)	13 (1989)
Economically active population (% of total)	42.6 (1988)
Illiteracy (% of total pop.)	95.2 (1991)
Available nutrition (daily calories per capita)	1658 (1989)
Energy consumption (10^6 tons coal equivalent)	1.2 (1987)

ported oil. Energy availability is nevertheless insufficient for industrial development, which has not progressed significantly in the last few decades.

Industry is limited to the processing of primary-sector products, especially cotton, which is processed in a few textile centers (Addis Ababa, Dire Dawa, Debra Berhan, and Bahir Dar) with some input of imported raw materials as well. Other activities include sugar refining, tobacco processing, brewing, food canning, and a paper mill. Domestic needs are also met by the steel mill at Akaki near Addis Ababa, and by a cement plant at Dire Dawa.

Commerce and communications. Ethiopia's balance of trade reflects the nation's state of severe underdevelopment, emphasizing its need for an industrial base. All capital goods and a vast range of finished products must be imported; the only exports are products derived from agriculture (hides and leather, oilseeds, sugar, and meat).

Communications and transportation facilities are grossly insufficient. Besides the old railroad from Addis Ababa to Djibouti there are fewer than 24,000 mi [39,000 km] of roads (1987), only a portion of them paved. East-west surface communications are especially inadequate, complicated by the natural barrier of the Galla trench. For its maritime transportation Ethiopia must use the ports of Aseb and Djibouti, located in Eritrea and Djibouti, respectively.

Historical and cultural profile. *The Aksum civilization.* The Classical writers used the name "Ethiopia" to describe all the known African lands south of the second cataract of the Nile; this therefore included eastern Sudan and the Ethiopian highlands. Toward the middle of the first millennium B.C., southern Arabian peoples migrated into the latter region and intermingled with the previous Cushitic inhabitants, becoming the ethnic foundation on which, several centuries later, rose the kingdom of Aksum, the first organized political and cultural manifestation in Ethiopia's history. From its inception, it created important artistic works such as stelai up to 100 ft [30 m] tall, stone thrones, and massive polygonal structures for secular and religious use.

The Aksum civilization reached its zenith in the 4th century A.D. under King Ezana, who conquered large tracts of land (including the kingdom of Meroë), and converted the inhabitants to Christianity. The spread of this religion had a decisive effect on Ethiopia's subsequent history, not just because the Ethiopian church embraced the Monophysite heresy and followed the patriarchate of Alexandria almost down to the present day (auton-

omy was not granted until 1959), but also because of the artistic expression inspired by Christianity. In architecture, this inspiration produced major monuments such as the monastery at Debra Damo and the monolithic churches of Lalibela; in literature, it was expressed in centuries of religious and courtly writings in the Gheez language, beginning with a translation of the Bible.

Paola Segre has described the ten monolithic churches built in about 1225 by Lalibela, a monarch of the Zaguè dynasty venerated as an Abyssinian saint:

An incredible complex, excavated entirely from living rock, constituting one of the most important cultural treasures anywhere on earth.... The churches are invisible; they are hidden in the ground, excavated into the plateau's covering of basaltic tufa. The buildings are cut out of blocks that are completely hollowed out inside and, unlike so many other rock-cut churches, isolated from the surrounding rock by trenches. They are also carved on the outside to represent a roof, facades, and walls, perforated by windows and doors. It is, in short, an entire church, completely dug out from a single block and still attached to the rock at the base. What is so astonishing about Lalibela, in addition to the number and size of the churches, is the richness of their architectural accompaniments: grottos, crypts, baptismal fonts, and underground passages from one group of complexes to another ...

The Ethiopian Middle Ages. The kingdom of Aksum changed as time passed, shifting its center of gravity southward under pressure from the Begia and Arabs who had occupied the Eritrean coasts, merging with the local Abyssinian populations, and becoming "Africanized" to a certain extent. This slow process crystallized into the kingdom of Ethiopia, established in 1270 by the "negus," or kings, of the Solomonic dynasty, so called because they traced their origins to Menelik, mythical son of Solomon and the Queen of Sheba. It is no accident that literature once again began to flourish in the 14th century with the *Kebra Nagast*, a poem about the dynasty's mythical founders that in time became the defining national epic.

In the period of the "Ethiopian Middle Ages" (13th–19th centuries), however, the religious element was the dominant cohesive factor within the kingdom of Ethiopia, which rebuffed repeated Muslim attacks and enlisted the aid of the Portuguese, who were interested in controlling the Red Sea coast. The tenacity of Ethiopia's defense of Christianity against the Islamic threat was echoed in its intransigent rejection of an attempt by the Jesuits (who had arrived in the wake of the Portuguese) to gain a foothold in Ethiopian territory and bring the country back into the arms of the Church of Rome. In the 17th century, when the negus Susenios converted to Catholicism, the event ignited such widespread xenophobic rebellion that the sovereign was forced to abdicate; in reaction, the Catholic faith was banned, the missionaries expelled, and relations with Europe severed.

The crisis that convulsed the country in the next two centuries derived not from external threats but from an intrinsic weakness: the negus' inability to assert their monarchical power over the feudal interests with their strongholds in the Semien, Tigre, Lasta, and Shoa regions. Not until the mid-19th century did the adventurer Ligg Kasa manage to subjugate all of Abyssinia; he had himself proclaimed emperor under the name of Theodore II and proceeded to rule as an absolute monarch, promoting the modernization of his country. Part of this process was the adoption of Amharic as the written language, a step that marked the decline of the ancient Gheez language and laid the foundation for a new Ethiopian literature.

Contemporary Ethiopia. Theodore II then resumed contacts with Europe, and the western powers did not waste this opportunity to establish their influence over eastern Africa. Alongside France and Great Britain, Italy in particular set its sights on Ethiopian territory; in 1885 it conquered Massawa on the Red Sea, and in 1889 obtained control over Eritrea under the treaty of Uccialli, in return recognizing the negus Menelik as emperor of Ethiopia. Menelik moved the capital to Addis Ababa, intending to continue the modernization programs of Theodore II with Italian aid; but Italian expansionist designs on Ethiopia itself changed his mind, and his victory at Adwa over the troops of General Baratieri (1896) allowed him to reaffirm his nation's independence. Forty years later that independence was called into question when fascist Italy resumed its expansionist policy. By then Haile Selassie I had acceded to the throne, and under his cautious reformist leadership the country became a constitutional monarchy. But Haile Selassie was unable to resist the Italian attack, and his appeals to the League of Nations were in vain: the second war between Italy and Ethiopia (1935–36) led to Ethiopia's annexation as part of Italian East Africa.

The colonial adventure ended after only five years, during World War II, when Ethiopia was occupied by Allied troops and its independence proclaimed. But the consequences of the colonial period continued to be felt for many years, since it had unleashed the divisive idea of Eritrean separatism. Following the collapse of the Italian colonial system, the northern region of Eritrea had taken a dim view of reunification with Ethiopia, with which it was first linked in a federation (1952), and ultimately annexed as a mere province (1962), until gaining independence in 1993.

Haile Selassie never found a solution to the Eritrean problem, and his cautious attempts at reform foundered on it. The situation was complicated by rising tensions within the ruling class; in 1974 a military coup deposed the aged monarch and brought to power the radical nationalist wing of the armed forces, headed by Colonel Mengistu Haile Mariam. The new regime nationalized land, banks, and insurance companies as part of a program of "Ethiopian socialism." In reality, what prevailed over the years were an authoritarian attitude and systematic violations of human rights. Mengistu proved incapable of responding to the nation's problems, aggravated by recurrent droughts, thus nourishing popular discontent and the armed struggles waged by liberation fronts in Eritrea (FPLE) and Tigre (FPLT). Further weakened by a conflict with Somalia over control of the Ogaden region (1977–84), and deprived of Soviet and Cuban aid, the regime put down an attempted coup in 1989 but emerged from it even more unstable and isolated; it did not survive the last assault mounted by independence fighters early in 1991, which forced Mengistu into exile.

DJIBOUTI

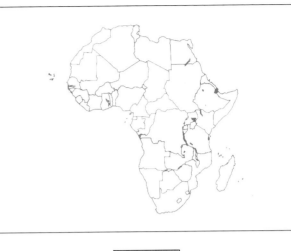

Geopolitical summary

Official name	Jumhuriya Jibuti / République de Djibouti
Area	8955 mi^2 [23,200 km^2]
Population	530,000 (1990 estimate)
Form of government	Parliamentary republic. The head of government and the Chamber of Deputies are elected by universal suffrage every 6 and 5 years, respectively.
Administrative structure	5 districts
Capital	Djibouti (pop. 290,000, 1988 estimate)
International relations	Member of UN, OAU, Arab League; associate member of EC
Official language	Arabic and French; Cushitic languages widely spoken
Religion	Predominantly Muslim; about 9000 Catholics
Currency	Djibouti franc

Natural environment. This, the smallest Eastern African nation, borders Eritrea to the north, Ethiopia to the northwest and southwest, and Somalia to the southeast, and is strategically located west of the Bab al-Mandab, the strait connecting the Red Sea with the Gulf of Aden and the Indian Ocean. Djibouti's territory includes the easternmost part of the Danakil tectonic depression, as well as a more depressed intermediate portion (Afar region) that expands toward the east, bordered on the north and south by mountains (largely made up of basaltic rocks produced by ancient volcanic activity) that occasionally exceed 6500 ft [2000 m] in altitude and enclose large plateau areas. Small active volcanic cones are present toward the Danakil region. The innermost part of the central depression is connected to the great Galla trench that bisects the Ethiopian highlands. The Awash river runs through it and empties at the Ethiopian border into Lake Abbe, which has no outlet to the sea. The land rises from the coast toward the interior in a series of stepped plains, and is cut into by the Gulf of Tadjoura, ringed with coral islets, on the southern edge of which is the capital, Djibouti.

Climate data

Location	Altitude (ft asl)	Average temp. (°F) January	Average temp. (°F) July	Average annual precip. (in.)
Djibouti	13	78	96	5.1

Conversion factors: 1 ft = 0.3 m; 1 in. = 25 mm; °C = (°F – 32) × 5/9

There is almost no surface hydrography. Wadis collect water from seasonal storms, but soon dry up. The Awash river ends at Lake Abbe, Djibouti's only fresh water body, while at the center of the country lies Lake Assal (over 300 ft [100 m] below sea level), the level of which is subject to severe seasonal variations. Intense evaporation causes salt deposits to form along its shores.

The climate is extremely arid, with very high average temperatures that drop below 86°F [30°C] only in winter. Absolute maximum temperatures can reach 113°F [45°C] in the summer, especially when the hot *khamsin* brings sandstorms from the nearby Arabian peninsula. The high relative humidity makes the heat even more unpleasant, but rainfall is extremely rare (about 5 in. [130 mm] a year at Djibouti), exceeding 20 in. [500 mm] only in the highlands, and even then with great variations from year to year. At elevations above 3300 ft [1000 m] temperatures become milder, but temperature swings also increase.

Typical steppe vegetation prevails in the lowland areas. The mountain slopes are bare and rocky, with scattered acacias and euphorbias, but above 2600 ft [800 m] there are woodlands with junipers, "candlestick" euphorbias, aloes, and umbrella-shaped acacias, as in Mt. Gonda National Park. The dum palm is a typical tree species. The increased moisture along wadis promotes denser vegetation, including fringing forests.

Population. Two ethnic groups make up most of the country's population: to the north the Afars, of Danakil stock; and to the south the Issas, of Somali origin. Djibouti is at the border between Ethiopia and Somalia, both of which have often claimed it; moreover, because of its location next to the Bab al-Mandab strait, it has always been exposed to influence from the nearby Arabian peninsula, from which Islam was brought to the country.

Fifteen percent of the present-day population is of Yemenite origin, concentrated almost exclusively in the capital. The Issas (47% of the population), are more numerous than the Afars (37%), but physical differences between the two peoples, both of

Administrative structure

Administrative units (districts)	Area (mi²)	Population (1990 estimate)
Ali Sabieh	926	–
Dikhil	2,779	–
Djibouti	232	–
Obock	2,200	–
Tadjoura	2,818	–
DJIBOUTI	**8,955**	**530,000**

Capital: Djibouti, pop. 290,000 (1988 estimate)

Conversion factor: 1 mi² = 2.59 km²

Socioeconomic data

Income (per capita, US$)	1250 (1990)
Population growth rate (% per year)	3.5 (1980–89)
Birth rate (annual, per 1,000 pop.)	19.9 (1989)
Mortality rate (annual, per 1,000 pop.)	17.7 (1989)
Life expectancy at birth (years)	45.4 (1989)
Urban population (% of total)	80 (1989)
Economically active pop. (% of total)	–
Illiteracy (% of total pop.)	41 (1987)
Available nutrition (daily calories per capita)	–
Energy consumption (10^6 tons coal equivalent)	0.12 (1987)

whom speak Cushitic dialects, are difficult to detect. There is a clearer distinction between the urban population and the nomadic herders in the interior. The annual population growth rate is high (3.5%), due partly to immigration; the population grew from 90,000 in 1966 to more than half a million in 1990. The capital, Djibouti, is home to more than half the inhabitants of this small country. Its fate hinges on its strategic location (exploited by the French colonists), and its function as a harbor and free port on the sea route from the Mediterranean to the Indian Ocean through the Suez Canal. Djibouti is also Ethiopia's most important outlet to the sea, thanks to a railroad built in 1929 linking it to Addis Ababa. The nation's second largest city is Tadjoura, on the northern shore of the gulf of the same name, a marshaling point for the caravans that once climbed through the Danakil Mountains to the Ethiopian highlands.

Economic summary. The economy is greatly influenced by the arid climate and lack of water, which make agriculture a precarious undertaking, except in the oases where there is some access to underground water. The principal activity in this sector is animal husbandry, especially of goats and sheep; this is declining, however, demonstrated by the fact that livestock holdings decreased slightly between 1960 and 1990 despite a fivefold increase in human population. The most important economic activities are those associated with commercial functions and other general service activities, in the capital and its port. Industry is practically nonexistent; the only mineral exported is salt.

Communications center on the capital's harbor and on the railway line to Addis Ababa, a true lifeline for this nation, which survives on the trade between Ethiopia and the rest of the world. There is an international airport at Ambouli, near Djibouti city.

Historical profile. This territory, not large but of pivotal commercial and strategic importance, was once called the *Côte française des Somalis* (French Somali Coast), then the French Territory of the Afars and Issas after the country's two major ethnic groups. The colonial administration, which took over power between 1859 and 1888, when the port of Djibouti was founded, exploited ethnic conflicts between the two groups for its own ends. France supported the Afars, who until recently favored maintaining ties with the European power, while the Issas tended toward union with Somalia, and organized the clandestine Somali Coast Liberation Front (FLCS). It was not until 1977 that independence was approved unanimously, and a republican form of government selected.

SOMALIA

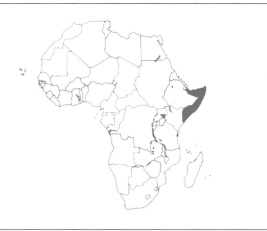

Geopolitical summary

Official name	Jamhuriyadda Dimugradiga Somaliya
Area	246,135 mi^2 [637,657 km^2]
Population	3,253,000 (1975 census); 5,074,000 (1980 estimate)
Form of government	Presidential republic on socialist model; the head of government and the popular assembly are elected every 7 years by universal suffrage
Administrative structure	16 regions
Capital	Mogadishu (pop. 750,000, 1989 estimate)
International relations	Member of UN, OAU, Arab League; associate member of EC
Official language	Somali; Arabic, Italian, and English are used administratively
Religion	Predominantly Muslim
Currency	Somali shilling

Natural environment. Somalia occupies the eastern end of the continent, extending along the peninsula called the Horn of Africa between the Gulf of Aden to the north and the Indian Ocean to the east, and bordered on the northwest by Djibouti, on the west by Ethiopia, and on the southwest by Kenya.

Geological structure and relief. Somalia constitutes the eastern fringe of the Ethiopian plateau, consisting of a series of tablelands sloping to the southeast, the natural continuation of the great highlands that rise to the east of the Galla–Danakil tectonic trench. The mountains along the coast of the Gulf of Aden are an extension of the raised rim forming the interface between the highlands and the Danakil depression, and contain the country's highest elevations (Surud Ad, 7898 ft [2408 m]), which drop down to the sea in steep cliffs all the way to rocky Ras Asir. From this mountainous rim, the highlands slope away toward the southeast, merging into the plateaus of the Ogaden (in Ethiopia) and Mijertins regions, which in turn descend to the

vast peneplain of Mudug and the coastal plain of Benadir, surrounded by long chains of sand dunes that block the passage of the Webi Shebelle, Somalia's largest river, as it flows to the ocean. Geologically, these lowlands constitute an ancient substructure of crystalline rocks that, as they sank, were invaded by the sea and blanketed with sediments of Cenozoic and Recent age, which were then overlain by alluvial deposits caused by surface erosion. Somalia can therefore be divided into two major natural regions: the north, characterized by highly incised highlands sloping toward the ocean; and the south, where the low tablelands fall away toward the wide coastal plains.

Hydrography. Somalia's hydrography is fairly simple; two large rivers, the Webi Shebelle and the Juba, flow down from the heart of the Ethiopian highlands and run almost parallel to one another through the Somali plains. The former, which carries more water, encounters lines of dunes near the shoreline that force it to flow parallel to the coast; then, much farther to the southwest, it joins the final stretch of the Juba in the coastal marshes. The Webi Shebelle is East Africa's largest river in terms of length and size of drainage basin. Its flow is not continuous throughout the year, however; like the Juba, it has two flood seasons corresponding to rainfall conditions on the Ethiopian highlands. The water of these rivers is nevertheless of fundamental importance to Somali agriculture. Thanks to their precious contribution, Somalia's "Mesopotamia" is the most fertile and the most densely populated region of the country. Other watercourses also descend to the Indian Ocean parallel to the two major rivers; they are highly torrential but contain no water for most of the year.

Climate. The climate is influenced by the proximity of the equator, which passes across the extreme southern tip of Somalia, and by the ocean, which moderates temperatures on the coast and induces a monsoon type of atmospheric circulation. Annual average temperatures at Mogadishu vary from 77–81°F [25–27°C], but can exceed 86°F [30°C] in the interior. In the northern highlands, temperatures are moderated by altitude. Rainfall is extremely sparse throughout the country, generally less than 20 in. [500 mm] per year, since the monsoons blow parallel to the coastline. As a result, the southwest monsoon, which blows during the summer and brings in maritime air masses, quickly loses its moisture and brings little rain; precipitation levels drop off rapidly toward the northeast. The dry but cool winter monsoon, on the other hand, which blows from the Asian continent, becomes weaker as it moves south and is barely perceptible at Mogadishu. Northern Somalia is extremely arid, with less than 8 in. [200 mm] of precipitation annually, decreasing to less than 4 in. [100 mm] in the narrow coastal strip along the Gulf of Aden. The southern part of the country lies between the 8 and 20 in. isohyets [200–500 mm], with the exception of the region between the two major rivers, where rainfall slightly exceeds 20 in. [500 mm].

Flora and fauna. The general aridity of the climate exerts a strong influence on vegetation. Savanna landscapes with xerophytic tree communities predominate, giving way in the highlands to typical steppe grasslands; tropical forest is limited to the wettest parts of southern Somalia along the two major rivers (fringing forests). Typical species include several aromatic plants such as frankincense and myrrh, along with cactuslike *Euphorbia* and gum-bearing species. The marshes along the southern coast, parallel to the lower course of the Webi Shebelle, are a typical moist environment.

Population. Ethnically, the Somalis are related to the Hamitic Galla peoples of southern Ethiopia, with whom they share certain Caucasoid traits—including facial features, tall stature, and brown skin color—and their language, which belongs to the Cushitic group (specifically lower Cushitic). Relatively long isolation has produced considerable uniformity in physical features, but foreign cultures have left their mark, notably Arab influences, which resulted in conversion of the country to Islam in the 7th century and the establishment of settlements of Arab minorities along the coast. The ancient Somalis lived predominantly as shepherds, and this way of life is reflected in the survival of tribal groupings (*qabilah*).

With the difficult environmental conditions and extensive exploitation of resources, population density is extremely low: a territory almost the size of Texas contained slightly more than 5 million inhabitants in 1980, with a density of barely 21 per mi^2 [8 per km^2]. But the high annual population growth rate (3%), which caused the population to double between 1961 and 1980, indicates that the number of Somalis now exceeds 6 million, despite the famines and wars that have tormented the country in recent years, culminating in the overthrow of the Siad Barre regime in January 1991. The most populous regions are those to the south, especially Benadir (where the capital city is located) and Lower Shebelle. The Northwest region also has areas of higher density around the cities of Hargeisa and Berbera.

The proportion of urban population is low (36%). Aside from Mogadishu (Xamar), which houses about 600,000 inhabitants, only four cities exceed 50,000 in population: these are Hargeisa, at the edge of the northern plateau (70,000); Kismayu, on the Indian Ocean; Berbera, the main port on the Gulf of Aden (65,000); and Merca, in the Lower Shebelle region (60,000).

Mogadishu, the ancient economic and religious center, began to decline in the 15th century as Portuguese commercial influence increased, but regained its pre-eminence when Italy acquired it from the ruler of Zanzibar, and made it the focus of its colonial hegemony over the country. The city owes much of its present appearance, and its role as the political and administrative capital of Somalia following independence (gained in 1960), to Italian domination. Since independence Mogadishu's population has increased fivefold, while the other cities have developed much more slowly or have in fact regressed, as is the case with Kismayu, in Lower Juba with an excellent harbor, that has suffered from its peripheral location. It is also worth mentioning the agricultural center of Jioher, established by Italians on the Webi Shebelle (under the original name of Villaggio Duca degli Abruzzi) in order to develop plantation agriculture. Attempts have been made since the Italian colonial period to reduce

Climate data

Location	Altitude (ft asl)	Average temp. (°F) January	Average temp. (°F) July	Average annual precip. (in.)
Mogadishu	16	79	78	16.9
Berbera	10	76	97	1.9

Conversion factors: 1 ft = 0.3 m; 1 in. = 25 mm; °C = (°F − 32) × 5/9

Administrative structure

Administrative unit (regions)	Area (mi^2)	Population (1980 estimate)
Bakol	10,422	148,724
Bari	27,020	222,287
Bay	15,054	450,986
Benadir	640	520,103
Galgudud	16,598	255,856
Gedo	12,352	235,061
Hiran	13,124	219,328
Lower Juba	23,546	272,368
Lower Shebelle	9,650	570,649
Middle Juba	8,878	147,810
Middle Shebelle	8,492	352,040
Mudug	27,020	311,230
Nagal	19,300	112,162
Northwest	17,370	654,990
Sanaag	20,844	216,539
Togder	15,826	383,867
SOMALIA	246,136	5,074,000

Capital: Mogadishu, pop. 750,000 (1989 estimate)
Major cities (1986 estimate): Hargeisa, pop. 70,000; Kismayu, pop. 70,000; Berbera, pop. 65,000; Merca, pop. 60,000.

Conversion factor: 1 mi^2 = 2.59 km^2

nomadic herding activities and encourage the population to settle permanently, but they have been successful only along the coast and in the more fertile farming areas.

Economic summary. The backwardness of the Somali economy is unequivocally evident from estimates of land utilization. Only 1.6% of the country's land area is arable or used for orchard crops, while 67.4% is occupied by meadows and pastures, and 14.2% by forests and woodlands; the remainder consists of unproductive, uncultivated land. The majority of the economically active population (almost 65%) is employed in farming and animal husbandry, with those engaged in nomadic and seminomadic herding clearly predominant. Permanent agriculture has taken hold only in the southern part of the country, especially in the region between the Webi Shebelle and the Juba, where Italian colonists also introduced cash crops such as sugar cane, bananas, cotton, peanuts, and citrus fruits. These activities were made possible by a system of irrigation canals, and by the creation of market towns, roads, and commercial infrastructures. Elsewhere the basic resources of the population are entrusted to a meager kind of subsistence agriculture based on the production of low-grade cereals (durum wheat, sorghum, and millet); animal holdings, which recovered after being devastated by drought, were again substantial by 1991 (5 million cattle, almost 7 million camels, 35 million sheep and goats). The government has devoted considerable attention to improving fishing, which until 1972 was practiced with very modest equipment and craft-based methods. In late 1974, drought and famine drove the population, most of them nomads, to move to the coast and engage in this activity, which was entirely new to them.

Socioeconomic data

Income (per capita, US$)	150 (1990)
Population growth rate (% per year)	3 (1980–89)
Birth rate (annual, per 1,000 pop.)	48 (1989)
Mortality rate (annual, per 1,000 pop.)	18 (1989)
Life expectancy at birth (years)	48 (1989)
Urban population (% of total)	36 (1989)
Economically active pop. (% of total)	40.1 (1988)
Illiteracy (% of total pop.)	45.2 (1991)
Available nutrition (daily calories per capita)	1736 (1989)
Energy consumption (10^6 tons coal equivalent)	0.41 (1987)

Italian agricultural aid continued during the decade of Italian trusteeship under UN auspices (1950–1960), and even after Somalia's independence in 1960, including support for infrastructure and certain food-processing industries (sugar refining, oil processing, canning) and primary processing sectors (cotton ginning, soap-making, distilleries). The Siad Barre regime, which came to power in 1969, failed in its efforts to promote a development plan. A high rate of population growth, internal conflicts, the unsuccessful attack against Ethiopia aimed at conquering the Ogaden, and most recently the civil war, have done perceptible damage to Somalia's civic and economic institutions. Added to this are a lack of energy and mineral resources (except for salt), and dependence on foreign sources for all consumer goods, with the result that economic aid from Italy and other countries has only temporarily alleviated some of the burden of external debt.

Internal communications routes in 1988 comprised little more than 13,600 mi [22,000 km] of roads (only 1500 mi [2500 km] paved); the railway from Jawhar to Mogadishu has been closed. Mogadishu is the location of the only international airport and the main maritime port, while there are other harbors at Berbera, Merca, and Kismayu. The mercantile fleet, with total tonnage of almost 2 million gross register tons in 1974, had declined to barely 17,010 t in 1990.

Historical and cultural profile. *Somalia and Islam.* The first firm dates in Somali history appear during the period of Muslim expansion. Beginning in the 7th century, commercial centers involved in trading slaves, gold, and ivory—Mogadishu, Merca, Brava, and Zeila—flourished along the Gulf of Aden and the Indian Ocean. In Mogadishu large mosques were erected, including those of Hamar wen, Arba-rukún, and Abd-al-Aziz. In the northwest corner of the country there arose first the nation of Ifat (13th century), then Adal (14th century); although the former was almost suffocated by the Christian kingdom of Ethiopia, the latter espoused holy war as its *raison d'être*, and in 1527 not only invaded Ethiopia but ravaged it for more than ten years. Only with Portuguese intervention were the Muslims driven out of Ethiopia, and this defeat, together with the invasion of the Galla people, caused the fall of the kingdom of Adal. A number of forces then took turns attempting to control the coasts of the Horn of Africa: Omani imams, Egyptians, French, British, and Italians.

The two Somalias. In the 1890s, it was the latter three European powers who ended up parceling out Somali territory. The French remained on its fringes, taking up a position in the northwest region corresponding to present-day Djibouti. Located immediately adjacent to this, around the cities of Berbera, Zeila, and Bulhar, was British Somaliland; Italian Somaliland gradually embraced the rest of the country. In the early 20th century both the British and Italian possessions took on a definite colonial configuration, at least after the defeat of a Somali insurrection headed by the religious leader Mohamed ibn Abdullah.

The presence on Somali territory of two rival colonial forces had a negative influence on the development of an independence movement, dividing and delaying local opposition. Moreover, during World War II Italy combined its Somali and Eritrean possessions with Ethiopia to form Italian East Africa, and used the Horn of Africa as a strategic base for attacks against the Allies; by 1940 they had also occupied British Somaliland. The situation very soon reversed itself, however, and the British got the better of the Italians; until 1949 the two Somalias were united under a British protectorate, a state of affairs that laid the groundwork for territorial integration and for the rise of a strong nationalist party, the Somali Youth League.

Independence. All through the 1950s, trustee administration of its former colony was entrusted to Italy; in 1960 this region merged with the portion controlled by Great Britain to create the Somali Republic. The parliamentary system instituted at independence remained in effect for a decade, and was dominated by the Somali Youth League, a patronage-ridden party with strongly conservative leanings.

In 1969 this regime was overthrown by a military junta led by General Mohamed Siad Barre, which espoused radical Arabism and socialism and ultimately created an openly Marxist political system. This allowed the "revolution" to move from nomadism to socialism.

Confronted with a nation of considerable diversity that had been weakened by the legacy of two colonial regimes, the military government undertook a massive reform program, involving agriculture, the educational system, even the creation of a homogeneous Somali literary language. One of the prime objectives was also to combat deeply entrenched tribal power structures. But over the course of two decades the nature of the regime changed: Marxism was abandoned, relations with the USSR were broken off (after Moscow supported Ethiopia during the skirmishes of 1977–78), and Siad Barre concentrated power in the hands of his own family and his own tribe, the Marehan, retaining it by systematic violations of human rights. Internal dissent steadily deepened; five guerrilla movements extended their power in both the north and south, ultimately controlling most of the country. Supported by wide popular consensus, they brought down Siad Barre's regime in January 1991. In the autumn of that year, the Somali capital was still the scene of violent clashes pitting two opposing factions: on one side the men of President Ali Mahdi Mohamed; on the other the rebel troops of General Mohammed Farah Aidid, leader of the United Somali Congress (CSU), which had assumed power after the overthrow of Siad Barre. These two members of the Hawiye clan—although Mahdi belonged to the Abgal subclan and Aidid to the Habar-Ghidir—fought for control of Mogadishu; Mahdi's forces in the CSU then overcame Aidid's military wing within the party. Ali Mahdi's government was not accepted throughout the country, however: the English-speaking northern region broke away, and the south remained in the hands of the Darod, Siad Barre's ethnic group.

EASTERN AFRICA

Images

1. *The village of Tadjoura, marshaling point for caravans to Ethiopia since the late 19th century, is hidden between the sea and the green, forested slopes of Mt. Gonda. Located amid palm trees, the little harbor is accessible only to small, lightweight boats, due to the presence of small coral islands and reefs made of ancient volcanic rocks, interrupted by semiconical natural deposits.*

2. *An Amhara shepherd and his flock on the Danakil plain, a vast arid region of Ethiopia. The country has such enormous animal holdings that despite the use of rather primitive methods and equipment, animal husbandry is the second most important economic activity in Ethiopia. It also supplies two export products, leather and hides.*

3. *Mangrove trees stand in the waters of the Red Sea along the coast of Ethiopia. Formed by the subsidence of a long trench between the Asian and African plates, the Red Sea acquired significance as a communication route between Europe and the Far East with the opening of the Suez Canal in 1869. Its waters, very warm and often containing red algae that gave it its name, help to moderate Ethiopia's climate, already tempered to some extent by a balance between the high elevations of the interior and the*

effects of tropical latitude.

4. *Lake Awasa, at an elevation of 5602 ft [1708 m] above sea level, is the second highest lake basin in the Galla trench, east of Mt. Badda (13,556 ft [4133 m]). Extending towards the north-northeast, the bottom of the trench gradually rises towards the greatest heights of the Somali plateau, and contains a number of lakes, all located at different elevations and separated from each other by a variety of volcanic phenomena.*

5. *The salt lake of Asale, located 380 ft [116 m] below sea level, occupies the largest depression in the northern part of the Salt Plain, near the eastern slopes of the Ethiopian highlands. The Salt Plain, an immense basin with an area of some 2000 mi² [5000 km²], is interrupted by basaltic dikes and faults in the volcanic rock. The floor of the basin is sandy, with extensive dune fields occasionally in the form of barkans, while its margins are terraced. The area is a true desert, with temperatures around 104°F [40°C] even in winter.*

6. *The falls of the Blue Nile. Hemmed into a narrow gorge through which the river runs for its entire course through Ethiopia, the falls face the "Portuguese bridge," the only one that crosses the river. The*

Blue Nile is the most important river in all of East Africa. It rises to an elevation of 10,234 ft [3120 m] in the heart of the Ethiopian highlands, in a marshy dell at the foot of Mt. Amedamit, and runs north to empty into Lake Tana. Emerging from the lake, it reaches the Sudanese plain, and after having traveled some 850 mi [1400 km], pours its waters—rich in fertile sediment—into those of the White Nile at Khartoum.

7. *An Amhara woman carries water in the Ethiopian highlands near Mt. Abune Yosef (13,743 ft [4190 m]). The Amhara are the major ethnic group in Ethiopia (approximately 2 million), and live mostly in the east central part of the highlands, in permanent villages where they practice agriculture and animal husbandry. The men wear a long shirt over tight white cotton trousers that fit snugly over the calf; women wear long, ankle-length tunics. The* mateb, *a kerchief of dark blue silk, is commonly worn to distinguish Christians from Muslims.*

8. *A woman of the Tigre people, a group of Ethiopic populations living predominantly in the region of the same name, between the city of Asmara and the Tekeze river. The Tigre nuclear family is based on a patriarchal model; each man*

uses the material and spiritual strength of his family to enhance and maintain his reputation and social status, with the primary objective of increasing the authority of his own house. Women play a decidedly secondary role, confined to conjugal duties and the performance of domestic tasks.

9. *A Danakil woman draped in the typical* futa, *a brightly colored cotton cloth that is tied around the shoulders but does not cover the neck, which is often adorned with showy necklaces. The Danakil, a Hamitic people who live in Djibouti and northern Ethiopia, are among the oldest inhabitants of East Africa. Their way of life is economically simple, based on nomadic stockraising and hunting. Their name for themselves is "Afar," meaning "free people"; the word also designates their language.*

10. *A Somali village along the banks of the Webi Shebelle. The coastal and riverside regions, where agriculture and fishing activities are concentrated, are characterized by permanent villages of huts called* mundul. *Where polygamy requires a separate dwelling for each wife, the head of the household and his sons use the largest hut, while the smaller ones are also used for livestock, for cooking, and—with the addition of a*

raised earth floor to protect against dampness—for grain.

11. *View of Asmara (Ethiopia). The capital of the Eritrean province is located at an elevation of 7698 ft [2347 m] on the northern slope of the Ethiopian highlands. Its climate is very healthy, and has encouraged a high population density. Asmara was originally a little village of huts surrounded by woodlands, the feudal seat of a ras. The arrival of the Italians in 1889 set off a process of explosive development in a modern style, with the construction of a large network of water and rail connections linking Asmara to other cities of the country.*

12. *The city of Djibouti. Capital of the republic of the same name, which became independent in 1977 after 115 years of French domination, it owes its own commercial prosperity to the harbor, a major international port of call for traffic through the Suez Canal, and to the railway linking it to Addis Ababa, its ultimate terminus. Alongside the administrative district, which sits high up on a coral platform, stretches the colorful native quarter of Bender Djedid, with its small mosques and typical terraced and arcaded houses.*

13. *The Parliament Building in Addis Ababa. The Ethiopian capital, whose name means "new flower" in Amharic, sits in a wide, verdant basin at an elevation of 8659 ft [2640 m], bordered by the Entotto mountains and the valley of the Awash river. Rapid population growth over the past few decades has also resulted in considerable urban development. The most modern part of the city, on the east bank of the Ghenfilwe torrent, is centered around the Imperial palace (ghebì), and the cathedral of St. George with its many Byzantine paintings. Other important buildings include the mausoleum of King Menelik (who founded the city in 1888), the theater, the university dedicated by Emperor Haile Selassie I, and the railroad sta-*

tion, terminus of the main line from Djibouti.

14. *A modern district of Addis Ababa, dubbed the "moral capital of Africa" at the summit meeting of African nations that met here in May of 1963. It is considered the main commercial, cultural, and industrial center of Ethiopia, and its continued development is also sustained by its fortunate geographical position at the focus of the country's major communication routes.*

15. *The old port of Mogadishu, guarded by the "Sultan's Lighthouse." The capital of Somalia is one of the oldest coastal centers, having first been settled in the 10th century A.D. Ever since then the port, the country's busiest, has been accessible only to flat-bottomed barges, due to the presence of shoals and coral outcroppings parallel to the shoreline, and the violence of the monsoon winds on the ocean.*

16. *Awasa National Park, near the greatest elevations of the highlands. Ethiopia is extraordinarily rich in animals of every type, not only because climatic conditions are favorable, but also because the country contains such a variety of highland, lowland, wet, and dry habitats. One notable species is a unique local quadruped called the warthog: with an enormous snout, tiny eyes and ears, long tusks, and hooves like a horse, it eats chickens and is hunted for its excellent meat.*

17. *Salt extraction, concentrated primarily in the salt pans of Mesewa and Aseb, is one of Eritrea's principal mineral resources. Subsurface minerals have not yet been completely exploited, although the platinum deposits at Bir Bir and gold mines at Adola (Kibre Mengist), Gojam, Beni Chargoul, and in Eritrea were discovered some time ago. One of the main obstacles to economic development and greater industrialization is a limited and inefficient network of road and rail communications; only air transport func-*

tions effectively.

18. *The market in Addis Ababa. Ethiopia's economy is based predominantly on agriculture and stockraising, which employ approximately 90% of the work force and generate 75% of the country's income. The most common non-subsistence agricultural products are coffee (which alone accounts for almost half of Ethiopia's exports), honey, and especially cereals.*

19. *The harbor of Djibouti, on the southern shore of the Gulf of Tadjoura, was built in 1888, four years before the city replaced Obock as the capital of French Somaliland. It is one of the principal outlets for goods from Ethiopia, and a well-equipped international port of call on the direct sea route from the Mediterranean to Asia and Africa via the Suez Canal, able to provide water, food, and fuel.*

20. *Remains of a prehistoric tomb. Ethiopia was originally inhabited by population groups that derived, linguistically and culturally, from two major stocks: Nilotic and Hamito-Semitic. These various tribes did not achieve unity until after the Arab invasion, which led to the foundation of the Christian kingdom of Axum, the region's first real governmental structure. From that time onward, the name "Ethiopia" was applied only to the lands inhabited by the Abyssinians of Cushite stock, who lived on the plateau and practiced agriculture and animal husbandry.*

21. *The city of Axum, former capital of the Axumite Empire, contains important ruins of churches and palaces, with inscriptions and painted decoration describing the conversion of the indigenous Ethiopian population to Christianity in about the 4th century A.D. The basalt stelai, which probably were funerary or commemorative monuments, reach heights of 108 ft [33 m] (as does the only obelisk still standing). They are surmounted by moon-shaped finials,*

and are covered with symbolic reliefs and inscriptions.

22. *The Ethiopian holy city of Lalibela, a large agglomeration of* tukuls *at the foot of Mt. Abune Yosef, is one of the most famous artistic and religious sites in East Africa. It comprises ten ancient monolithic churches, carved out of the living rock by Egyptian Copts who had arrived as refugees from Islam. According to tradition, the churches were constructed by order of King Zagwe, of the dynasty of the same name, who is venerated by the Abyssinians as a saint of the Coptic church. The church of St. George, shown in the photograph, differs from the other religious buildings around it by having the architectonic form of a Latin cross that carries over onto the outer walls; the symbol is repeated in the carvings on the roof.*

23. *Coptic art—the Arabic term was applied by the Muslims to the ancient inhabitants of Egypt—flourished from the 4th century A.D., and survived until the mid-7th century, using every conceivable artistic form. The Monophysite heresy propounding the single nature of God—which caused the Egyptian Coptic church to be declared heretical by the Council of Chalcedon in 451 A.D.—had a considerable influence on artistic expression: a characteristic motif is the image of Christ enthroned as the single God, surrounded by armies of mounted saints and anchorites, symbols of the local Christian society.*

24. *The castles of Gondar (Ethiopia). Capital of the mountainous Begemdir region, the city reached the peak of its splendor in the 17th century, when Emperor Faliside transformed it from a simple military camp into the capital of the Abyssinian kingdom. During this period, the Negus and their successors adorned their city with lavish castles in the European style and numerous churches, with the assistance of Portuguese craftsmen and laborers.*

8

9

10
11→

WESTERN AFRICA

The medieval Arab geographers gave the name *bilad es-sudan* to that vast stretch of Africa extending between the desolation of the Sahara and the lush equatorial forest, bathed by the waters of the Atlantic along the long and indented coast of the Gulf of Guinea, and bounded on the east by the Ethiopian highlands. The expression literally means "land of the blacks," and from it derived the modern term "black Africa," indicating not just a physical context, but the incredible ethnic and cultural panorama, the melding and intermingling of races, languages, religious beliefs, and customs that make Africa one of the most fascinating and mysterious of all human horizons, intensely permeated with an awareness of nature and the idea of the supernatural. And it is precisely in its relationship with the natural environment that the ethnic panorama of sub-Saharan Africa reveals its great variety.

In terms of habitat, the region can indeed be regarded as a transitional area between dry Africa (the Sahara) and wet Africa (the Congo basin). From the shores of Lake Chad to the Gulf of Biafra, one moves from an arid, steppelike environment to the dense rain forests that march directly into the ocean amid the inextricable tangle of the mangrove swamps.

Between these two extremes is a considerable variety of plant communities: scrubby steppes with typical thorny plants and sparse acacias; rolling savannas with green meadows of tall grasses or scattered trees (from the leafy acacia to the majestic baobab); and fringing forests along the river, outliers of the dense evergreen rain forest that covers the central portion of the continent.

The changes in plant life occur typically in bands running east-west. South of the Sahel, the subarid band fringing the Sahara desert extends the domain of the savanna, with trees or shrubs depending on how much moisture the environment receives. This is the kingdom of the grasses, and thus an environment particularly suitable for agriculture, although such activities are often impeded by the presence of laterite hardpan (*bowal*). Favorable factors include the seasonality of the climate: precipitation tends to coincide with passage of the sun through the zenith, so that the two periods of greatest rainfall (which occur during summer in the Northern Hemisphere) tend to merge into a single one near the tropics, but become more clearly differentiated toward the equator. Among the tree species of the savanna (which in more southerly areas takes on the character of a genuine parklike forest with flourishing thorny mimosas) are such typical giants as the baobabs, with their enormous trunk that bulges out like a gigantic bottle, and the acacias with their wide leafy crown. Approximately at the Congo basin or the shores of the Gulf of Guinea, where precipitation tends to spread over a single half-year period and increase in intensity (above 60 in. [1500 mm] per year), is the realm of the "tropophilic" forest, characterized by the alternation of dry and wet seasons, which has all the attributes of a monsoon forest including a high percentage of deciduous trees (including the typical *Malvaceae*), as well as a few xerophytic species. Where the impact of the monsoons is more pronounced, precipitation can be as high as 78 in. [2000 mm] annually, and it often exceeds 156 in. [4000 mm] along the coasts of Sierra Leone and Liberia, in the Niger delta, and in the adjacent Gulf of Biafra (and on the slopes of Mt. Cameroon). Here the tropical forest achieves its greatest lushness, developing a fringe of mangroves that constitutes a true shoreline forest, 25–35 ft [8–10 m] tall, in the narrow intertidal zone; typical species that flourish here include true mangroves with their stiltlike roots and budding plantlets hanging from the branches, like the avicennia whose leaves are covered with saline secretions. The seasonal nature of rainfall along the northern shore of the Gulf of Guinea (a name that has also supplied a widely used term for this entire region, often referred to as "Guinean Africa") is governed by the southeast trade winds that, during the summer half of the year, veer to the northeast once they cross the equator and discharge all their moisture on the coastal mountains, from the Fouta Djallon to Mt. Cameroon. In the winter months, however, in addition to suffering from a lack of rain, much of western Africa is lashed by the *harmattan*, a hot, dry Saharan wind that raises clouds of dust and greatly aggravates the dryness of the air. The topography also presents few obstacles to the winds and allows very clearly defined bioclimatic bands to develop, since there are few abrupt changes in elevation; the prevalent landforms are large depressions created by lakes or rivers, like those of the Niger or Lake Chad.

The generally flat topography also results from a geological substrate of substantial age; from the banks of the Senegal to the Benue stretch vast outcroppings of the Precambrian basement rock which forms the framework of the entire continent. The few traces of ancient Paleozoic (Caledonian) folding have been practically leveled by hundreds of millions of years of erosion, while a few rare fractures have allowed magma to rise and form occasionally active volcanic structures. The only exception is the substantial alignment running from the Adamawa Mountains (at the border between Nigeria and Cameroon) to the island of São Tomé, passing through the great mass of Mt. Cameroon (13,350 ft [4070 m]). Lastly, the depressions are covered with sediments of continental origin (sands and clays), while the monotony of the morphological landscape is relieved here and there by river terraces or tectonic escarpments.

Along with the entire range of intertropical biogeographic environments, western Africa is distinguished by the presence of all the principal traditional ways of life that characterize human activity, from hunting and gathering to pastoral nomadism to fishing to a wide range of agricultural systems, each tightly linked to individual ethnic and social groupings and strongly determined by environmental constraints. The original ethnic and racial substrate, consisting of numerous Sudanese Negroid groups (Fulbe, Mandingo, Hausa, Songhai, Ashante, Yoruba, Ibo, and many more), have been joined over the centuries by Berbers and Arabs and then by the Europeans themselves, with their colonial regimes that often had profound effects on economic, social, and cultural development. In particular, as the old animist traditions declined, Islam on the one hand and European culture on the other can be considered the basis for the present-day development of the many nations (whose existence is a legacy of the previous colonial geopolitical structure) into which Black Africa is now divided. Beneath the ethnic and political mosaic that characterizes it, however, one can still discern a geographical articulation that is once again logically based on specific environmental conditions. These include a southern strip along the Gulf of Guinea, rich in mineral deposits and hydroelectric power capabilities; the inland plains at the center of the region, with a savanna environment favorable to an agricultural and pastoral economy; and lastly a northern Sahelian band more suited for raising livestock.

In strictly economic terms, it is the first two zones, intensively exploited during the period of European colonial domination, that still guide the development of certain countries, all of them characterized by substantial population growth rates (on the order of 3–4% per year) which in some cases have produced true urbanization. The nations making up this part of Africa (Senegal, Gambia, Guinea-Bissau, Guinea, Sierra Leone, Liberia, Ivory Coast, Burkina Faso, Ghana, Togo, Benin, Nigeria, and Cape Verde, since it is at the same latitude as Senegal), in fact contain numerous cities with more than 500,000 inhabitants, and six with more than a million. This area of close to 1 million mi^2 [3 million km^2] (one tenth the size of the entire continent) is home to a population of some 170 million (two-thirds Nigerian), with a population density very much greater than the average for the African continent.

BENIN

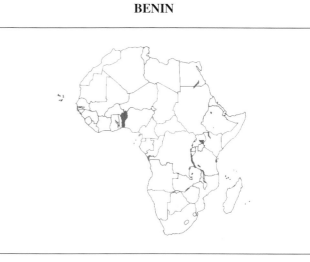

Geopolitical summary

Official name	République du Bénin
Area	43,472 mi^2 [112,622 km^2]
Population	3,338,240 (1979 estimate); 4,308,000 (1987 estimate)
Form of government	Presidential republic. The head of government, who also heads the executive branch, is elected by a Revolutionary People's Assembly for a term of five years; the same term applies to the Assembly.
Administrative structure	6 provinces divided into 84 districts
Capital	Porto-Novo (pop.164,000, 1984 estimate)
International relations	Member of UN, OAU, *Conseil de l'entente*; associate member of EC
Official language	French; Bantu and Sudanese languages also widely spoken
Religion	Christian (Catholic) 18%; Muslim 15%; the remainder of the population holds animist beliefs
Currency	CFA franc

Natural environment. Benin faces the Gulf of Guinea to the south along a coastline of only 75 mi [125 km], and is bordered on the west by Togo, on the northwest by Burkina Faso, on the northeast by Niger, and on the east by Nigeria.

Southern Benin consists of a narrow alluvial plain fringed with lagoons, lakes, and swamps, built up from recent (Cretaceous and Eocene) sediments and alluvium which rest, toward the interior, on the crystalline Precambrian basement; this extends north to the Niger valley, where it is again covered by recent sediments. In the rest of the country, the geomorphological framework consists of a large central plateau at an average elevation of 1600 ft [500 m], the northwestern edge of which rises slightly to form the Atakora range (2102 ft [641 m]), a direct continuation of the adjacent pre-Paleozoic Togo highlands.

Climate data

Location	Altitude (ft asl)	Average temp. (°F) January	Average temp. (°F) July	Average annual precip. (in.)
Porto-Novo	–	82	78	50
Cotonou	13	81	77	51.7
Natitingou	1,509	80	77	52.3

Conversion factors: 1 ft = 0.3 m; 1 in. = 25 mm; °C = (°F – 32) × 5/9

The hydrographic system consists of numerous rivers that flow into the Niger in the northern part of the country (except the Pendjari, a subtributary of the Volta), and to the south empty directly into the Atlantic. The largest of these is the Ouémé, which rises in the Atakora Mountains and, after a course of 280 mi [450 km], flows into the ocean. Other rivers of significance include the Mono (which marks the southernmost part of the border with Togo), the Couffo, and the Zou.

The climate reflects the country's geographical position and the absence of any significant high ground. In the southern part it is equatorial, with average temperatures in the hottest month (March) of around 82°F [28°C] (with minimal variation throughout the year) and rainfall of about 60 in. [1500 mm]. Toward the interior the equatorial features fade away, as temperature swings become more pronounced and precipitation decreases. In the northern regions the climate becomes hotter and drier, although rainfall still exceeds 30 in. [800 mm] per year.

The rainforest is limited to thick mangroves on the coastal plain; the dominant plant communities in the southern parts of the plateau are wooded savanna and fringing forests along the rivers. Toward the north the savanna thins out progressively, ultimately becoming a steppe with short grasses. The wild fauna is rich and varied: numerous hippopotami and crocodiles live in the lakes and marshes of Mono province to the southwest, and the natural environment is protected by two national parks to the north, Pendjari and W (the latter is named after a W-shaped series of bends in the Niger river).

Population. The population of Benin, concentrated predominantly in the southern regions, comprises a large number of ethnic groups (more than 60), who are often at odds with each other. The

Administrative structure

Administrative unit (provinces)	Area (mi²)	Population (1987 estimate)
Atakora	12,043	622,000
Atlantique	1,244	909,000
Borgou	19,686	630,000
Mono	1,467	610,000
Ouémé	1,814	806,000
Zou	7,218	731,000
BENIN	43,472	4,308,000

Capital: Porto-Novo, pop. 164,000 (1984 estimate)
Principal cities: Cotonou, pop. 478,000 (1984 estimate); Abomey, pop. 54,418 (1979 census); Natitingou, pop. 50,800 (1979 census).

Conversion factor: 1 mi² = 2.59 km²

most important in the southern provinces are the Fon (39%), the Yoruba, originally from Nigeria, and the Adja; the north is populated by Sudanese groups, such as the seminomadic Peul herders in the Sahel; the Bariba, with their Paleonegritic characteristics and feudal organization; and the Somba in the Atakora region. Along the coast live the "lake people": Aizos, Pedah, and Mina.

After independence, a period of political instability ensued, resulting from ethnic rivalries and the economic disparity between the richer south and the poorer north. In 1972 a radical social and economic reform program was initiated by the new socialist government. Continuing internal difficulties and international mistrust, especially of France and Nigeria, generated by new alliances and programs aimed at nationalizing foreign companies, have ultimately led to an erosion in government *dirigisme*.

The principal city is Cotonou, which has an international airport and an active port at the mouth of the Nokoué lagoon. The political capital is Porto-Novo, located on the lagoon of the same name; inhabited mostly by Yoruba, it was founded in the 16th century as a regional center of the Dahomey kingdom. Abomey, the historic city *par excellence*, is located in Zou province. A few miles from Cotonou are the evocative villages of Ganvié, Soava, So-Tchanhoué, and Honedo, built on pilings in a lake; they are unique in Africa.

Economic summary. Although the reforms of 1972 were followed by relative stability, Benin has remained a very poor country in economic terms. The economy is based almost exclusively on agriculture, which employs 44% of the labor force. The best land is on the alluvial plain and is worked on a plantation basis, growing mostly coconut and oil palms, the major sources for the country's agricultural exports. The other export crops—coffee, cocoa, peanuts, and cotton—are of lesser importance. Crops grown for domestic consumption include manioc, sweet potatoes, and rice in the southern areas, and cereals in the savanna regions. The present government has devoted considerable attention to these subsistence crops, organizing labor into state cooperatives. Animal husbandry, practiced mostly in the savanna, involves 951,000 head of cattle (1990), a similar number of sheep and goats, and 714,000 swine, which are raised in the south. Fish are abundant, and are caught both for the domestic market and for export. Forest covers 33% of Benin's territory, and produces valuable hardwoods (teak, mahogany, iroko).

Petroleum, a resource that supplements a low level of hydroelectric power production, has been extracted since 1983; there are no other mineral resources. The industrial sector is poorly developed, comprising mostly vegetable-oil processing factories and cotton ginning plants, along with a few cement factories, breweries, and beverage bottling plants. There is little in the way of an engineering sector. The industrial area is centered around the coastal cities of Cotonou and Porto-Novo.

The country's trade deficit is large and growing. Export products include primarily palm oil, peanuts, oilseeds, cotton, and cocoa. There is still a considerable "underground" trade between Benin and Nigeria, carried on largely by people with a shared Yoruba background who live near the border between the two countries. Communications routes are modest, with 4600 mi [7445 km] of highway in 1985 (750 mi [1200 km] paved), and 359 mi [579 km] of railroad. The port of Cotonou is becoming more and more closely linked with traffic to and from Lagos.

Socioeconomic data

Income (per capita, US$)	421 (1990)
Population growth rate (% per year)	3.2 (1980–89)
Birth rate (annual, per 1,000 pop.)	46 (1989)
Mortality rate (annual, per 1,000 pop.)	15 (1989)
Life expectancy at birth (years)	51 (1989)
Urban population (% of total)	37 (1989)
Economically active population (% of total)	47.1 (1988)
Illiteracy (% of total pop.)	73 (1991)
Available nutrition (daily calories per capita)	2145 (1989)
Energy consumption (10^6 tons coal equivalent)	0.19 (1987)

Historical and cultural profile. The ancient kingdom of Benin occupied the Guinean region west of the Niger delta, and thus included only the margins of the territory which has been known by that name since 1975. A number of kingdoms arose in this area beginning in the 12th century, in particular the kingdom of Dahomey, with its capital Abomey. Famous for its military organization and armies, which included female warriors, it engaged in continual struggles with neighboring states until it became a vassal of the Yoruba kingdom of Oyo in 1712. This subjugation ended after a century under the leadership of King Gezo, who expanded the nation to its greatest extent.

An account of this kingdom and its capital has survived, written by a Dutch traveler who visited the coastal area in 1602:

> The houses of this city are arranged in an orderly fashion, like the houses in Holland. The city appears to be very large, with one great unpaved street perhaps seven times as wide as Warmoes Street in Amsterdam; ... The king of Benin has many soldiers, and courtiers, and people who come to the palace riding on cattle ... one also sees many slaves who carry water, yams, and palm wine, which they say is for the king....

In the second half of the 19th century the first trade and friendship treaties were signed with the French, to whom the Cotonou coast was ceded. Relations with the new arrivals deteriorated in only a few decades, however, and by the turn of the century the French had sent several military expeditions to occupy the country.

Initially an autonomous colony, in 1902 Dahomey became part of French West Africa, ruled by a lieutenant governor residing at Porto-Novo. Full independence was granted in 1960 along with several other neighboring states; the president of the republic was Hubert Maga, who was overthrown just three years later in the first of a series of coups led by military commanders. In 1972 another "putsch" installed a military regime (with socialist leanings) under Mathieu Kérékou that proved more durable than its predecessors. Eight years later the government became civilian: Kérékou was elected president of the republic, and was re-elected to the post in 1984.

A rich artistic style flourished around the royal court in the former capital, Abomey, and the palace still contains impressive examples of clay polychrome bas-reliefs ornamenting the exterior. Yoruba influences are evident in the large painted wooden statues depicting ruler figures, and in evocative, stylized iron statues depicting the war god Gu.

BURKINA FASO

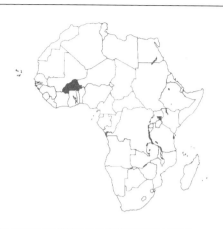

Geopolitical summary

Official name	Burkina Faso
Area	105,841 mi² [274,200 km²]
Population	7,976,019 (1985 census); 8,760,000 (1990 estimate)
Form of government	Military government, with a National Revolutionary Council headed by a single person as head of government and of the executive
Administrative structure	30 provinces
Capital	Ouagadougou (pop. 442,223, 1985 census)
International relations	Member of UN, OAU; associate member of EC
Official language	French; Bantu languages very common
Religion	Predominantly animist; also Muslim (30%) and Catholic (4%)
Currency	CFA franc

Natural environment. Burkina Faso extends between the basins of the Niger and Comoé rivers, and is bordered on the north by Mali, on the east by Niger, and on the south by Benin, Togo, Ghana, and Ivory Coast. The territory consists of a broad stretch of crystalline basement (consisting of granite, quartzite, schist, etc.) which crops out at the surface and is only partly covered by recent alluvium and laterite incrustations. Average elevation is about 1000 ft [300 m], and the tablelands that characterize the country's morphology are occasionally interrupted by the rounded profile of isolated residual hills a couple of thousand feet high, such as the Banfora massif (2404 ft [733 m]) in the southwestern corner of the country.

The hydrographic system comprises three different basins: the first in the eastern section, with minor rivers that flow into the Niger; the second to the south, with the three branches of the

Climate data

Location	Altitude (ft asl)	Average temp. (°F) January	Average temp. (°F) July	Average annual precip. (in.)
Ouagadougou	1,010	76	83	34.9

Conversion factors: 1 ft = 0.3 m; 1 in. = 25 mm; °C = (°F – 32) × 5/9

Volta (Black, Red, and White); and the third in the western region, represented by the Comoé and its tributaries. It is also possible to identify, in the northernmost part of the country, an endorheic area in which water disperses into marshes and evaporates during the dry season. All these rivers are governed by a rainfall pattern that is closely linked to climate. The temporary and violent nature of the precipitation, with winter floods and totally dry summers, makes watercourses largely useless for either navigation or the production of electricity.

The country's geographical position in the Sahel belt gives it an arid, steppe-like climate in the northern and eastern sections, and a humid tropical one in the southwest. Temperatures are high (annual average 81°F [27°C]), with considerable seasonal temperature swings. In winter the dry *harmattan* blows from the Sahara, while in summer (June–September), the Atlantic monsoon brings more than 43 in. [1100 mm] of rain each year to the central and southern parts of the country, and lower levels of precipitation (slightly above 20 in. [500 mm]) in a shorter season (July–September) to the more arid northern regions.

The predominant plant community consists of wide expanses of steppe, with rare baobabs and thorny scrub consisting of xerophytic plants; in the wetter southern regions, wooded savanna and parklike forests predominate. Burkina Faso also contains part of the W National Park (Benin), and a short distance away the small Arly National Park adjacent to Benin's Pendjari reserve.

Population. Half the population consists of Mossi, the ethnic group with the most ancient cultural traditions. They are sedentary Sahel farmers living in mud huts with straw roofs, and have retained animist cults and rituals associated with their former kingdom. Other groups include the Gurunsi (210,000), Lobi, Bobo, and the Senufo, southern farmers. To the north live the Peul and Tuareg, Muslim nomads who herd animals.

Burkina Faso's population is currently growing at an annual rate of 2.6% despite a mortality rate of 1.8%. Population density is 83 per mi^2 [32 per km^2], with the population concentrated largely in the central, southern, and western areas between the capital and the major cities; the northern desert area and several unhealthy river regions are largely unpopulated.

Ouagadougou, formerly the capital of the Mossi empire, performs administrative functions and is an agricultural and pastoral market center. The city has developed especially quickly since the construction of a railroad linking it to Abidjan in Ivory Coast. Bobo-Dioulasso takes its name from the two ethnic groups in the region, the Bobo and the Dioula; until 1960 it was the nation's largest city, and is linked by rail to the capital. Other important centers include Koudougou, Ouahigouya, and Banfora.

Burkina Faso is in absolute terms one of the poorest countries in the world, with per capita income of US$370 per year and an 81.8% illiteracy rate. The already precarious living conditions are aggravated by the droughts which periodically devastate the entire region, and by endemic disease. The country therefore receives substantial economic aid from international organizations (World Bank Group, FAO, etc.), and from the governmental support and development programs of the major industrialized countries.

Economic summary. Burkina Faso's economy is based essentially on agriculture and animal husbandry (which employ

Administrative structure

Administrative unit (provinces)	Area (mi^2)	Population (1985 census)	Administrative unit (provinces)	Area (mi^2)	Population (1985 census)
Bam	1,551	164,263	Passoré	1,574	225,115
Bazéga	2,051	306,976	Poni	3,999	234,501
Bougouriba	2,736	221,522	Sanguié	1,994	218,289
Boulgou	3,487	403,358	Sanmatenga	3,556	368,365
Boulkiemdé	1,597	363,594	Séno	5,201	230,043
Comoé	7,100	250,510	Sissili	5,302	246,844
Ganzourgou	1,578	196,006	Soum	5,153	190,464
Gnagna	3,320	229,249	Sourou	3,662	267,770
Gourma	10,273	294,123	Tapoa	5,705	159,121
Houet	6,358	585,031	Yatenga	4,745	537,205
Kadiogo	451	459,138	Zoundwéogo	1,333	155,142
Kénédougou	3,207	139,722			
Kossi	5,086	330,413	**BURKINA FASO**	105,841	7,976,019
Kouritenga	628	197,027			
Mouhoun	4,031	289,213			
Nahouri	1,483	105,273			
Namentenga	2,993	198,798			
Oubritenga	1,811	303,299			
Oudalam	3,878	105,715			

Capital: Ouagadougou, pop. 442,223 (1985 census)

Major cities (1985 census): Bobo-Dioulasso, pop. 231,162; Koudougou, pop. 51,670; Ouahigouya, pop. 38,604; Banfora, pop. 35,204; Kaya, pop. 25,779.

Conversion factor: 1 mi^2 = 2.59 km^2

75% of the working population), and is one of the most backward on the continent. Only one-tenth of the land area is used for cultivation; a third is meadows and pastures, a quarter is occupied by forests, and a further third is uncultivated and unproductive. Subsistence crops, which comprise low-grade cereals (millet, sorghum, corn), tubers (manioc and sweet potatoes) and oil-seed plants (peanuts and sesame), are not produced in sufficient quantity to meet internal demand. The only products exported are peanuts, cotton, and sesame. The government is promoting cultivation of sugar cane, rice, and sweet potatoes. Animal husbandry, practiced on a nomadic basis by herders in the Sahel, provides 50% of total exports. Animal holdings consist of 2.9 million head of cattle (1990), 8.85 million sheep and goats, 520,000 horses and donkeys, 5000 camels, and a large number of domestic fowl. Forest products are of low quality and used exclusively for internal consumption, meeting demands for construction material and fuel.

Mineral resources are of some significance, especially manganese, which is present in the northern part of the country (Tambao), and to a lesser extent gold (now being mined again at Boura) and phosphates (at Tilemsi). Exploitation of these minerals is hampered by limited equipment and the lack of an adequate transportation system; these two factors result in prices that are not competitive on the world market. The industrial sector is small and concentrated in the major cities, employing only 12% of the working population and yielding 24% of gross national product. The predominant industries are food processing, textiles, leather, and consumer goods such as beer, sugar, and cigarettes. Small mechanical and chemical complexes also exist.

The external trade balance is decidedly negative, due to massive importation of energy sources, food, and capital equipment, which are not compensated for by the low level of exports of animals, cotton, hides, and leather. The sparse communications network comprises 341 mi [550 km] (1988) of railroad, a road system covering 6,963 mi [11,231 km] (1983), and the two international airports at Ouagadougou and Bobo-Dioulasso.

Historical and cultural profile. The region that is now Burkina Faso was once the site of three major governmental entities, namely the ancient empires founded by the Mossi after 1000 A.D. around Ouagadougou, Yatenga, and Fada N'Gourma. These empires withstood attacks from Mali, and for several centuries remained immune to outside influences, including Islam.

What changed these empires and their subsidiary principalities were the Europeans, who sent out a series of expeditions at the end of the last century to explore the upper basin of the Volta. The first was led by the Scotsman Mungo Park, who was followed by Heinrich Barth from Germany and Louis-Gustave Binger from France. These scientific expeditions were succeeded by military ones and, in 1896, in the heat of the battle against Samoury Touré, French troops entered Ouagadougou and received its surrender. The Yatenga empire also capitulated to the French, and it is certainly possible that the triumph of the Europeans was abetted by disputes among the various tribes. Some groups put up particularly tough resistance to colonial penetration; these included the Lobi, settled south of the basin of the Black Volta, who presented the French with considerable problems and earned the reputation of "rebellious anarchists," as evidenced by this reconstruction made in 1981 by a team of anthropologists studying this ethnic group:

The difficulties began almost immediately, as soon as the French demanded certain services from the natives (labor, transport of materials, provision of food, etc.). These difficulties intensified between 1901 and 1902 when the administrative center was transferred to Gaoua…. Not only did the population refuse to participate in construction work on the emplacements (which were ultimately built by workers sent from Diébougou), but they turned down all requests to provide foodstuffs. In December of 1901 the inhabitants of the village of Poni revolted, and like wildfire the spirit of rebellion spread through the entire country. Despite their superior weaponry, the French found themselves confronted by a hard, exhausting struggle…. The tactics of the Lobi warriors were aimed primarily at disorganizing the troops and creating conditions for hand-to-hand fighting, at which they excelled; their knowledge of the terrain also worked in their favor. The Lobi used all sorts of traps and tricks to achieve their aims: ambushes in the forest or from ditches, sowing panic among infantry and horses with deafening shouts and horn blasts, releasing swarms of smoke-maddened bees among the troops, and planting poisoned arrows along trails…. Soothsayers were consulted before each attack or act of defiance. The purpose was to kill the white man's soul with sacrifices and libations offered to deceased ancestors.

The territory became a colony in 1919 under the name of Upper Volta; after World War II it became a French Overseas Territory, then a republic associated with the French Community (1958), and finally an independent state in 1960.

In the years that followed, despite a constant stream of civilian and military governments, power remained essentially in the hands of a pro-French oligarchy. Only in 1983, following a coup d'état led by Colonel T. Sankara, was a decisive break made with the colonial past. In 1984 Upper Volta changed its name to Burkina Faso, which in the Mossi language means "land of the upright and just." Despite this hopeful name, the country has suffered some episodes of real tension in recent years.

Over the centuries the Mossi people have produced typical "wando" masks, large objects consisting of a central element shaped like an antelope's muzzle, topped by a horn which extends into an anthropomorphic figure; the overall result is balanced and expressive. Also very evocative and infused with emotional power are the Bobo masks, which perpetuate the image of an ancestor or totemic animal. This ancient people is also famous for its clay bas-reliefs decorated with figures or geometric elements, designed exclusively as hut decorations.

Socioeconomic data

Income (per capita, US$)	370 (1990)
Population growth rate (% per year)	2.6 (1980–89)
Birth rate (annual, per 1,000 pop.)	47 (1989)
Mortality rate (annual, per 1,000 pop.)	18 (1989)
Life expectancy at birth (years)	48 (1989)
Urban population (% of total)	9 (1989)
Economically active population (% of total)	53.5 (1988)
Illiteracy (% of total pop.)	81.8 (1991)
Available nutrition (daily calories per capita)	2061 (1989)
Energy consumption (10^6 tons coal equivalent)	0.22 (1987)

CAPE VERDE

Geopolitical summary

Official name	República de Cabo Verde
Area	1,557 mi^2 [4033 km^2]
Population	295,703 (1980 census); 369,000 (1990 estimate)
Form of government	Constitutional republic with a National Assembly elected every five years by universal suffrage; the Assembly elects the head of state, who nominates the government and prime minister.
Administrative structure	2 districts, divided into 14 councils
Capital	Praia (pop. 61,797, 1990 estimate)
International relations	Member of UN, OAU; associate member of EC
Official language	Portuguese; Crioulo dialect widely spoken
Religion	Catholic
Currency	Cape Verde escudo

Natural environment. The Cape Verde archipelago, located in the Atlantic Ocean approximately 370 mi [600 km] to the west of the promontory of the same name, consists of ten principal islands arranged in an arc and divided into two groups: the Sopravento to the north, with Santo Antão, São Vicente, Santa Luzia, São Nicolau, Sal, and Boa Vista; and the Sottovento group to the south, with Fogo, São Tiago, Brava, and Maio. These islands of volcanic origin constitute an isolated structure, now almost completely extinct, between the Mid-Atlantic Ridge and the African landmass. Only the volcano on the island of Fogo is still active (most recent eruption in 1951), and reaches a height of 9279 ft [2829 m]. All the others are rather rugged, with elevations that are often substantial on some islands (6491 ft [1979 m] on Santo Antão, 4566 ft [1392 m] on São Tiago), or more softened on others that have been worn down by erosion.

The climate is of the dry tropical type, influenced by the northeast boreal monsoon that blows from the Tropic of Cancer toward the equator; this cool but dry wind brings no rain. Average temperatures are fairly stable, ranging between 72 and 77°F [22–25°C]. The presence of the cold Canary Current means that evaporation is reduced, so that annual rainfall levels are no more than 8–12 in. [200–300 mm] (4.8 in. [123 mm] at Praia); precipitation is irregular, and is followed by periods of drought that on some islands can last several years. The islands have no surface watercourses, and the potentially fertile volcanic soil suffers from lack of water and the intense deforestation that has been practiced for centuries.

The vegetation is consequently very poor, reduced to a succession of steppe-like heaths composed largely of shrubs and grasses, with thorny acacias and a few palms. A reforestation project sponsored by the FAO has been initiated.

Population. With a population of 369,000, Cape Verde is one of the smallest states on the continent. In the past, population growth was limited by a series of disasters resulting from epidemics, volcanic eruptions, and famine; the mortality rate is now lower while birth rate has remained high, yielding a natural growth rate of 3.2%. The population density of 236 per mi^2 [91 per km^2] is in actuality much higher, since the population is concentrated on the larger islands, on the coast, and in the interior up to an elevation of 1600 ft [500 m]. The most populous southern island is São Tiago, with Praia and three other cities, along with excellent farmland; the major northern island is São Vicente, with the port of Mindelo. Another center is Espargos on Sal island, an important air base.

The population of the Cape Verde archipelago consists of descendants of Sudanese-Guineans belonging to the Felup and Papel tribes, and Balante from Guinea-Bissau. Once the islands had become an important port of call, the Portuguese also populated it with Guinean slaves, prisoners, Portuguese convicts, and sailors, producing over the centuries a highly diverse population, most of whom are of mixed race.

Emigration has traditionally been high: about 550,000 of Cape Verde's inhabitants (more than one and a half times the total current population) have been forced to emigrate overseas to survive, and much of the nation's earnings comes from their remittances in hard currency and from international aid. The disastrous economic situation is evident especially in the lack of food (83.9% of the territory is uncultivated and unproductive), necessitating frequent distributions of free food by the government. Less negative aspects are seen in education, where the illiteracy rate of 52.6% is well below the average for western African nations, and in average life expectancy (63 years).

Economic summary. Cape Verde's economy is based largely on agriculture (even though only 9.7% of its land is used for this

Climate data

Location	Altitude (ft asl)	Average temp. (°F) January	Average temp. (°F) July	Average annual precip. (in.)
Praia	89	72	77	4.8

Conversion factors: 1 ft = 0.3 m; 1 in. = 25 mm; °C = (°F − 32) × 5/9

Administrative structure

Administrative unit (districts)	Area (mi²)	Population (1980 census)
Sopravento	861	107,685
Sottovento	696	188,018
CAPE VERDE	1,557	295,703

Capital: Praia, pop. 61,707 (1990 estimate)
Major cities (1980 census): Mindelo, pop. 36,746.

Conversion factor: 1 mi² = 2.59 km²

purpose) and on fishing, which is still practiced with primitive methods. About 67% of GNP derives from service-sector activities, principally port fees from international transportation.

The principal crops are corn, sweet potatoes, tropical fruits, coconuts, sugar cane, beans, castor beans, and manioc, but they meet only approximately 5% of local needs.

After independence, the state nationalized all food-related businesses and developed irrigation in order to improve the agricultural production of bananas, sugar cane, coffee, and peanuts. There were also plans to develop fishing (mostly for cod, tuna, shellfish, and coral), with the construction of a freezing plant at São Vicente to can the fish, some of which is exported. Animal husbandry, mostly involving goats, is extremely limited in scope. Forests cover only 0.2% of the area of Cape Verde; a major reforestation effort is under way to combat the erosion caused by generations of woodcutting, involving the planting of more than a million trees.

What little industry there is—employing only 20,000 people (16% of the work force)—is closely associated with agricultural activities: processing of food products, salt production on the islands of Sal and Maio, fish canning, cement plants, and electrical energy. The maritime and air-traffic service functions associated with international commerce (at the ports of Mindelo and Porto Grande, and the international airport at Espargos) have recently lost their significance as the need for stopovers by international fleets has diminished. Internal communications are provided by sea and by 1395 mi [2250 km] of roads (1984). Service-sector workers (39% of the work force) are employed predominantly in public administration. The balance of payments is heavily negative: exports of fish, salt, and bananas account for only 5% of the value of imports. Economic plans are aimed at developing tour-

Socioeconomic data

Income (per capita, US$)	890 (1990)
Population growth rate (% per year)	3.2 (1980–89)
Birth rate (annual, per 1,000 pop.)	32.1 (1989)
Mortality rate (annual, per 1,000 pop.)	7.7 (1989)
Life expectancy at birth (years)	63 (1989)
Urban population (% of total)	33.1 (1989)
Economically active population (% of total)	37.1 (1988)
Illiteracy (% of total pop.)	52.6 (1991)
Available nutrition (daily calories per capita)	2550 (1989)
Energy consumption (10⁶ tons coal equivalent)	0.02 (1987)

ism (beach resorts and cruises), communication routes (both for internal access and between islands), the fishing industry, and energy, utilizing geothermal and wind sources.

Historical profile. The first Europeans to set foot on the Cape Verde islands, in 1456, were two Italian navigators—Alvise da Ca' da Mosto and Antoniotto Usodimare—in the service of King Henry of Portugal. Ca' da Mosto left this record of the expedition:

> We marveled greatly because we did not know that there was any land thereabouts. And when two men were sent aloft they discovered two large islands.... But then, hearing of these four islands that I had found, others arrived here and it was they who discovered these ten islands, both large and small, uninhabited, nothing being found on them other than doves and strange sorts of birds, and great numbers of fish.

Exploration was completed in the next few years by Antonio da Noli and Diego Gomez, revealing an uninhabited archipelago rich in watercourses and covered with vegetation, amply justifying the name "Green Cape." The Portuguese were quick to realize the possibilities offered by such a location, and in about 1470 the first colonists began to arrive from the mother country: the islands soon became one of the most important Portuguese trade outposts, especially for the slave trade. Populated in the course of time by highly mixed communities, they were subjected to intense colonial exploitation, and improper use of their resources ultimately caused profound changes in the environmental equilibrium: crop yields steadily declined, the soil dried out, and the country slid into increasingly dire poverty. By the late 17th century, this situation had generated a flow of emigrants that in time reduced Cape Verde to nothing more than a reservoir of labor for other Portuguese colonies. From the mid-19th century to the recent past, this flow became a true exodus: it has been calculated that between 1968 and 1974, 8000 Cape Verdeans left their country each year.

The Cape Verde islands, which from an administrative point of view were associated with Portuguese Guinea as an "overseas province," shared the historical vicissitudes of that colony. The movement toward independence was led by the African Party for the Independence of Guinea and Cape Verde (PAIGC), which worked from bases on the mainland but included many leaders of Cape Verdean origin. After the fall of the Salazar regime in Portugal (1974), the province of Cape Verde gained independence along with the other "overseas provinces." In a referendum in 1975, the population of the archipelago voted for union with Guinea-Bissau, while retaining basic reciprocal autonomy. The link between the two nations was broken, however, in 1980, when a coup in Guinea-Bissau brought to power João Bernardo Vieira, a leader hostile to the considerable number of Cape Verdeans within his country's ruling class. The PAIGC was then dissolved, replaced by the African Party for the Independence of Cape Verde (PAICV). Aristides Pereira, president of the republic since 1975, was re-elected to his post in 1986.

IVORY COAST

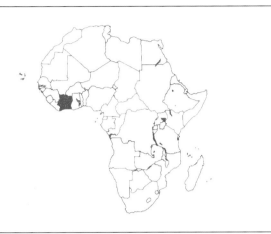

Geopolitical summary

Official name	République de Côte-d'Ivoire
Area	124,471 mi^2 [322,463 km^2]
Population	6,897,301 (1975 census); 12,100,000 (1990 estimate)
Form of government	Presidential republic, with head of government and National Assembly elected every five years by universal suffrage
Administrative structure	34 departments
Capital	Abidjan (pop. 2,534,000, 1985 estimate); Yamoussoukro, under construction (pop. 120,000, 1983 estimate)
International relations	Member of UN, OAU, *Conseil de l'entente*; associate member of EC
Official language	French; Sudanese languages widely spoken
Religion	Animist 37%, Muslim 34%, Catholic 22%, Protestant 5%
Currency	CFA franc

Natural environment. Ivory Coast, which is bordered on the west by Guinea and Liberia, on the north by Mali and Burkina Faso, and on the east and south by Ghana and the Atlantic Ocean (Gulf of Guinea), consists of a large peneplain sloping slightly from north to south, modeled into the Precambrian basement rock that extends without a break from the Atlantic coast to the upper Volta and the Niger, where it is covered by Paleozoic sediments and where the land rises to elevations of more than 1600–2600 ft [500–800 m]. Along the coast is a narrow band of more recent sediments (Cretaceous and Tertiary). Higher elevations are found in the west central region, such as the Man massif (4300 ft [1311 m]), which culminates farther west, at the border with Liberia and Guinea, in the Nimba Mountains (5747 ft [1752 m]). The alluvial coastal plain forms a hinterland some

100–125 mi [150–200 km] wide, dotted with marshes and characterized by wide river meanders. The 340 mi [550 km] of coastline is high and rocky west of Fresco, while the longer eastern portion is a succession of broad lagoons defined by a long chain of sandbars.

Numerous rivers run almost parallel to one another from north to south, interrupted by many rapids and falls which make them unusable for navigation. The longest is the Sassandra (680 mi [1100 km]), now restrained by dams and hydroelectric plants at Buyo and Soubré. Next longest is the Bandama, which forms the great artificial lake of Kossou, a valuable energy and fishing resource. To the northwest, the watercourses run in the opposite direction and feed into the Niger basin; these include the Sankarani, which defines a short stretch of the frontier with Guinea, and the Bagoé. To the east, water flows into the Volta basin through the branch of the Black Volta that constitutes the border between Ivory Coast and Ghana.

The subequatorial climate is greatly influenced by geographical position, which determines rainfall volume, length of the rainy season, and temperature swings. In the southern half of the country precipitation ranges from 80–100 in. [2000–2500 mm], occurring over a long, uninterrupted season 8 months long that reaches a maximum in May–June; the dry season is short, and concentrated in January and February.

Average temperatures in the south are about 79°F [26°C] with little variation; average rainfall is 82 in. [2100 mm] per year. The moist southwest winds gradually lose their effect as they move toward the north, giving way to the dry Saharan *harmattan*. Precipitation decreases considerably (47 in. [1200 mm]) and the rainy season becomes shorter, coinciding with the sun's changing positions in relation to the zenith. Temperature swings increase: the average minimum temperature drops to 68–77°F [20–25°C] during the rainy season, and average maximums rise considerably during dry periods (95–105°F [35–40°C]).

The coast, with its hot, wet climate and fairly fertile alluvial soils, is covered by a dense rainforest containing numerous valuable hardwoods (mahogany, ebony, rubber) that have nevertheless become gradually less abundant due to human exploitation. On the less fertile granite soils of the central regions, on the western highlands, and on the eastern clay soils, the rain forest gives way to wooded savanna dominated by enormous baobab and leafy acacia trees.

The rich savanna fauna, both herbivores (elephants, zebras, antelopes) and carnivores (lions, hyenas, leopards), is protected in Komoé, Tai, Marahoué, and Asagny National Parks. Marine flora and fauna are protected in the Esotilé islands park in Aby lagoon, while Banco National Park near Abidjan preserves part of the virgin rainforest.

Population. The 60 ethnic groups living in Ivory Coast are largely of Sudanese stock, divided into many tribes including the

Climate data

Location	Altitude (ft asl)	Average temp. (°F) January	Average temp. (°F) July	Average annual precip. (in.)
Abidjan	79	81	75	82
Bouaké	1,194	81	77	47
Conversion factors: 1 ft = 0.3 m; 1 in. = 25 mm; °C = (°F – 32) × 5/9				

Administrative structure

Administrative unit (departments)	Area (mi²)	Population (1975 census)	Administrative unit (departments)	Area (mi²)	Population (1975 census)
Abengourou	2,663	175,522	Katiola	3,636	76,256
Abidjan	5,481	1,661,841	Korhogo	4,825	276,299
Aboisso	2,413	146,551	Lakota	1,054	76,821
Adzopé	2,019	160,931	Man	2,721	273,826
Agboville	1,486	139,269	Mankono	4,115	81,322
Biankouma	1,911	74,916	Odienné	8,184	123,561
Bondoukou	6,381	295,368	Oumé	926	85,369
Bongouanou	2,150	218,812	Sassandra	6,767	121,712
Bouaflé	2,189	168,091	Séguéla	4,339	75,847
Bouaké	9,187	804,955	Soubré	3,192	73,687
Bouna	8,287	78,041	Tingréla	849	36,061
Boundiali	2,972	96,884	Touba	3,339	77,446
Dabakala	3,733	55,636	Zuénoula	1,092	99,996
Daloa	4,478	266,757			
Danané	1,776	117,173	**IVORY COAST**	124,471	6,897,301
Dimbokro	3,293	255,817			
Divo	3,057	201,932			
Ferkésédougou	7,373	91,516			
Gagnoa	1,737	171,444			
Guiglo	5,462	135,808			
Issia	1,386	101,828			

Capital: Abidjan, pop. 951,216 (1975 census); pop. 2,534,000 (1985 estimate); new capital Yamoussoukro under construction: pop. 120,000 (1984 estimate).

Principal cities (1975 census): Bouaké, pop. 175,264; Daloa, pop. 60,837; Man, pop. 50,288.

Conversion factor: 1 mi² = 2.59 km²

Agni-Ashanti and the Baule. The Mandingo (Malinke), Senufo, Culango, and Lobi are agriculturists and live in the northern savanna regions; except for the Culango and Lobi, who are animists, they are predominantly Muslim. The Beké, Dida, Krue, Guro, and Yacouba populations live in the western regions. The central and southern areas are inhabited by lakeside dwellers who live in stilt houses and practice primitive fishing methods; the Akan, Agni, and Baule, who practice plantation agriculture, have become the wealthy leaders of the country.

When it became independent in 1960, Ivory Coast had some 3 million inhabitants (including 15,000 Europeans); this number has quadrupled in thirty years (to a population of 12.1 million in 1990, with a density of 96 per mi² [37 per km²]) as a result of a 4% population growth rate and intensive immigration from Burkina Faso, Mali, and Ghana. The highest density, which was once found in the north where the people practiced traditional agriculture and animal husbandry, is now recorded on the coast, around Abidjan and on the road to Bouaké. Two thirds of the population live in rural villages, with circular clay huts and thatched roofs, grouped around the "meeting house."

The capital, Abidjan, with 2.5 million inhabitants (up from 17,000 in 1936), is located on the peninsula at the end of Ebrié lagoon, and is connected to the Gulf of Guinea by the Vridi canal, in operation since 1950: its inland port handles more than 6.6 million t of traffic per year. In addition to the Treichville and Plateau districts that bear the imprint of French colonial architecture, and the more recent neighborhood called "Zone Quatre," the city comprises Petit Bassam island with its harbor and industrial complex, and several residential areas. Yamoussoukro, the official capital since 1983, is south of Lake Kossou, and still has no real administrative functions. Other cities are Bouaké, Daloa, Man, and Korhogo.

With a per capita income of US$730 and an illiteracy rate of 46.2%, Ivory Coast can consider itself a fairly modern nation, capable of facing the more serious problems that still afflict most African states. Consisting largely of a young population (50% below twenty years of age), most of them educated thanks to the spread of elementary instruction into every village via television, the country also embodies the contradictions arising from rapid modernization and the crisis of traditional society.

Economic summary. The colonial economy was based on exploitation of the coastal region: cocoa and coffee plantations were established, and forced labor was used. A railroad from Abidjan to Bobo-Dioulasso (1934) was built in order to extend peanut and cotton cultivation to the interior savanna regions. It was in fact on the cocoa plantations, in 1944, that the first stirrings of the anticolonial movement arose, led by a physician named Félix Houphouët-Boigny. The ultimate result was independence in 1960, and the institution of a free-trade economy combined with a long-term state-controlled policy. The nation continued to develop on the basis of plantation crops, which were expanded and diversified with the technological assistance of outside specialists, including Europeans. A "stabilization fund" was created, which today protects the country from fluctuations in international prices and allows it to accumulate capital for investment in other economic sectors and in social services (schools, hospitals, roads, etc.), which are among the best in Africa. The corporate structure, characterized by small and medium-sized companies (only 6% of companies are large), ensures broad profit distribution and fairly widespread prosperity.

Agriculture is the most important economic activity, employing 59% of the work force and producing 29% of GNP, although the land area in use is limited (7.5%). Subsistence crops (millet, corn, sweet potatoes, manioc, and rice) cover most of the demand, but the country's real wealth consists of its plantation products grown on the southern coastal plain, such as cocoa (of which Ivory Coast is the world's leading producer, with 781,000 t in 1991), coffee (264,000 t), oil and coconut palms, rubber, bananas, pineapples, and cotton. Animal husbandry is of little significance except in the north central region; the fishing industry is highly productive and well equipped. The forests, covering 23.7% of the nation's territory, produce large quantities of timber, including construction lumber and valuable hardwoods. Massive and indiscriminate logging in the last few decades, which had caused some depletion of the forest, prompted the government to impose substantial limitations on lumber exports.

The country has no significant mineral resources: there are a few diamond and iron deposits at Klahoyo, but they remain unexploited for lack of available energy. Petroleum deposits were recently discovered offshore at Bélier and near Grand Bassam. Massive amounts of electrical energy are produced, covering 50% of demand and generated with huge dams on the Bandama river (at Kossou), and the Bia, Comoé, and Cavally.

Industry is of relatively minor but growing importance, with 10% of the work force employed in manufacturing activities, contributing 17% of GNP. Particular emphasis is being placed on iron-and-steel and chemical sectors: a large refinery is in operation at Vridi (near Abidjan), and there are sugar processing plants at Ferkessédougou, Borotou, Sérébou, Katiola, and Zuénoula. The most highly developed activities are those related to agriculture and food processing, such as oil extraction, coffee processing (Port San Pedro), brewing, and tobacco, in addition to wood products (Bassam and Abidjan), textiles (Bouaké), cement, and light manufacturing (Abidjan and San Pedro).

Despite a generally healthy balance of trade, the crisis of world overproduction of cocoa, uncertainties about coffee price trends, and lastly its level of debt to international institutions have placed Ivory Coast, like many other nations, in a difficult situation, fueling social and political tensions. Principal exports are cocoa (20%), coffee (20%), timber (16%), and bananas; imports consist of petroleum and its derivatives (15%), machinery (12%), and transportation equipment (10%). The country's primary trade partners are France, the U.S., the Netherlands, and Italy.

Tourism, with 300,000 visitors per year (1990), is a very recent source of income. The country has one of Africa's best road systems, with 33,316 mi [53,736 km] of all-weather highways (1984), a railroad system (730 mi [1177 km] in 1986), modern and efficient ports such as Abidjan, Tabou, and the spe-cialized lumber-handling harbor at San Pédro, and two main airports at Port Bouët (Abidjan) and Bouaké.

Historical profile. By the 5th century B.C., Carthaginian and Phoenician ships had already explored the littoral of what is now Ivory Coast. Undoubtedly the area's interior regions fell into the orbit of the great empires of Ghana, Mali, and Songhai, which dominated the history of western Africa until the 16th century. The last of these empires, the Songhai, declined as the Europeans advanced: first the Portuguese with their commercial outposts and Catholic missionaries, then the French with their slave traders and occupation troops. Between 1838 and 1842, France imposed a series of treaties on local tribal chieftains and secured control of the region; it was declared a colony in 1893, stirring up fierce opposition among the population led by Samoury Touré.

Incorporated into French West Africa, Ivory Coast experienced a sudden economic boom based on the introduction of monocultures (coffee and cocoa) and the creation of proper communications. On the French side, however, investments were meager and forced labor was frequently used. After World War II, the push for independence was led by the African Democratic Assembly (RDA) founded by Félix Houphouët-Boigny, which moderated its initial anticolonialist fervor in favor of a more conciliatory approach, and ultimately aligned itself for the most part with De Gaulle's policies. Independence was achieved gradually: a framework law was adopted in 1956; semiautonomy within the French Community was granted in 1958; and full sovereignty was attained in 1960. The new nation, guided by the single party of Houphouët-Boigny, was from then on the leading representative of the moderate, liberal, and pro-Western attitude that prevailed in several African nations. This orientation brought a generous influx of European capital and ignited another phase of rapid development. Despite the economic stability achieved by Houphouët-Boigny, who was repeatedly and lawfully re-elected as president, this policy has raised increasing concern, especially in intellectual circles, about the penetration of European attitudes into every aspect of social life.

Culture. The most interesting traditions are those of the Baule, a group that moved in the 18th century from the Ashanti kingdom to what is now Ivory Coast, and became assimilated with the Guro and Senufo. The Baule instituted a kind of religious revolution, replacing the natives' animism with belief in a host of well-defined divinities descended from the sky god Nyamye and the earth goddess Assye. Artistic expression was profoundly affected by the change: the primitive and occasionally violent forms of some African styles are absent here, replaced by serene, refined depictions of the deities. Wood carvings, masks, bronze objects, and even everyday utensils bear witness to the self-assured aesthetic of the Baule people. The other tribes in this area also produce ritual masks, often (as in the case of the Senufo) of anthropomorphic figures; these objects are produced under the auspices of the secret societies that control religious life. Literary expression in the local languages is through an oral tradition, with themes that include the legends surrounding Aura Poku, the queen who brought the Baule to Ivory Coast. A written literature in French has developed since the beginning of the 20th century; early works include the theatrical sketches of Barnard Daidié, foreshadowing Senghor and the theme of "Negritude."

Socioeconomic data	
Income (per capita, US$)	730 (1990)
Population growth rate (% per year)	4.1 (1980–89)
Birth rate (annual, per 1,000 pop.)	50 (1989)
Mortality rate (annual, per 1,000 pop.)	14 (1989)
Life expectancy at birth (years)	53 (1989)
Urban population (% of total)	40 (1989)
Economically active population (% of total)	37.9 (1988)
Illiteracy (% of total pop.)	46.2 (1991)
Available nutrition (daily calories per capita)	2365 (1989)
Energy consumption (10^6 tons coal equivalent)	2.4 (1987)

THE GAMBIA

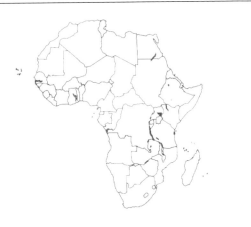

Geopolitical summary	
Official name	Republic of The Gambia
Area	4360 mi² [11,295 km²]
Population	687,817 (1983 census); 875,000 (1990 estimate)
Form of government	Presidential republic with a House of Representatives that performs legislative functions; the head of government is elected by universal suffrage.
Administrative structure	35 districts combined into 6 Area Councils, plus the capital (City Council)
Capital	Banjul (pop. 44,188, 1983 census)
International relations	Member of UN, OAU, Commonwealth; associate member of EC; combined with Senegal in the Senegambia confederation
Official language	English; Bantu dialects widely used
Religion	Predominantly Muslim (89%); remainder animist (9%), Protestant, and Catholic
Currency	Dalasi

Natural environment. The Gambia is bordered on the north, east, and south by Senegal, and faces the Atlantic Ocean along a short stretch of coastline to the west. The territory is flat and uniform, consisting of a narrow strip on either side of the Gambia river, an average of 30 mi [50 km] wide and extending approximately 185 mi [300 km] inland. There are only isolated areas of rolling terrain, and the land is often marshy; it consists of recent river sediments, muddy toward the coast, and older sandy soils well suited for the cultivation of peanuts. The mouth of the Gambia river is a well-protected estuary with a narrow neck only 2 mi [3 km] wide; this moderates the tidal flows that are nevertheless quite substantial (penetrating 90 mi [150 km] inland during the rainy season, and much farther during the dry season).

The climate of this small country is subequatorial, with a short, very rainy summer and a long dry season extending from November to May. Precipitation exceeds 40 in. [1000 mm] per year, and average monthly temperatures range between 77°F [25°C] in winter and 82°F [28°C] in summer.

The vegetation is characterized by the typical association of mangroves along the coast, and a dense fringing forest toward the estuary and near the river, with various species of palms. Forest covers 17% of the territory. Toward the country's southern border, the vegetation is already that of the rainforest, well-endowed with tall trees (mahogany) and shrubs (bamboo); toward the interior, this merges into a tall-grass savanna. The acacia is the typical tree of the region, together with the inevitable baobab.

Population. The most numerous ethnic group is the Mandingo, an agricultural people who are also skilled traders and artisans (especially weavers). The Wolof, the second-largest group in the country, also retain their African customs and traditions, despite efforts by the government to transmit European culture via a widespread school system. Other minor groups include the Fulbe, Jola, Serahuli, and the Aku, few in numbers but highly Anglicized and established in most positions of power.

The population is distributed quite uniformly throughout the country, with an average density of 212 per mi² [82 per km²]. Rural settlement in scattered villages is the norm (about 80% of the population), while urbanization is concentrated almost exclusively in the capital, Banjul (almost 50,000 inhabitants). Founded in 1816 by the British under the name of Bathurst (which it retained until 1974), on a promontory on the left bank of the Gambia river estuary, it still presents the typical appearance of a British colonial city, and performs the functions of an administrative capital, large city, and the nation's principal port.

Mario Appelius described the city as follows in a report dating from 1926:

> *Finally you are ashore, and immediately you recognize the typical British colonial city, a carbon copy of the same barracks, the same commercial buildings, the same streets, the same Scottish villas ... all the commercial and colonial bureaucratic trappings of the Empire, all of them reproduced at Bathurst on a smaller scale ... but always cast from the same mold, cut from the same cloth, governed by a single custom, which is the great law of the Empire. You see them on the streets and in the bars ... the same petty officials with Kitchener or Cromer mustaches, the same "made in England" shoes, the same "made in England" hats, the "made in England" tennis rackets, and the ladies and the schoolmistresses most especially "made in England"!*

Social conditions are characteristic of a very poor country: GNP per capita is US$260 (1990), one of the lowest figures in the world; illiteracy afflicts 72.8% of the people, and the population growth rate is approximately 2.8% per year, with infant mortality at 14.3% annually.

Climate data				
Location	**Altitude (ft asl)**	**Average temp. (°F) January**	**Average temp. (°F) July**	**Average annual precip. (in.)**
Banjul	7	73	80	50.5
Conversion factors: 1 ft = 0.3 m; 1 in. = 25 mm; °C = (°F − 32) × 5/9				

Administrative structure

Administrative unit (councils)	Area (mi²)	Population (1983 census)
Kombo Saint Mary	29	101,504
Lower River	625	55,263
MacCarty Island	1,117	126,004
North Bank	871	112,225
Upper River	799	111,388
Western	681	137,245
Banjul (city)	5	44,188
GAMBIA	4,126(*)	687,817

(*) Excluding 234 mi² of internal waters

Capital: Banjul, pop. 44,188 (1983 census)
Principal cities (1983 census): Serekunda, pop. 68,433; Bakau, pop. 19,309; Brikama, 19,584; Sukuta, pop. 7,227.

Conversion factor: 1 mi² = 2.59 km²

Economic summary. The country's economy is decidedly traditional, based on agriculture (which produces 35% of GNP) and propped up by aid from Western countries. The most common agricultural products are peanuts, developed by the British during the colonial period, coconuts, palm oil, rice, and cotton, the latter increasing rapidly in terms of both quantity and cultivated area. Animals are raised exclusively for family consumption, while fishing, although limited by a small fleet, is also practiced for export; forest products such as mahogany, bamboo, and *Borassus* palm are also exported.

There are no mineral resources; industry (11% of GNP) is linked to peanut processing, and energy sources are based on hydroelectric exploitation of the river.

The nation's real resource is tourism, which contributes 9% of GNP; 200,000 foreign tourists arrive each year, bringing US$300 million in income. The country's capital is the terminus for several cruise lines, and boasts a number of large international-class hotels including a floating hotel ship anchored in the harbor. Tourists come primarily from Europe, but they have been joined by many Americans looking for their African roots along the Gambia river. There is also a flourishing contraband trade with Senegal based on the fact that the two countries are located in different monetary areas: Gambia is within the sterling sphere of influence, and Senegal falls within the orbit of the French franc.

Socioeconomic data

Income (per capita, US$)	260 (1990)
Population growth rate (% per year)	2.8 (1980–89)
Birth rate (annual, per 1,000 pop.)	47.4 (1989)
Mortality rate (annual, per 1,000 pop.)	21.4 (1989)
Life expectancy at birth (years)	49.9 (1989)
Urban population (% of total)	21.5 (1989)
Economically active population (% of total)	46.1 (1988)
Illiteracy (% of total pop.)	72.8 (1991)
Available nutrition (daily calories per capita)	2339 (1989)
Energy consumption (10^6 tons coal equivalent)	0.09 (1987)

The services sector provides 54% of GNP and is based on foreign trade, dominated by exports of peanut oil (50%), palm nuts, fish, and hides. Internal communications are provided principally by 150 mi [240 km] of navigable river, but also by 1911 mi [3083 km] of highways (1983), only half of them usable throughout the year. The harbor at Banjul and the airport at Yundum provide links to the outside world.

Historical profile. The historic development of The Gambia is unusual, since it constitutes a tiny enclave of English colonization and language within French-speaking Africa. The Gambia river basin was discovered by Italians in 1445, explored by the Portuguese and then by English gold prospectors and French traders; in the 17th and 18th centuries, it was caught up in rivalries between Britain and France for control of the Senegal region.

Early in the 19th century, the British established a city on Saint Mary island and called it Bathurst in honor of the current minister for the colonies. During this same period, the territory became a stable part of Britain's possessions in western Africa, along with the Gold Coast and Sierra Leone, whose governor was also responsible for The Gambia. In the course of the century, The Gambia acquired its own particular identity; the hinterland was governed as a protectorate, while the region around Bathurst (the capital, now known as Banjul) was a colony.

Having achieved internal autonomy in 1963, the country was given full independence two years later as part of the British Commonwealth. In reality, independence was political but not economic, and Great Britain's supporting role in this arena remained critical. Despite the failure of an attempt at integration with Senegal, special agreements have been made between the two states in the areas of foreign policy, defense, and security. Twice thereafter, in 1980 and 1981, Senegalese troops were called in by President Jawara to re-establish domestic order, threatened first by a destabilization plan masterminded by Libya, and then by a leftist insurrection.

In the latter instance, following a deterioration in the economic situation and a series of increasingly unpopular government actions, the Revolutionary Socialist Labor Party attempted to gain power by force, proclaiming a Marxist-Leninist government. The party freed prisoners and gave them weapons, unleashing a wave of looting and killing in the capital; at the request of President Jawara, the Senegalese government then sent in troops who were welcomed as liberators by the population. The Senegalese reestablished order, and between 1982 and 1989 The Gambia and Senegal joined together in a kind of confederation called Senegambia. Under this arrangement, a contingent of some 1300 Senegalese soldiers was permanently assigned to The Gambia (which gave up its own armed forces except for a reduced police force). Over the next few years, this "irritating" presence became a source more of tension than of cohesion. Problems were also evident in the economic area, and relations between the two states deteriorated to the point of crisis in 1989. The final break came when Jawara requested an end to the practice by which Senegal held the permanent presidency of the confederation, asking instead for a system that allowed each head of government to act as president in alternate years. Faced with this request, the Senegalese leader Abdou Diouf recalled his soldiers and declared the dissolution of Senegambia.

GHANA

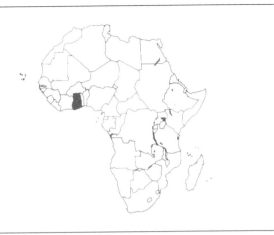

Geopolitical summary

Official name	Republic of Ghana
Area	92,076 mi^2 [238,538 km^2]
Population	12,296,081 (1984 census); 14,925,000 (1990 estimate)
Form of government	Military government, headed by a Provisional National Defense Council
Administrative structure	10 regions, subdivided into 110 districts
Capital	Accra (pop. 867,459, 1984 census)
International relations	Member of UN, OAU, Commonwealth; associate member of EC
Official language	English; Sudanese dialects widely spoken
Religion	Christian 52% (including approximately 1 million Catholics); animist 35%; Muslim 13%
Currency	Cedi

Natural environment. Ghana is bordered on the west by Ivory Coast, on the north by Burkina Faso, on the east by Togo, and faces the Atlantic Ocean (Gulf of Guinea) to the south.

Geological structure and relief. The territory is approximately rectangular in shape with the long side oriented north–south; its morphology is extremely varied, governed both by the lithological structure and by hydrology. The principal areas of higher ground, with the exception of the Kwahu plateau (which barely exceeds 2300 ft [700 m] in elevation) that is aligned northwest–southeast in the southern part of the country, are located predominantly at the periphery of the country, thus defining a large interior which contains the lower course of the Volta with its two principal branches, the White Volta and Black Volta, the latter constituting much of Ghana's western border. Geologically speaking, the area occupied by the Volta basin (and largely submerged beneath the huge lake of the same name created in the early 1970s) is formed of sedimentary rocks (sandstone, lime-

stone, conglomerate) of Lower Paleozoic age (Cambrian and Silurian), which partly cover the older crystalline basement rock. The latter crops out directly in the eastern peripheral highlands (Fazao and Togo mountains, elevation less than 3300 ft [1000 m]), and in the entire southwestern section of the country, including the Kwahu plateau already mentioned, and the adjoining Ashanti plateau. Outcrops of sedimentary rocks, including more recent ones, are also found along the coastal strip, generally compacted and made slightly rolling by the succeeding lines of dunes that block the many river mouths, especially the great Volta delta (about 60 mi [100 km] across), creating numerous lagoons. Lastly, just to the southeast of Kumasi is the little lake called Bosumtwi located inside a characteristic caldera, the remains of a volcanic structure that was still active in Quaternary times.

Hydrography. As already mentioned, Ghana's territory consists largely of the Volta basin with its two branches: the White Volta, which rises to the north near the border between Burkina Faso and Mali; and the Black Volta, which rises to the east at an elevation of 2404 ft [733 m], near Bobo Dioulasso. The river is 1000 mi [1600 km] long, and has been dammed, 60 mi [100 km] from its mouth, by the Akosombo dam constructed by Italian contractors. This created an artificial lake with an area of 3280 mi^2 [8500 km^2], one of the world's largest, with many complex branches along the course of the two principal rivers and their tributaries (Sene, Nasia, Daka, Oti, and Wawa). In addition to its principal function—as a source of electrical energy for the entire country and for export—the dam also plays a critical role in flow control, preventing the devastating floods that occur elsewhere during the rainy season. The Bia and Tano rivers flow down from the southern Ashanti and Kwahu plateaus and empty into the Aby lagoon at the border with Ivory Coast; the Ankobra and Pra rise in the same area, and drain the southwestern region of the country.

Climate. Ghana's climate can be considered entirely typical of the countries that face the Gulf of Guinea: it is very hot and humid, with subequatorial characteristics on the coast and decreased precipitation toward the interior, where the temperatures are always very high. Maximum rainfall occurs at Axim on the western coast (97 in. [2488 mm] and 151 days of rain per year), distributed into two seasons: the "great rains" from April to May and the "lesser rains" from September to October, with an average maximum temperature of 85°F [29.3°C] and average minimum of 74°F [23.3°C] in the months of March and October. On the eastern coast the wet southwestern winds have a more marginal effect, resulting in much lower levels of precipitation (48 in. [1230 mm] at Accra), with slightly higher average maximum temperatures (86°F [29.9°C]; average minimums are 73°F [22.5°C]). In the interior, at Tamale, precipitation is concentrated in a single season due to the closeness of the peak zenithal rain and is still above 40 in. [1000 mm] (45 in. [1163 mm]), while maximum temperatures are even higher (91°F [33°C]). In the north-central regions, however, the climate is of the Sudanese type, with a short rainy season and a long, hot dry season.

Flora and fauna. Along the coast and in the rainier areas of the plateaus, vegetation is characterized by rain forest, including valuable tropical woods such as ebony, mahogany (including Gabon, wawa, and sapele mahogany), guarea citrus, and mangroves, especially throughout the Volta delta. Toward the interior the forest thins out, forming "fringes" along rivers; it ultimately

Climate data

Location	Altitude (ft asl)	Average temp. (°F) January	Average temp. (°F) July	Average annual precip. (in.)
Accra	190	80	77	48.0
Kumasi	1014	77	76	55.0
Tamale	689	91	72	45.4

Conversion factors: 1 ft = 0.3 m; 1 in. = 25 mm; °C = (°F – 32) × 5/9

becomes a wooded savanna in the north-central part of the country, with expansive steppes rich in xerophytic plants. The great size of the forest (covering 35.6% of the nation's area) and its variable density offer optimum habitats for Ghana's abundant wild fauna. The most common large herbivores are elephants, buffaloes, antelopes, and gazelles, while the most prevalent carnivores are lions, leopards, cheetahs, and hyenas. Hippopotami and crocodiles live along the rivers, while huge numbers of insect species (some of them harmful to humans and animals, like the tsetse fly) infest the area. This enormous natural wealth is protected, however, only in two large parks: the Kujani nature reserve on a large peninsula on the western shore of Lake Volta, and the Mole nature reserve in the savanna on the left bank of the Black Volta, near the borders with Ivory Coast and Burkina Faso.

Population. The ethnic and cultural picture is dominated by the traditional prestige of the Ashanti, who ruled the region with a powerful kingdom from the 17th to the 19th centuries, and resisted colonization with tenacity and military skill. They belong to the Sudanese Akan group, are the most numerous ethnic group in the country (53% of the population), and live in the southern regions; the north is inhabited by the Mola-Dagbani (16%) and the east by the Ewe (12%), the dominant group in Togo; the prevalent group in the southwest is the Ga-Adangme (7.6%). Relations among the various ethnic groups are not entirely peaceful; disputes flared up in 1981 between two smaller tribes (Konkomba and Manumba), resulting in about a thousand deaths. European influence in the region, which goes back many years, has resulted in extensive conversion to Christianity (52% of the population, including 8% Catholic), 35% animist, and 13% Muslims, largely in the north.

Demographic structure and dynamics. In the colonial period the area within Ghana's current borders, defined by lengthy negotiation among the European colonial powers, was densely settled by comparison with the rest of Africa (78 per mi^2 [30 per km^2]), and boasted a more developed and better-educated population. The British therefore adopted a colonial policy of "indirect rule," which in practice left most matters of local government in native hands. Semi-independence already existed by 1925; this autonomy was extended further in 1946, and culminated in absolute sovereignty in 1957.

The rate of population increase (approximately 1.6% per year) resulted from a previous situation of equilibrium that was completely upset, after independence, by a drop in mortality rate to 1.3%, bringing the annual increase to a record level of 3.4% and raising the population to 14,925,000 (1990 estimate), with a density of 161 per mi^2 [62 per km^2]. The highest population density is found in the central and southern regions of the country: Ashanti, Brong-Ahafo, Western, Central, and Eastern. The urban

population here exceeds 30%, and the urban structure is represented by the three major cities of Accra, Sekondi-Takoradi, and Kumasi. There is a considerable imbalance by comparison with the northern regions, in which a low population density is perpetuated by the small number of cities (Tamale and Bolgatanga). Rural settlement patterns are still traditional, with scattered villages of round or rectangular straw huts with thatched roofs. The cities present the appearance of ultramodern centers (Sekondi-Takoradi and Tema) ringed by miserable shantytowns; some, however, have retained the air of a pleasant colonial capital (Kumasi). The capital, Accra, located on a stretch of coastline, was reconstructed after several episodes of destruction resulting from fire (1894) and epidemics (1907) and now comprises a highly modern administrative center designed by European and American architects, containing banks, the Supreme Court, the Central Library, and the University of Ghana.

Social conditions. The improvement in general social conditions that followed independence has radically changed the nation's demographic structure, reducing mortality rate, raising the natural rate of population increase, and extending the average life expectancy. Ghana is therefore a very young country, with 53% of the population less than 20 years old: this mass of young people is exerting pressure on the countryside (which they have been abandoning in large numbers) and on the cities, which are more and more oppressed by a huge unemployed underclass. GNP per capita is approximately US$390, which is higher than in many western African countries but still cannot conceal a difficult social situation, indicated by a deficient diet (2209 calories per day per person), and a high rate of illiteracy (39.6%).

Economic summary. Ghana's economy illustrates perfectly the economic dependency and political insecurity that result from monoculture. As the world's leading producer of cocoa (30% of total production) in the 1960s it was able to afford the costs of both independence and an ambitious policy of Panafricanism. The drop in the price of cocoa on the world market

Administrative structure

Administrative unit (regions)	Area (mi^2)	Population (1984 census)
Ashanti	9,415	2,090,100
Brong-Ahafo	15,269	1,206,608
Central	3,793	1,142,335
Eastern	7,711	1,680,890
Greater Accra	1,001	1,431,099
Northern	27,168	164,583
Upper Eastern	3,413	772,744
Upper Western	7,132	438,008
Volta	7,941	1,211,907
Western	9,234	1,157,807
GHANA	92,076	12,296,081

Capital: Accra, pop. 867,459 (1984 census)
Major cities (1984 census): Kumasi, pop. 488,991; Sekondi-Takoradi, pop. 175,352; Tamale, pop. 168,091; Bolgatanga, pop. 142,003.

Conversion factor: 1 mi^2 = 2.59 km^2

in 1966 thwarted the ambitions of Ghana's leader, Kwame Nkrumah: toppled from his position as "champion" of the oppressed peoples, he became the scapegoat for galloping inflation, enormous debt, the reluctance of wealthy countries to invest, and difficulties in basic subsistence. The situation returned to normal after a few years, but in 1972 there was another fall in cocoa prices and another coup, followed by further normalization and more changes in 1982. At the moment, a Council for National Renewal is maintaining political stability and promoting economic development (with considerable government guidance), aimed at self-sufficiency in food, increased agricultural productivity, and the creation of a national industrial base. In addition to cocoa production, Ghana's economy is based on the presence of substantial mineral deposits, and on the vigorous industrialization that occurred right after World War II, developed with particular speed in the first years of independence, and is now back on solid ground. Nationalization of the economy remains the cornerstone of economic organization, and the country's third-world "anti-imperialism" has mellowed into a more pragmatic neutralist approach that welcomes foreign investment.

Agriculture and livestock. Agriculture is of predominant importance to the nation's economy, contributing 37% of GNP and employing 21% of the work force. In terms of the problem of food self-sufficiency, the situation is complicated by the small area dedicated to cultivation (4.8%) and the low productivity of traditional crops such as corn, millet, sorghum, sweet potatoes, and manioc. Exceptions include the rapid development of rice and sugar cane production, and a general improvement in rural incomes achieved by means of farm cooperatives. But the most important production sector involves plantation crops (6.6% of the land area is devoted to permanent crops), dominated by cocoa production. For many years the world leader, Ghana has suffered not only from price fluctuations and from the consequences of ill-advised internal politics, but also from aging plantation stock, depletion of the soil, drought, and the fires that in 1983 destroyed 15% of the plantations. In 1991 it was the world's third-largest producer (after Ivory Coast and Brazil) with 12% of production. Other plantation crops include peanuts, coconut and oil palms, and new products such as pineapples, lemons, and tomatoes. Fishing and animal husbandry are primarily of domestic significance, while the forests, covering 34.1% of the territory and containing many valuable hardwoods, provide material both for internal use and for export.

Energy resources and industry. The richest energy resource is hydroelectric potential, specifically the enormous capacity created by the dam at Akosombo, which accounts for almost all of Ghana's energy production (4.75 billion kWh in 1988). Traces of natural gas and petroleum have also been discovered 30 km from Half Assini, but the nation's traditional mineral wealth is based on extraction of gold (29,183 lb [13,265 kg] in 1989) in the Western and Central regions; production has fluctuated greatly since 1965, a year in which diamonds were also extracted at three times the current rate from a mine near Accra. The reasons for this trend include obsolescent equipment, increased smuggling, buyouts of multinational corporations, and greater involvement by the state in mineral extraction in order to establish an industrial base. Small amounts of bauxite are extracted at Awaso in the Western region and are used to produce aluminum; manganese ore is mined at Nsuta near Tarkwa, also in the Western region.

Industry has declined slightly in recent years, and now contributes 30% of gross domestic product and employs 8% of the work force. Large factories are producing at only 30% of capacity due to the changes in goals that have followed Ghana's various economic crises (decreasing cocoa prices and rising petroleum costs), and an economic policy that has slowed down the activities of multinational corporations. The industrial base comprises a large aluminum plant at Tema, a chemical and pharmaceutical plant at Kwabenya (near Accra), and a petroleum refinery and cement plant also at Tema; light industry includes the textile, wood, and food-products sectors (tobacco, beer, sugar) at Takoradi. Industrial installations are located in a coastal region centered around Tema, Accra, and Sekondi-Takoradi, an adjacent mining district defined by Nsuta, Tarkwa, and Prestea, and another mining area farther inland to the south of Kumasi, comprising Obuasi (gold), Akwatia (diamonds), and Awaso (bauxite). The great power plant and dam at Akosombo stand alone, linked to the focal points of Tema and Accra.

Commerce and communications. The balance of trade is essentially neutral, and reflects a development policy focusing on the country's true strengths and aimed at deemphasizing exports of cocoa (25%), gold (10%), and diamonds (0.3%) in favor of a wider range of processed products such as wood (0.7%) and manganese (0.6%). Imports include manufactured goods and machinery (50%), fuels and chemical products (30%), and foodstuffs (11.6%). Communications routes are among the most efficient on the continent. The road network is ubiquitous, with all-weather capability over 17,500 mi [28,300 km] (1985) of the system; rail lines (591 mi [953 km]) connect the urbanized Tema–Accra–Kumasi triangle with Sekondi-Takoradi, providing links between the three industrial and mining areas and the three major ports. The international airport at Accra/Kotoba, together with smaller fields at Tamale, Takoradi, Sunyani, and Kumasi, rounds out the nation's transportation system.

Historical profile. ***The ancient kingdom.*** Ghana was the most ancient and glorious black empire of precolonial Africa. Founded around the 4th century A.D. by Berbers, it took its name from its capital Ghana (or Ganata), which was located near present-day Koumbi-Saleh in Mauritania. According to Arab sources, twenty kings had reigned in Ghana before the prophet Muhammad was even born. These kings bore the name *Kaya-Maga*, "masters of gold," since gold was the country's wealth. The economic prosperity of this region is remarked upon by a 14th-century Arab chronicle written by an unknown traveler:

Socioeconomic data

Income (per capita, US$)	390 (1990)
Population growth rate (% per year)	3.4 (1980–89)
Birth rate (annual, per 1,000 pop.)	45 (1989)
Mortality rate (annual, per 1,000 pop.)	13 (1989)
Life expectancy at birth (years)	55 (1989)
Urban population (% of total)	33 (1989)
Economically active population (% of total)	37.1 (1988)
Illiteracy (% of total pop.)	39.6 (1991)
Available nutrition (daily calories per capita)	2209 (1989)
Energy consumption (10^6 tons coal equivalent)	1.86 (1987)

The prince of Ghana who reigned in the time of Muhammad had a thousand horses in his stables. If a horse died, it had to be replaced before sunset … Every animal slept on a real mattress, and was tethered only by a silken ribbon about its neck and leg. Each one had three people to care for it, seated nearby: one provided food, the other drink, and the third kept the animal clean.

The fame of the kingdom of Ghana, its army, its matriarchally descended dynasties, and its magnificent barbarian court, was kept alive through the centuries all over western Africa; some of these traditions were inherited by the later kingdom of Mali.

The colonial period. After it achieved independence, the former British colony of Gold Coast changed its own name to Ghana, although its territory did not entirely correspond to the Sudanese region in which the ancient empire of that name had developed between the 8th and 13th centuries. The name "Gold Coast," moreover, seemed too closely linked to a colonial past that was better forgotten: it recalled the primary reason that had drawn so many European adventurers and traders. The first to gain control of the gold trade had been the Portuguese, who landed in the area in 1471 and within a decade had established fortified bases all along the coast. Fifty years later, however, the Portuguese were already being confronted by Spaniards, French, Dutch, and English. By the mid-16th century, the slave trade was proving to be much more profitable than gold, and first the Dutch, then the British distinguished themselves in this new activity. In the 19th century, Great Britain made even greater strides in exploiting the region, shifting from the slave trade to actual colonial conquest. This effort, however, encountered the resistance of the Ashanti, the people who had gathered together the earlier Akan groups and created in the interior a huge and prosperous state based on trade and agriculture. The wars between the British and the Ashanti lasted the entire century, and this fierce people were not even subdued by the destruction of their capital, Kumasi, which was burned down in 1874.

Only at the beginning of this century could the British say that they had established control over the entire region, which was then gradually unified and subjected to the typically English colonial system of "indirect rule." The institution that acquired increasing significance was the Legislative Council; operating in the southern region of the colony by the mid-1800s and consisting of British civil servants and unofficial representatives elected by the local population, in 1946 it finally welcomed the Ashanti as well, so that the local representatives found themselves in the majority.

Independence. As far as the Europeans were concerned, the country appeared to be on the road to a slow and gradual acquisition of autonomy, but these predictions were confounded after World War II by the advent of Kwame Nkrumah, with his radical slogan "self-government now." Nkrumah's party raced along the road toward independence, which was achieved in 1957 by constitutional means, without violence, and with strong support from a mobilized population. Having achieved independence so quickly, Ghana became the ideal for the entire African continent, and Nkrumah's decisions—in favor of socialism, anti-imperialism, and Panafricanism—were regarded as paradigms. In the years that followed, however, this approach became more and more frequently belied by facts and tarnished by authoritarian relapses. By 1966, therefore, when Nkrumah was overthrown by a military junta, the popular consensus that he initially enjoyed had already been eroded.

In a message to the Popular Party of the Overseas Conventions, published in *Africa and the World* in 1968, Nkrumah exhorted the militants to bring the party back to power, and to fight against the neocolonialist intentions of the new ruling group:

> *The beginning of 1966 found Ghana poised for a breakthrough in her national economy. As Ghana's economic progress gained momentum so did the imperialist neo-colonialist intrigues and subversion increase. Our experienced Party, however, was equal to the tasks and the challenges. Ghana became an inspiring spearhead of the African revolution.*
>
> *The situation of Ghana since February 24, 1966 has been a chronicle of mass unemployment, sadness, shame, incompetence, and chaos. Proud Ghanaians now walk about with heads bowed down in shame.*
>
> *Ghana is sold out and is in the grip of neo-colonialists and their lackeys. Everything is in continuing collapse and tribalism has raised its head again.…*
>
> *For us, the militants of the Party, to sit down doing nothing and watch twenty years hard work with concrete achievements destroyed by a short-sighted and corrupt few, is infamy and downright cowardice.*
>
> *Ghana must be saved. The return to power of the Party and its popular government is the only way to save Ghana now. And this cannot be done constitutionally.*
>
> *I have closely and carefully followed the collapsing trends in Ghana. I have deliberately kept silent all this time because I know the time has not yet come for me to say anything. We shall soon have a great deal to say—very soon.*
>
> *Our militants are everywhere in Ghana. For them, too, now is not the time for words. Now is a time for action.*

In the years that followed, Ghana was subjected to an alternation of dictatorial regimes and fleeting restorations of democracy, in a context of instability that increasingly weakened the nation.

Culture. The art that flourished in the Ashanti kingdom was rich and unique, both in its popular manifestations (like the wooden depictions of the "fertility doll") and in the "official" styles. Before the arrival of the English, Kumasi was famous for its gold objects, from necklaces to scale weights to the royal stove that also served as a sacrificial altar.

Also unique is Ghana's literary heritage, given written expression in the nation's major languages using the "Africa" alphabet, derived with certain modifications from Roman letters. Every language has its own original literature, and since the early 20th century there have been many works, both prose and poetry, that have contributed to the creation of a national awareness. In addition to authors who write in the languages of their own ethnic groups, there are others who use English, such as Michael F. Dei Anang, Gilbert A. Sam, and Ayi Kwei Armah.

GUINEA

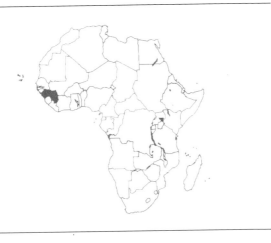

Geopolitical summary

Official name	République de Guinée
Area	94,901 mi^2 [245,857 km^2]
Population	5,781,014 (1983 census); 6,876,000 (1990 estimate)
Form of government	Presidential republic with military leadership
Administrative structure	29 provinces combined into 4 major ethnic regions, plus the capital
Capital	Conakry (pop. 705, 280; 1983 census)
International relations	Member of UN, OAU; associate member of EC
Official language	French; Sudanese languages widely spoken
Religion	Muslim 85%; animist 5%; remainder Catholic
Currency	Guinea franc

Natural environment. Guinea is bordered on the north by Guinea-Bissau, Senegal, and Mali, on the east by Mali and Ivory Coast, on the south by Liberia and Sierra Leone, and faces the Atlantic Ocean to the west.

Geological structure and relief. The territory is extremely varied, with topography extending up to considerable elevations and a hydrographic and morphological organization that results in very distinct regional entities. Facing the Atlantic is a sandy, marshy coastal plain, cut into by deep embayments and fringed with numerous small islands. The coastal plain extends inland for some 25–50 mi [40–80 km], with slightly rolling topography modeled onto early Paleozoic rocks, to the foot of the Fouta Djallon massif (4674 ft [1425 m]). This feature, located entirely on Guinean territory, is an important element of the largest mountain complex in western Africa, together with the Loma Mountains (6389 ft [1948 m]) in neighboring Sierra Leone and the Nimba Mountains (5747 ft [1752 m]) that straddle the border

with Liberia and Ivory Coast. The Fouta Djallon is a substantial massif made up of Paleozoic (Ordovician) sandstones covering a crystalline basement (granite and dolerite); its slightly fractured subhorizontal strata often form huge plateaus dissected by deep valleys. Other rocks, such as volcanic dolerites and metamorphic schists, constitute the skeleton of the hills that overlook Conakry from the hinterland and reach elevations of more than 3300 ft [1000 m] (Kankan and Kakoulima mountains). The entire eastern half of Guinea consists of a series of highlands with elevations of between 1000 and 1600 ft [300–500 m], the framework of which consists of pre-Paleozoic crystalline rocks: this is in fact an outcropping of the ancient African continental foundation, dissected by numerous watercourses that flow to the northeast into the upper Niger, which rises in this region. Lastly, the entire southeastern sector of Guinea is occupied by a massive alignment of mountains oriented northwest–southeast, with peaks often exceeding 5000 ft [1500 m], including the high ground of the Nimba Mountains. The Archeozoic African shield, consisting of intrusive igneous rocks, crops out here as well; its western slope plunges directly toward the Atlantic.

Hydrography. The mountains of Guinea constitute the principal hydrographic node of western Africa, the source of the Senegal and the Gambia to the north, the Niger and the Volta to the east, and many smaller rivers which descend parallel to one another to the southwest, directly to the ocean. Of these, the ones which flow directly toward the Guinean coast (the Kogon, Fatala, and Kolente) and those that flow through Liberian territory are characterized by extreme variability, with high flow rates during the rainy season. These rivers are short and interrupted by frequent cataracts, and have a characteristic estuary mouth that is often very deep. The Niger, which rises in Guinea on the eastern slopes of the Loma Mountains, receives water from numerous tributaries in a wide depression that becomes swampy in the rainy season and in which the river is already navigable.

Climate. The extremely intense rainfall (averaging more than 80 in. [2000 mm] annually) and the country's latitude give Guinea's climate a definitely equatorial character, but it is still characterized by two seasons, with different features on the coast and in the interior. Near the ocean, rainfall reaches levels among the highest in the world (168 in. [4295 mm] at Conakry), with a long wet season from April to November governed by winds from the southwest (the austral southeast trade wind, which changes direction as it crosses the equator), and more than 40 in. [1000 mm] of rain each month in July and August. Temperatures are always very high, around 81°F [27°C] as the annual average (80°F [26.6°C] at Conakry), with minor temperature swings. The winter (December–March) is entirely dry, and dominated by the *harmattan*, the hot northeast wind from the Sahara. Toward the interior the climate becomes less humid, with a shorter rainy sea-

Climate data

Location	Altitude (ft asl)	Average temp. (°F) January	Average temp. (°F) July	Average annual precip. (in.)
Conakry	23	80	77	167.5
Kankan	1,246	75	82	66.1
Labé	3,451	69	77	66.7
Conversion factors: 1 ft = 0.3 m; 1 in. = 25 mm; °C = (°F − 32) × 5/9				

Administrative structure

Administrative unit (regions)	Area (mi²)	Population (1983 census)
Guinée-Forestière	22,127	1,246,747
Guinée-Maritime	16,976	1,147,301
Haute-Guinée	35,719	1,086,679
Moyenne-Guinée	19,960	1,595,007
Conakry (city)	119	705,280
GUINEA	94,901	5,781,014

Capital: Conakry, pop. 705,280 (1983 census)
Major cities (1983 census): Kankan, pop. 88,760; Labé, pop. 65,439; Kindia, pop. 55,904; Nzérékoré, pop. 23,000.

Conversion factor: 1 mi² = 2.59 km²

son and reduced rainfall—66 in. [1695 mm] at Kankan, in the eastern part of the country on the Niger floodplain, and 67 in. [1710 mm] at Labé in the Fouta Djallon at an elevation of 3451 ft [1052 m]. Temperatures are milder at higher elevations, but increase considerably in the interior (82°F [28°C] in July at Kankan), with the dry season predominating, echoing the Sudanese type of climate found further inland.

Flora and fauna. In the southern regions on the border with Liberia, Guinea's plant life consists of the typical evergreen rain forest that is also found along a narrow coastal strip accompanied by mangrove swamps. Toward the interior, and farther north, the forest thins out and becomes a densely wooded savanna; this, together with tall grasses, also covers the northernmost part of the country (Fouta Djallon). In the upper Niger basin, this community is succeeded by a more open savanna. The transition from forest to savanna also occurs with increasing elevation, as is evident in the highland areas (Fouta Djallon, Loma, Nimba, etc.).

Population. The population consists of a number of ethnic groups: the most numerous are the Fulbe (40%), Muslim nomads who have now settled permanently and live in the mountainous Fouta Djallon area; the Malinke (25%), also Muslim, a sedentary farming group living in the easternmost reaches of the upper Niger basin; and the Muslim Sousou (11%) who inhabit the southern regions along the border with Sierra Leone. All of these populations are of Sudanese stock from northern regions, and have supplanted the region's original forest-dwelling Bantu peoples. Minor groups include the Kissi, Guerze, Kpelle, Toma, Kouranko, and Baga; all are animists (5%) and have contributed to the country's process of revival.

At the time of independence in 1958, the country contained 3 million inhabitants, but with a 2.5% growth rate the population rose to 6,876,000 in 1990. Population density is 70 per mi² [28 per km²] on average, but at least twice as high on the coast and along the Conakry–Kankan railroad line that leads into the center of the country. The population is 75% rural, scattered among 4500 traditional villages with their circular straw huts; this pattern is the result of migration from the interior, and the mining and plantation activities of the French colonial period. The cities are all in the colonial style: the largest, Conakry, was established by the French in 1884 and became the capital in 1958. The old section is located on Tumbo island, and contains

not only buildings in the European style, but also a few native dwellings made of dried mud. The new section stretches along the Kaloum peninsula, with a grid plan and large modern buildings. Around the periphery is a third, very poor section of wretched shantytowns. Mario Appelius visited Conakry in the early 1930s, and described it as follows:

> *Conakry is a great garden floating on the sea. Not a single cement edifice disturbs the sinuous windings of the shoreline.... The rare houses that face the sea are concealed by baobabs and mangroves. One sees a window here, a fragment of terrace, farther off the corner of a roof, a few little houses that look more like toys than real dwellings. The eye is captured by a green lushness that rises and falls.... You wonder if you are faced with an artificial park or an almost virgin forest; but the last thing you would think of is a city. This, nonetheless, is the capital of Guinea.*

Other important cities are Labé, Nzérékoré to the south near the Liberian border, and Kindia near the capital.

For many years, the economic and political system was based on a planned economy and a single party, which absorbed rural underemployment and resulted in a large work force (46% of the population). The low per capita income (US$480) indicates poverty, which is also confirmed by a deficient diet (2042 available calories per day) and high infant mortality (14%), and a life expectancy of only 43 years.

Economic summary. Guinea, one of the richest colonial countries, provided Europe with raw materials in accordance with a very precise speculative timetable. First of all the French exploited rubber cultivation to the fullest, then developed banana cultivation (making the region the world's fourth-largest producer) and extended coffee production. After World War II they moved on to extraction of mineral products, built the railroads linking the interior with the coast, and developed the harbor at Conakry. It was in fact the mine workers who established a strong union movement which called into question the entire colonial system; the result was independence and the establishment of a continually expanding production system.

The basis of the economy is agriculture, which is being more and more highly diversified to improve self-sufficiency in food. The principal objective is to develop industry in order to eliminate dependence on imports. The production process is organized into production cooperatives and agricultural work brigades, while the mining companies are supported by both government and foreign capital.

Socioeconomic data

Income (per capita, US$)	480 (1990)
Population growth rate (% per year)	2.5 (1980–89)
Birth rate (annual, per 1,000 pop.)	48 (1989)
Mortality rate (annual, per 1,000 pop.)	21 (1989)
Life expectancy at birth (years)	43 (1989)
Urban population (% of total)	25 (1989)
Economically active population (% of total)	46 (1988)
Illiteracy (% of total pop.)	71.7 (1991)
Available nutrition (daily calories per capita)	2042 (1989)
Energy consumption (10^6 tons coal equivalent)	0.5 (1987)

With only 2.9% of its land used for cultivation, Guinea's agriculture cannot satisfy the nutritional needs of its population. Both overall production and carbohydrate levels are insufficient, but the output of subsistence foods (rice, corn, sorghum) is continually increasing, and traditional crops such as manioc and sweet potatoes are still grown. Export crops, such as bananas and coffee, are now undergoing a crisis, while cultivation of pineapples, tropical oils (palm and peanut), sisal, citrus, tobacco, and cocoa is being developed. Woodlands and forests cover 59.5% of the nation, and produce moderate quantities of construction timber, hardwoods, and rubber.

Animal husbandry constitutes an important resource, especially in the mountainous northern regions, while fishing is practiced by companies financed by both government and Japanese capital, and concentrates largely on sardine production.

Mining activities represent the most profitable and dynamic sector of the economy. Guinea's deposits of bauxite—with an extremely high aluminum content, which is extracted in quantities equivalent to 16.1% of world production—constitute 30% of the entire world reserves of this ore (second only to Australia). The most important mineral basins are those at Fria, Boké, and Kindia; the Gaoual region is located farther north in the Fouta Djallon Mountains. Iron is the other great mineral resource, found on the Kaloum peninsula near Conakry, and in the much larger and more recently discovered deposits in the Nimba Mountains. Ore is transported directly to the ocean by a railroad, entirely on Liberian territory, that connects these mines to the port of Buchanan. Other resources include diamonds (extracted at Macenta near Kankan) and uranium.

Although Guinea has Africa's largest aluminum extraction plant (at Fria), and its industry produces 35% of GNP, this sector is still not highly developed. Only 13% of the work force is employed in industry, which includes factories processing tobacco at Conakry, tea at Macenta, peanuts at Dabola, and sugar at Madimoula.

The balance of trade, which was comfortably positive in the past (with exports producing more than twice the value of imports), is now approaching parity because of factors that include an increase in imports of capital goods. Exports consist primarily of bauxite, alumina, and diamonds, along with agricultural products such as bananas, coffee, peanuts, pineapples, coconuts, palm oil, and rice, 21% destined for EC countries and 9% for the U.S. Imports consist of textile products and clothing (14.3%), transportation equipment (12.1%), petroleum products (7.2%) and foodstuffs (5.8%). The country is accessible by sea through the two ports of Conakry and Kamsar, and by river on the Niger, which is navigable within Guinea. Air traffic passes through the international airport at Conakry/Gbessa, although the airfields at Kankan, Labé, and Nzérékoré also handle a small amount. The highway system consists of 18,047 mi [29,108 km] of roads usable only in the dry season, 8300 mi [13,400 km] of which are paved (1986); the railroad network in 1983 consisted of three main lines: Conakry–Kankan (410 mi [662 km]), Conakry–Fria (97 mi [156 km]), and Kamsar–Sangaredyi (83 mi [134 km]).

Historical profile. According to legend, the Guinean coast was visited by the voyage of pharaoh Necho II in the 7th century B.C., then two centuries later by Carthaginian explorers. The first European awareness came much later, in the 14th and 15th centuries, when the territory was divided into little local chiefdoms dependent to various degrees on the great Sudanese kingdoms. Resistance by the natives was not enough to prevent the region from being almost depopulated, first by the Portuguese and then by the French, who deported slaves to Brazil and the Caribbean. After the era of slavery ended in the 19th century, the age of colonialism began; once a few treaties had been signed with tribal leaders, the French were the first to attempt to occupy the area, although they encountered more obstacles than they had expected. From 1822 to 1898, colonial troops faced one of the most obstinate resistance movements in the entire history of sub-Saharan Africa. The fight was led by Samoury Touré, today regarded not only as a national hero, but as a champion of the struggle against oppression even beyond Guinea's borders. At the turn of the century the city of Conakry was founded, capital of the colony that very shortly became part of French West Africa.

After decades of apparent tranquillity, the first independence movements began to stir in the post-World War II period. They were headed by the Democratic Party of Guinea (PDG), which became radicalized under the leadership of Sékou Touré. As a result, in the referendum promoted by De Gaulle in 1958 in order to establish the French Community of Nations, Guinea was the only country in francophone Africa to vote for full independence. A republic was duly proclaimed, and Sékou Touré served as its president until 1984, the year of his death. Deprived of technical and financial aid from France and the United States, the country turned to the Soviet bloc, while domestically it selected a socialist model for development, attempting with dubious success to move directly from a tribal to a socialist society. After Touré's death, a bloody coup brought to power a Military Committee for National Redressment (CMRN), which dissolved the single political party, released political prisoners, and restored labor rights and freedom of the press.

Culture. In the coastal region, where animist cults have survived Islamization, artistic production is mediated even today by the secret society called Simo, which determines both form and content. The themes passed down by tradition include the fertility goddess Nimba and the serpent Banjonyi, depicted in masks and wooden statues.

As far as literature is concerned, the Fulbe have a unique written tradition. Expansive poems called "kasydy" (from the Arabic *qasida*), devoted to religious or historical themes, are very popular. A shorter form is represented by the "jerfi," poetic compositions found throughout Guinea which deal with love, war, disputes, human relations, animals, and good and evil spirits. There are also jerfi which have no subject, in which the words are connected solely on the basis of phonetic free association. Of course the oral literary tradition common to African countries is present everywhere in Guinea, and here it is extraordinarily rich, consisting of cycles of legends passed on by the "griots" from father to son. In the last few decades, this rich tradition has served as the foundation for a new literature, written especially in French and including the works of famous authors such as the poet and playwright Keyta Fodeba, considered the father of contemporary African theater. Novelists include not only Nene Kchali, Camara Laye, and William Sassine, but especially Djibril Tamsir Niane, author of *Soundiata*, a work that addresses the deeds of the legendary king of Mali as told by the "griots."

GUINEA-BISSAU

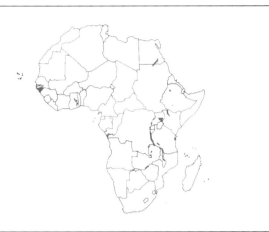

Geopolitical summary

Official name	República da Guiné-Bissau
Area	13,944 mi² [36,125 km²]
Population	767,739 (1979 census); 966,000 (1990 estimate)
Form of government	Presidential republic with a National Popular Assembly consisting of a single party
Administrative structure	3 provinces and 9 regions, including the capital
Capital	Bissau (pop. 125,000, 1988 estimate)
International relations	Member of UN, OAU; associate member of EC
Official language	Portuguese; Creole and Sudanese languages widely spoken
Religion	Animist 65%; Muslim 30%; remainder Catholic
Currency	Guinea peso

Natural environment. Guinea-Bissau is bordered on the north by Senegal, on the south and east by Guinea, and faces the Atlantic Ocean to the west.

The territory consists of a wide stretch of low alluvial plain between the Fouta Djallon highlands (in Guinea) and the ocean, traversed by numerous rivers (Cacheu, Mansôa, Geba, Corubal) whose deep estuaries make for an extremely jagged coastline. Subsidence in geologically recent times has also resulted in submerging of the shoreline, forming the Bijagós islands just offshore. The land rises slightly toward the interior of the country, with slight undulations in the terrain (Bafatá and Gabú hills) modeled onto extremely ancient (Paleozoic) sediments that are partly covered by Tertiary formations.

The climate is predominantly subequatorial, with considerable rainfall (55–100 in. [1400–2600 mm] per year) and extremely lush vegetation; it is characterized by one long, very wet season lasting from June to October, with precipitation reaching maximum levels in July and August and frequent flooding. In winter the *harmattan* wind from the Sahara, blowing from the northeast, increases both aridity and temperature (with high temperatures in April and May exceeding 86°F [30°C]).

The vegetation consists of a dense forest cover that is especially thick along the coast, where mangroves are prevalent, and along riverbanks. The interior is dominated by wooded savanna communities, with umbrella acacias and large baobabs.

Population. Small and thinly settled, the country underwent profound changes in population during the struggle for independence. The countryside became unsafe because of the war and was depopulated, increasing the flow of people to the cities and creating desperately poor shantytowns. The present government, embodied in the National Popular Assembly and inspired by a cautious, pragmatic socialism and a policy of nonalignment in foreign policy, is now providing incentives to return to the land. It was in fact the long war of independence that led to the organization of parties supported by the various ethnic groups. The Balante (25% of the population)—an economically backward and animist society descended from Paleo-African groups, living in the central part of the country—supported the socialist PAIGC (African Party for the Independence of Guinea and the Cape Verde Islands), which led to independence. The Mandingo, on the other hand—Sudanese, Muslim, more advanced, and subdivided into various groups of which the largest is the Malinke (12%)—supported FLING (Front for the Struggle for National Independence of Guinea), which received assistance from Senegal. The Peul (22%), well-off Muslims living in the eastern region, supported the Portuguese, thus drawing the hostility of the promoters of independence. At present power is firmly in the hands of PAIGC, which is emphasizing free trade in an effort to attract foreign investments that might perhaps improve the miserable social conditions into which the country is sliding: per capita GNP is low (US$155), the illiteracy rate is 63.5%, and the diet provides little more than 2500 calories per day.

Economic summary. The economic situation reflects backwardness, due to the tremendous importance of agriculture in generating GNP (60%), and especially the high rate of employment in this sector (70% of the work force). In addition, the principal crops are still plantation products grown for export, requiring massive imports of foodstuffs (41% of total imports). When imports of capital equipment (30%), energy products, and other consumer goods are added, the result is an appalling deficit in the balance of trade, with an external debt equal to six times GNP, which in 1990 was barely US$150 million.

Agriculture favors cash crops for export, such as coconuts, copra, and peanuts, although rice is now becoming a more important presence on the small amount of land dedicated to cultivation and orchard crops (9.3% of the country). Elements of the

Climate data

Location	Altitude (ft asl)	Average temp. (°F) January	Average temp. (°F) July	Average annual precip. (in.)
Bissau	66	76	80	76.1

Conversion factors: 1 ft = 0.3 m; 1 in. = 25 mm; °C = (°F – 32) × 5/9

Administrative structure

Administrative unit (regions)	Area (mi²)	Population (1979 census)
Bafatá	2,309	116,032
Biombo	324	56,463
Bolama-Bijagós	1,013	25,473
Cacheu	1,998	130,227
Gabú	3,532	104,315
Oio	2,086	135,114
Quinara	1,211	35,532
Tombali	1,442	55,099
Bissau (city)	30	109,214
GUINEA-BISSAU	13,944	767,739

Capital: Bissau, pop. 125,000 (1988 estimate)
Major cities (1979 census): Bafatá, pop. 13,429; Gabú, pop. 7,803; Mansoa, pop. 5,390; Cantchungo, pop. 4,965.

Conversion factor: 1 mi² = 2.59 km²

current agricultural development policy include promotion of subsistence crops, a regionally based administrative structure, nationalization of land, and organization of labor on a cooperative basis. Products from the forest (which covers 29.6% of the land area) include a small quantity of timber, some of it exported; in the absence of suitable financial resources that could be used for industrial investments such as boats, freezers, etc., rubber production and fishing in the rich territorial waters (which is being heavily emphasized) are being partly exploited by joint-venture foreign companies.

Development programs aimed at reducing dependence on foreign sources are not being helped by a lack of energy and mineral resources (although recent exploration suggests the presence of bauxite and petroleum), and a low level of hydroelectric production. The industrial sector, which consists only of small food-processing establishments, employs only 11% of the work force and produces a meager 9% of GNP.

The ratio between imports and exports is five to one. Goods arrive primarily from Portugal (20%), and from Germany, Sweden, and the United States (10% each), while most exports are sent to Angola (35%), Spain (18%), Portugal (18%), and Great Britain (10%). The communications system consists of 3136 mi [5058 km] of roads (775 mi [1250 km] paved in 1982),

Socioeconomic data

Income (per capita, US$)	155 (1990)
Population growth rate (% per year)	2.7 (1985–90)
Birth rate (annual, per 1,000 pop.)	42.9 (1989)
Mortality rate (annual, per 1,000 pop.)	23 (1989)
Life expectancy at birth (years)	43.1 (1989)
Urban population (% of total)	29.3 (1989)
Economically active population (% of total)	47.1 (1988)
Illiteracy (% of total pop.)	63.5 (1991)
Available nutrition (daily calories per capita)	2543 (1989)
Energy consumption (10^6 tons coal equivalent)	0.06 (1987)

the harbor at Bissau, which is not equipped for large oceangoing vessels, and the city's international airport.

Historical profile. From the 12th to the 14th century, the territory that is now Guinea-Bissau was part of the empire of Mali, and declined along with it. The Portuguese appeared in the 15th century, limiting themselves at first to trade with the natives and the establishment of forts along the coast. In 1610, a private company in Lisbon acquired the coastal strip of Bolama island, and then expanded inland to increase the slave trade: Portugal obtained a monopoly on this extremely lucrative business, although for two centuries it was threatened by the activities of French and English companies, and only achieved uncontested supremacy again in the 1800s. Until 1879 the territory was administered by the government of the Cape Verde islands; it then became for all intents and purposes a Portuguese colony, and its borders with the surrounding territory of French West Africa were ratified in 1886. The Portuguese occupation campaign encountered enormous difficulties and was not really completed until 1915, with the exception of a few areas where armed resistance was never subdued.

The Portuguese were unable to consolidate their rule by simply applying the usual method of forced labor: strong-arm tactics only increased popular discontent, and the nationalist movement finally came out into the open in 1956, when Amilcar Cabral established the African Party for the Independence of Guinea and Cape Verde (PAIGC). This party was widely supported by workers, especially stevedores. Strikes were organized, including one in 1959 at the port of Pidgiguiti at Bissau, which resulted in a massacre. As a result, PAIGC changed its tactics, emphasizing clandestine action and guerrilla warfare in the countryside, as outlined by the plan of action adopted at that time:

1. Without delay mobilize and organize the peasant masses who will be, as experience shows, the main force in the struggle for national liberation.
2. Strengthen our organization in the towns but keep it clandestine, avoiding all demonstrations.
3. Develop and reinforce unity around the Party of the Africans of all ethnic groups, origins and social strata.
4. Prepare as many cadres as possible, either inside the country or abroad, for political leadership and the successful development of our struggle.
5. Mobilize emigrés in neighboring territories so as to draw them into the liberation struggle and the future of our people.

In 1961 the transition was made to an armed struggle, which progressed almost unopposed: within five years PAIGC already controlled two thirds of the country, and in 1972 formal recognition was obtained from the United Nations. Although Amilcar Cabral was assassinated in 1973 by African killers working for the Portuguese secret police, PAIGC went ahead and proclaimed the independence of Guinea-Bissau; its status was recognized a year later by Portugal, where in the meantime the last vestiges of the Salazar regime had been swept away. The country's first president was Luis Cabral, brother of the murdered leader; he was overthrown in 1980 by João Bernardo Vieira, in a coup that caused the breakup of the party and terminated the process of unification with Cape Verde.

LIBERIA

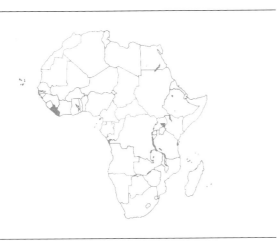

Geopolitical summary

Official name	Republic of Liberia
Area	42,988 mi^2 [111,369 km^2]
Population	2,101,628 (1984 census); 2,436,000 (1988 estimate)
Form of government	Presidential republic with a National Assembly consisting of a Senate and House of Representatives
Administrative structure	11 counties and 2 territories
Capital	Monrovia (pop. 425,000, 1984 census)
International relations	Member of UN, OAU; associate member of EC
Official language	English; Sudanese languages also used
Religion	Christians (65%, including 78,000 Catholics); animists (20%); Muslims (15%)
Currency	Liberian dollar

Natural environment. Liberia is bordered on the east by Ivory Coast, on the north by Guinea, on the northwest by Sierra Leone, and faces the Atlantic Ocean to the south.

The territory consists predominantly of an extensive plateau, at an elevation of between 650–1600 ft [200–500 m], at the base of the mountain range that runs northwest–southeast from the Loma Mountains (Sierra Leone) to the Nimba Mountains in southern Guinea. Shaped on a foundation consisting essentially of Precambrian crystalline rock, and dissected by numerous watercourses which descend parallel to one another to the ocean, the plateau slopes gently down to a narrow coastal plain. Toward the interior the terrain becomes more varied, often exceeding elevations of 2000–2300 ft [600–700 m] (Bong Mountains, 2116 ft [645 m], northeast of Monrovia), and even reaching beyond 3300 ft [1000 m] (Mt. Wativi, the highest peak in Liberia, 4530 ft [1381 m], in the northernmost part of the coun-

try) to culminate at 5747 ft [1752 m] in the Nimba Mountains on the border with Guinea and Ivory Coast. The coastal plain, some 30 mi [50 km] wide, is dotted with large, mangrove-choked salty lagoons. The coastline is 355 mi [570 km] long, straight, low, and sandy except for the high, rocky area of Capes Mount and Mesurado; despite the numerous estuaries, there are no natural ports because of shallow water. The hydrographic system is fairly dense, with numerous watercourses that are short but with a high flow rate and a pluvial flow pattern. The longest rivers, such as the Lofa, St. Paul, St. John, and Cavally, rise in the mountains of Guinea, while others descend directly to the sea from the lower hills of Liberia and from the Loma Mountains; these include the Morro, Mano, Cestos, Sanguin, and Sino. They are not navigable because of numerous rapids and falls, and the presence of sandbars at their mouths.

The climate of Liberia is predominantly subequatorial: hot and humid, as is typical of coastal areas along the Gulf of Guinea. Average annual temperatures are very high, with minimal temperature variations. Rainfall is extremely abundant, exceeding 120 in. [3000 mm] per year (the actual figure for Monrovia is 156 in. [4000 mm], with a relative humidity of 85% and 169 days of rain each year), falling from May to October in a long rainy season dominated by the southwest trade wind. The winter, from November to April, is a drier period characterized by the prevailing desert *harmattan*, which blows from the Sahara toward the Gulf of Guinea. The uniformly high temperature and plentiful rainfall have led to the development of the dense evergreen forest that covers a third of the nation, rich in many valuable hardwoods such as mahogany, ebony, and rosewood, accompanied by a thick understory. Toward the north the forest becomes sparser, marking the transition to a wooded savanna with its typical baobab, shea, acacia, and oil palm trees. Mangroves are common near the coast, especially in swampy areas.

Population. The most ancient settlers were peoples arriving from the Niger basin; the Europeans, who were familiar with the coastal area from their trade in spices and slaves, never established settlements here because of the difficult environmental conditions. At the beginning of the 19th century, the United States founded a colony of freed slaves in this region (as had been done in Sierra Leone by the English Quakers), which was called Liberia and was the first free republic on the continent. In reality it was a "black colony," in which almost the entire population suffered under the tyranny of the tiny minority that governed them in close consultation with the United States.

Although they represent only 4% of the population, the Americo-Liberians (or "freedmen"), who live mostly in the capital or on the coast, had always retained economic and political power until the military coup in 1980. The native population, belonging to numerous tribes and still largely nonurban, live pre-

Climate data

Location	Altitude (ft asl)	Average temp. (°F) January	Average temp. (°F) July	Average annual precip. (in.)
Monrovia	75	79	76	156

Conversion factors: 1 ft = 0.3 m; 1 in. = 25 mm; °C = (°F − 32) × 5/9

Administrative structure

Administrative unit	Area (mi²)	Population (1984 census)
Counties		
Bong	3,126	255,813
Grand Bassa	3,381	159,648
Grand Cape Mount	2,249	79,322
Grand Gedeh	6,573	102,810
Grand Kru	–	–
Lofa	7,473	247,641
Margibi	1,260	97,992
Maryland	2,065.	132,058
Montserrado	1,058	544,878
Nimba	4,649	313,050
Sinoe	3,958	64,147
Territories		
Bomi	755	66,420
River Cess	1,693	37,849
LIBERIA	38,239(*)	2,101,628

(*) Excluding 4,749 mi² of internal waters.

Capital: Monrovia, pop. 425,000 (1984 census)
Major cities: Yekepa, pop. 14,189 (1974 census); Tubmanburg, pop. 14,089 (1974 census); Gbarnga, pop. 10,860 (1980 estimate)

Conversion factor: 1 mi² = 2.59 km²

dominantly in the interior. They can be divided into the Kwa-speaking peoples living in the southern part of the country, including coastal tribes such as the Bassa, Grebo, Gbe, and Kru; and those that speak Mande, which include the Vai on the coast, the Kpelle in the Nimba Mountains area, and the Dan and Mano in the interior near the border with Ivory Coast.

At present, 45% of the population is urbanized and lives in Monrovia or in towns that have grown up alongside plantations and mines. The area immediately around the capital has the highest population density (almost 520 per mi² [200 per km²]), while the average density of 57 per mi² [22 per km²] drops to less than 15–18 per mi² [6–7 per km²] in the interior. The capital is the only real city, named in honor of the American president James Monroe who supported the foundation of this village of former slaves. Located in southern Liberia at the mouth of the St. Paul river, near the promontory of Cape Mesurado, it is the rainiest reporting station in western Africa. A cultural and administrative center, Monrovia is the site of a university and various institutions of higher learning, along with libraries and museums. Its port, among the best-equipped on this coast, handles all foreign trade (assisted by its status as a free-trade zone). The city is connected by two rail lines to the mining centers of Bong Town in the Bong Mountains and Bomi Hills on the border with Sierra Leone, and is served by two nearby airfields: one at Sinkor, and the Roberts International Airport near the headquarters of Firestone, an American multinational rubber company. Other population centers are Buchanan, Yekepa, Tubmanburg, Harper, Greenville, and Robertsport. Despite the existence of a modern education and health-care system, 77.6% of the population is illiterate, life expectancy is 54 years, and half the national income is concentrated in the hands of 5% of the population.

Economic summary. The character of the present-day economy was defined during the thirty-year term of President William Tubman (1943–1971), who established a free-trade policy and turned over many mining and agricultural concessions to large American corporations, reducing the country to a state of genuine economic subjugation. A precedent was set in 1926, when almost a million acres [400,000 ha] of coastal land was granted to the Firestone Plantations Company for rubber plantations (85% of the population utilizes only 3% of the territory). A change in course occurred in 1980, with a coup d'état that overthrew the Americo-Liberian-dominated government and brought to power a group of military men representing the interests of the native population. In the fiscal arena, they increased ship registration fees; in agriculture, new land was brought under cultivation and agricultural cooperatives were promoted; and in industry, local initiatives were encouraged.

Agriculture and animal husbandry, practiced on 54.5% of the country's land area, employ 55% of the working population. The most important food crop is rice, grown by the dry method in rotation with other cereals; the technology is extremely simple, as described in the following excerpt from an anthropological study of the Kpelle tribe in central Liberia:

> *The farming cycle begins in February or March, toward the end of the dry season. A plot of land that has lain fallow for at least seven years is cleared of brush by the men, usually working in one of several types of cooperative work groups called kuu. Trees such as the citrus and palm, which are economically useful, and certain cottonwood trees, which are thought to have supernatural powers, as well as stumps of other trees are left standing. After the cut foliage dries, the entire area is burned, then cleared and burned again, and left to stand until the rains arrive to soak the ashes into the soil and loosen the soil for planting.*
>
> *At this point the women take over and plant the seed rice, scratching the surface of the soil with short-handled hoes and planting the seeds in the furrows. From the time the rice is planted until the harvest, some four to five months later, the family spends a good deal of time on the farm.... A fence of sticks held together by vines is built around the farm to help keep out small foraging animals, and traps are made to catch those animals that bypass the fence. Children, who participate in earlier portions of the work according to their strength, have primary responsibility for protecting the ripening crop from the ubiquitous rice birds. Sticks and slingshots are the main weapons in the child's arsenal. In November and December the whole family harvests, dries, and stores the rice.*

Production of manioc, sweet potatoes, and tropical fruits is also important. Crops grown for export include primarily cocoa, coffee, coconuts, peanuts, citrus, and sugar cane. The most important product is rubber, of which Liberia is Africa's leading producer, and eighth largest in the world. Fishing generates a limited quantity of salable product, while animal husbandry is important only within the domestic economy. Forests and woodlands (15.8% of the territory) produce a large volume of hardwood for export (5% of total exports).

Liberia is rich in mineral resources, including iron, gold, diamonds, bauxite, and manganese. The ferrous minerals hematite and magnetite are extracted in the Nimba Mountains, in the

Socioeconomic data

Income (per capita, US$)	250 (1990)
Population growth rate (% per year)	3.2 (1980–89)
Birth rate (annual, per 1,000 pop.)	44 (1989)
Mortality rate (annual, per 1,000 pop.)	14 (1989)
Life expectancy at birth (years)	54 (1989)
Urban population (% of total)	45 (1989)
Economically active population (% of total)	37.6 (1988)
Illiteracy (% of total pop.)	77.6 (1991)
Available nutrition (daily calories per capita)	2270 (1989)
Energy consumption (10^6 tons coal equivalent)	0.34 (1987)

Bong Mountains in the central part of the country, and, until the deposit was exhausted in 1977, at Bomi Hills on the Lofa river. The record production figure of 17.5 million t in 1971 has now been cut almost in half, however, due to declining international demand; Liberia nevertheless still extracts a substantial part of world production. Gold and diamonds are the country's traditional resources, while manganese and bauxite (extracted at Kolahun in the north) are the most recently exploited. Industrial activity is of considerable significance, accounting for 30% of GNP and employing 25% of the work force. The major sectors are iron and steel (iron ore is processed at Africa's largest such plant, at Buchanan), rubber processing at the Harbel complex, and a petroleum refinery and other chemical plants at Monrovia. Light manufacturing is of minor importance, operating on a craft basis and dealing with traditional local products such as sugar, beer, cement, and lumber.

The balance of trade is positive, thanks to exports of iron ore (63% of the total), rubber (19%), wood products (5%), diamonds (4.6%), and coffee, and to controls on imports, which are used both for subsistence (food 22%, manufactured goods 13%) and for development (machinery 25%, fuels 27%, chemical products 7%). A customs union called the Mano River Union has recently been established by Liberia, Sierra Leone, and Guinea. Transportation is characterized by the nominal presence of the world's largest merchant fleet sailing under the Liberian "flag of convenience": ships of every nation sail under Liberian registration thanks to low fees and the complete freedom accorded to foreign vessels. The principal ports—Monrovia, Buchanan, Greenville, and Harper—handle mostly mining and commercial traffic. The rail system in 1985 included 304 mi [490 km] of mine railroads, while 6389 mi [10,305 km] of highways were available for road traffic (1984). Air transport uses the Roberts International Airport (near the Firestone plantation), and the domestic field at James Springs Payne.

Historical profile. *An ideal of liberty.* In the 11th century, the territory of present-day Liberia became part of the great empire of Ghana. This stretch of coastline facing the Gulf of Guinea was called the "Pepper Coast" by the Portuguese navigators who arrived here in the 15th century; they were followed by the Dutch, the French, and the British, who began to pillage the African tribes for commercial gain. At the beginning of the 19th century, the Pepper Coast embarked on a course different from that of neighboring countries, one that spared it the European colonial aggression suffered by the rest of the continent. In 1816,

a group of American blacks, freed by a private abolitionist society, landed here. They obtained from several tribal chiefs the coastal region where they then established Monrovia, named in honor of the American president James Monroe; under the auspices of the American Colonization Society, and despite the hostility of the local population, 6,000 freed slaves came to this region in the next forty years. In 1847 they created the independent nation of Liberia, complete with a constitution based on the American model; ten years later this state merged with a similar settlement founded in 1833 at Cape Palmas, and the country took on its present configuration as its borders were defined by treaties with the British and the French. Sheltered from European colonial expansionism, the new nation remained in some respects on the sidelines of the historical processes that would give rise to modern Africa.

Liberty denied. American protection also had perceptible effects on both politics and the economy, since the exercise of power was reserved for the black elite of American origin. The indigenous population, excluded from public life, found itself discriminated against by measures such as the adoption of English as the official language and the imposition of Protestant Christianity. The situation remained the same into this century, with increasing economic dependence and social tensions. Between the 1920s and 1940s, the United States appropriated for itself the right to exploit certain major Liberian resources, such as rubber and iron ore, through companies such as Firestone and the Liberia Mining Company. Internally, the corruption of the ruling class, the privileges of the big landowners and urban middle class, unspoken popular discontent, and harsh repression of strikes and demonstrations all increased. And the process of national recovery was certainly not helped along by the fact that after World War II Liberia became one of the most sought-after "tax havens" for foreign merchant ships, which could avoid taxation in their respective countries simply by flying the Liberian flag.

This context of gradual national deterioration may explain the initial success enjoyed by the regime of Sergeant Samuel K. Doe and his Council of Popular Redemption, who came to power in 1980 in a bloody coup that kicked out the corrupt political class. But after his first "revolutionary," pro-Soviet statements, Doe failed to live up to expectations: he replaced the power of the American elite with a dictatorship run by himself and other members of his Krahn tribe. In the economic and financial arena, he did nothing but tighten the bonds tying his country to Washington. In a few years the situation came to a head and Liberia became a powder keg, which finally exploded with an insurrection led by the National Patriotic Front and the violent death of Doe in 1990.

Culture. Inhabited by ethnic groups that have very tenuous cultural links, Liberia does not exhibit any homogeneous artistic forms. Although the production of bronze objects, typical of the Dan-Ngere groups, may be considered an established practice, more recent forms include wall painting, the masks made by the various secret societies, and statuettes associated with the ancestor cult. Literature has its own written tradition in the Vai language, in which a collection of *Historic legends and folklore of the Grebo tribe* appeared in 1957. In recent decades several English-speaking literary groups have come into being in Monrovia, and the poet Roland Dempster has established a Liberian writers' association.

NIGERIA

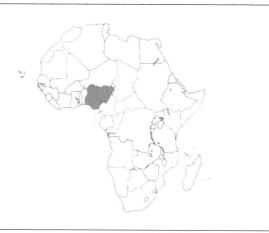

Geopolitical summary

Official name	Federal Republic of Nigeria
Area	356,574 mi² [923,768 km²]
Population	88,514,501 (1991 census)
Form of government	Military government controlled by an Armed Forces Ruling Council whose president is also the head of government
Administrative structure	30 federal states administered by a military governor, plus the federal capital territory
Capital	Abuja (pop. 378,671, 1991 census)
International relations	Member of UN, OAU, Commonwealth; associate member of EC
Official language	English; Sudanese languages also used
Religion	Muslim 45%; Christian 38% (including over 3 million Catholics), and animist
Currency	Naira

Natural environment. Nigeria is bordered on the east by Cameroon and Chad, on the north by Niger, on the west by Benin, and faces the Atlantic to the south.

Geological structure and relief. Nigeria's territory consists of three major sections, defined by the lower course of the Niger and by its principal left-bank tributary, the Benue. North of these watercourses is the north central region of the country, consisting of a large plateau cut into Paleozoic crystalline rocks, dominated by Mt. Sara (5543 ft [1690 m]), the highest point of the Bauchi massif (Jos Plateau) and the site of intense volcanic activity in the Mesozoic era. These highlands are bounded by steep cliffs and cut into by deep gorges on the southern and eastern slopes, but decline gently toward the north and west; rivers branch out radially in all directions throughout the plateau.

South of the Niger and the Benue, the Nigerian plateau falls away even more and is marked by numerous rivers flowing into the Atlantic; ultimately it gives way to a broad alluvial coastal plain more than 185 mi [300 km] wide, which includes the extensive Niger delta. This feature (12,000 mi² [30,000 km²] in total area), which is continually being extended by the constant addition of sediments, divides the Gulf of Guinea into two smaller parts: the Bight of Benin to the west and the Bight of Biafra to the east. With its many lagoons and marshes, the delta, like the rest of the coastal fringe, is a difficult environment for human beings to utilize. To the east of the Benue, Nigerian territory is defined by the complex spine of the Cameroon Mountains, running southwest-northeast, which rise to elevations of more than 6500 ft [2000 m] (the highest peak is Dimlang at 6698 ft [2042 m]). The northeastern part of the country, drained by the Komadugu basin, slopes away toward the endorheic depression of Lake Chad (787 ft [240 m] above sea level).

Hydrography. The well-developed hydrographic system is dominated by the great basin of the Niger, the last 700–750 mi [1100-1200 km] of whose course lies within Nigeria. Its extremely high flow rate (28,000 ft³ [800 m³] per sec on average) is governed by seasonal variations in precipitation in the section upstream of the artificial Kainji reservoir. Downstream from Jebba, the river runs northwest–southeast in a wide tectonic fissure that makes it navigable; after joining with its great tributary, the Benue (which extends for more than 500 mi [800 km] within Nigerian territory), with its more steady flow, it then runs south. The remainder of Nigerian territory is drained by the endorheic basin of Lake Chad, fed by the various Nigerian tributaries of the Komadugu river which rises on the Bauchi plateau and runs from southwest to northeast.

Climate. Climatic conditions in Nigeria are highly diverse: in summer, warm, wet air masses from the sea (the southwest monsoons that result when the southeast trade winds continue across the equator) are pulled in by a low-pressure area that forms in the Sahara; in winter, a dry wind (the *harmattan*) blows from the Sahara. This pattern produces three general climatic regions. On the coast and in the Niger and Benue valleys, the prevalent climate is tropical with monsoon features, with minimal annual temperature changes and an annual average temperature of about 77–79°F [25–26°C] (average temperatures at Lagos are 81°F [27.2°C] in January and 77°F [25°C] in July). Precipitation is extremely abundant, reaching 148 in. [3800 mm] along the Niger delta (91 in. [2324 mm] at Lagos), with maximum values from June to September (more than 16 in. [400 mm] in July alone) and minimums in January (2–4 in. [50–100 mm]). The influence of the monsoon weakens in the north central regions, and the effects produced by high pressure in the Sahara become

Climate data

Location	Altitude (ft asl)	Average temp. (°F) January	Average temp. (°F) July	Average annual precip. (in.)
Lagos	10	81	77	90.6
Ibadan	751	80	76	43.7
Kano	1,607	70	79	33.9
Port Harcourt	49	79	77	97.3

Conversion factors: 1 ft = 0.3 m; 1 in. = 25 mm; °C = (°F − 32) × 5/9

more perceptible: temperature swings increase, winter temperatures are lower, and there are two identifiable rainy seasons (corresponding to the periods when the sun passes the zenith) in quick succession, followed by a long dry season. This is a typical savanna climate. The rainy seasons become increasingly short (53 days of rain per year at Kano) and the amount of rainfall decreases, to as little as 20 in. [500 mm] in the northernmost region of the Sahel, which is characterized by a generally arid climate. Seasonal changes in temperature become even greater: temperatures can reach 104°F [40°C] in the summer months, and drop to less than 68°F [20°C] in winter.

Flora and fauna. Latitude, climate, and hydrography all have significant effects on Nigeria's vegetation and fauna, which are therefore extremely varied. Mangrove communities are common in the Niger delta and in coastal lagoons, while the southern plains are covered with a rain forest containing many tall trees such as mahogany, obeche, abura, and rubber. The forest reaches farther north along the rivers, creating the "fringing forest" with its characteristic mahogany, kapok, raffia, and wild coffee, and a lush understory. On the inland plateaus, as far as a line between Lake Kainji and the Shebshi Mountains, is a thick, densely wooded wet savanna (parkland). Even farther north this becomes more open, with occasional baobabs, ultimately thinning out into the Sahelian steppe, the typical home of nomadic herders. As in many regions of Africa, the originally rich wild fauna has been greatly reduced by intensive agricultural activity and relentless hunting. Two large reserves have been established for protection of the fauna at Borgu on the Niger and Yankari on the Bauchi plateau.

Population. The British wisely decided not to allow the purchase of land by foreigners, thus leaving the plantations in the hands of the native people. Although this did forestall racial conflicts between whites and blacks, it did not prevent ethnic disputes among the local populations, stimulated in various ways by environmental conditions. The advantages associated with the highly profitable products of the cocoa plantations, managed by the Yoruba populations in the southwest region, and of the oil-palm areas cultivated by the Ibo in the southeast, stood in contrast to the lower incomes earned by the Hausa and Fulbe peoples from the cultivation of cotton and peanuts in the country's northern provinces. These economic and ethnic disparities, which were aggravated by religious differences, resulted in the formation of the first ethnically based parties, struggles for supremacy, and separatist thrusts.

The Yoruba, heirs to a rich civilization and constituting 20% of the population, represent the most dynamic and highly developed Sudanese ethnic group in Nigeria. Their primacy dates back to the Portuguese colonial period (15th century), when they built kingdoms based on the trade in slaves, whom they obtained in raids on various other tribes, especially the Ibo. The latter, who constitute another large semi-Bantu ethnic group (17% of Nigeria's population), are a sedentary farming people, mostly Catholic, with their cultural focus on their traditional capital, Enugu. After the abortive secession of Biafra, which led to civil war (1967-70) and to devastation of the region, the importance of the Ibo within the nation has diminished.

The Muslim Hausa, traditionally the dominant group in both quantitative and cultural terms (21% of the population), have developed a dynamic focus of settlement centered around Kano in northern Nigeria; the Fulbe (9%), nomadic herders, are the largest of the minor groups, together with the Kanuri, Jos, Ibidio, Tiv, Iganu, and Edo. From a cultural point of view, the only unifying language is English; the local languages are fragmented into some 300 dialects. The prevailing religion is Islam, which is practiced in the southern and western regions; Christianity is concentrated in the eastern federal states, formerly the components of Biafra, while animist cults predominate in central Nigeria.

Administrative structure

Administrative unit (states)	Area (mi²)	Population (1991 census)	Administrative unit (states)	Area (mi²)	Population (1991 census)
Abi	–	2,298,978	Niger	25,104	2,482,367
Adamawa	35,277	2,124,049	Ogun	6,470	2,338,570
Akwa Ibom	2,733	2,395,736	Ondo	8,090	3,884,485
Anambra	6,823	2,767,903	Osun	–	2,203,016
Bauchi	24,938	4,294,413	Oyo	14,554	3,488,789
Benue	17,437	2,780,398	Plateau	22,400	3,283,704
Borno	44,930	2,596,589	Rivers	8,434	3,983,857
Cross River	7,780	1,865,604	Sokoto	39,579	4,392,391
Delta	–	2,570,181	Taraba	–	1,480,590
Edo	13,703	2,159,848	Yobe	–	1,411,481
Enugu	6,823	3,161,295	Abuja		
Imo	4,574	2,485,499	(Federal Capital Territory)	2,824	378,671
Jigawa	–	2,829,929			
Kaduna	17,776	3,969,252	**NIGERIA**	356,574	88,514,501
Kano	16,708	5,632,040			
Katsina	9,338	3,878,344			
Kebbi	–	2,062,226			
Kogi	–	2,099,046			
Kwara	25,811	1,566,469			
Lagos	1,291	5,685,781			

Capital: Abuja, pop. 378,671 (1991 census)

Major cities (1988 estimate): Lagos, pop. 1,243,000; Ibadan, pop. 1,172,000; Oyo, pop. 1,000,000; Ogbomosho, pop. 597,500; Kano, pop. 551,800; Oshogbo, pop. 390,400; Ilorin, pop. 389,500; Abeokuta, pop. 349,800; Port Harcourt, pop. 335,600.

Conversion factor: 1 mi² = 2.59 km²

Population structure and dynamics. Nigeria is the most populous nation in Africa (17% of the continent's total population), with an extremely high growth rate (3.4%) that has allowed the population to double from 1965 to the present (45% is now under the age of 15). This population is almost exclusively autochthonous; Europeans constitute a tiny minority.

One peculiarity of Nigeria is the existence, in the southwest, of the only example of an urban settlement pattern in equatorial Africa prior to European colonization; the population rate here is higher than the national average (35%), and among the highest in Africa, with many cities of more than 200,000 inhabitants. Lagos, founded by the Portuguese on an island at the mouth of a lagoon outlet, now extends over adjacent islands and onto the mainland, is a huge conurbation with a population of over 1.6 million. It developed on the basis of commercial, administrative, and port-related activities, which produced a chaotic metropolis seething with activity, as reported by Mario Appelius back in 1933:

> *All the races of Nigeria are represented in Lagos, and because the mercantile district is the heart of the city, I am in the midst of a living, talking, movie.... Lagos does not have a single market, but a whole series of specialized markets, ranging from ordinary fish and carrot markets to more theatrical ones for textiles, perfumes ... amulets, terracottas, jewelry, and carpets. Every one of them is crowded; the crowds are black, of course, but very diverse because this black crowd is enlivened by the variety of its clothing, the infinity of its nakedness, the eternity of its conversations, the deafening clamor of voices, the insignificance of the articles that are for sale ... the ubiquitous tam-tams, and ultimately by the intrusion into human affairs of innumerable animals: rams, sheep, camels, chickens, geese, monkeys, and parrots.*

In an attempt to solve at least some of the serious problems afflicting the city, a program is underway to lift certain administrative burdens by building a new capital, Abuja, located on federal territory at the center of the country. Ibadan is the second most important Nigerian city, located 75 mi [120 km] from the coast. Other major centers include Ilesha, Ilorin, Ogbomosho, Oshogbo, and Abeokuta.

Also densely populated is the southeast (250–500 per mi^2 [100–200 per km^2]). Inhabited by the Ibo, in a scattered settlement pattern characterized mostly by small villages of straw and mud huts or corrugated metal shacks, it includes only a few cities of colonial date. Port Harcourt has developed on an eastern branch of the Niger delta, as the terminus of the railroad that links the coast to the mineral-rich basin of Jos, and to the city of Calabar in a lagoon on the Cameroon border.

The north central region is the least urbanized, with only two cities of more than 200,000 inhabitants: Kano and Zaria. Peripherally located in northwestern Nigeria is Sokoto, an agricultural and commercial center on the river of the same name.

The oldest cities are Ife, founded around the 7th century; Ibadan, site of one of Africa's most prestigious universities; Ilesha; and Oyo. Kano, which flourished thanks to the commercial traffic on a major trans-Sahara caravan route that passes through it, is a famous historic and cultural center. The walls surrounding the city enclose the market and the renowned dyeing industry. Characteristics shared by all these cities are the walled citadel with the royal residence, an immensely lively market, and an architectural "split personality," with both traditional single-storied houses and modern buildings used for administrative, cultural, and business purposes.

Social conditions. With a per capita income of US$270 and an annual economic growth rate of less than 2% (compared with population growth of 3.4%), Nigeria is a developing country that runs the risk of rapid impoverishment. Although its infant mortality rate is relatively low (12.2% in 1990), the average life expectancy for Nigerians is 51 years. There have been some successes in education in recent years, with extraordinary development of schools and universities, but the illiteracy rate is still 57.6%. About 65% of the population is still engaged in agricultural activities, mostly on plantations, while the traditional economy survives only in the poorest northern Sahel regions, and on the eastern slopes of the Bauchi Mountains. The plantation agriculture system itself is creating serious problems of food supply for a continually increasing population. The "Feed the Nation" campaign, aimed at raising production of subsistence crops, has been in operation since 1976 in an effort to resolve this problem.

Economic summary. The British consolidated the region's two principal activities—plantation agriculture and mineral extraction on the Jos plateau—by constructing a railroad (1901) to bring down to Lagos' harbor the products of Ibadan's cocoa plantations, and the mineral products of the Jos plateau. Later extended to Kano (1911), this rail line also provided a link between the coast and the northern areas where cotton and peanuts were grown. At independence in 1960 Nigeria was a prosperous nation, with a rich agricultural base oriented toward exports and organized into small peasant holdings, good mineral resources, and the beginnings of some industry. The discovery of large, very high-quality petroleum deposits in the eastern lobe of the Niger delta (Port Harcourt) revolutionized the country's entire economic and social situation. When world oil prices rose in 1973, ambitious plans were laid. One of the most challenging was the construction of a new capital, Abuja, in the center of the country; others included modernization of the rail system with construction of another 600 mi [1000 km] of rail line, and a kind of "green revolution" that would guarantee self-sufficiency in food and reduce economic and regional inequities. But this enormous expenditure of foreign funds and local public money on investments that were not always productive, along with ethnic conflicts, ended up corrupting the entire nation; private citizens grew rich, the bureaucratic apparatus deteriorated, and crime increased, especially in the cities. The fall in oil prices in 1983, together with a progressive and ill-considered de-emphasis on

Socioeconomic data	
Income (per capita, US$)	270 (1990)
Population growth rate (% per year)	3.4 (1980–89)
Birth rate (annual, per 1,000 pop.)	47 (1989)
Mortality rate (annual, per 1,000 pop.)	15 (1989)
Life expectancy at birth (years)	51 (1989)
Urban population (% of total)	35 (1989)
Economically active population (% of total)	37.5 (1989)
Illiteracy (% of total pop.)	57.6 (1991)
Available nutrition (daily calories per capita)	2039 (1989)
Energy consumption (10^6 tons coal equivalent)	16.9 (1987)

many crops grown for export (oil palms, peanuts, etc.), produced a severe decline in national earnings, and made it impossible to meet obligations that had been incurred with foreign financial institutions. This already precarious situation was aggravated by a succession of coups that did even further damage to Nigeria's international credibility. The grand development projects were therefore abandoned, and runaway inflation caused increasing unemployment and labor unrest.

Agriculture and livestock. Agriculture plays a fundamental role in the Nigerian economy, employing about 47% of the population, contributing 32% of GNP, and supplying the domestic market. Approximately a third of the nation's territory is devoted to cultivated or orchard crops, and one-fifth is used for pasturage; forests now make up only one-sixth. The principal agricultural crops are plantation-grown and intended for export, including oil palm (363,000 t of nuts and 990,000 t of oil in 1991), cocoa (126,500 t in 1991), peanuts, and cotton; minor crops include bananas, rubber, ginger, tobacco, soybeans, and sugar cane. The important traditional crops grown for food are millet, manioc (which produces the cassava flour used for everyday nutrition), and sorghum (5,280,000 t in 1991), which is especially common in the northern part of the country. The predominant crops in the south are sweet potatoes (which are also exported) and rice, which is now being produced in ever larger quantities. Wheat, however, must all be imported, and consumption of this grain is constantly increasing. However, even traditional agriculture seems to be undergoing a process in which new products are being introduced (sweet potatoes, coconuts, tomatoes), cultivation is being extended to satisfy domestic demand (especially for rice), and farms are being modernized. Despite the healthy state of the production system, Nigerian agriculture is still at a disadvantage because of the emphasis placed on petroleum, and is not managing to meet fully the demand created by rapid population growth, which in turn fuels what has now become chronic urban unemployment.

Livestock holdings are enormous and varied, especially in the north, with 14.5 million head of cattle, 24 million sheep, and 36 million goats (1991). Fishing on a modest scale is practiced along the Niger, and the catch is used entirely for internal consumption. The forest supplies exports of hardwoods, especially mahogany, and softwoods, in quantities equivalent to 4% of GNP.

Energy resources and industry. Nigeria's principal mineral resource is petroleum; extraction reached a peak in 1979 with 720 million bbl [115 million metric t] of crude (presently 541 million bbl [86 million metric t]); natural gas production is still low. The oil fields, discovered in 1966, are located on either side of the Niger delta and extend offshore onto the islands and into the ocean, and are connected to the refineries at Port Harcourt and Warri (the latter constructed by the Italian state-owned petroleum conglomerate ENI in 1978). A large refinery was recently built at Kaduna, fed by a 1750-mi [2800-km] pipeline that links the coast with Kano and Maiduguri in the northeastern corner of the country. The other mineral resources are concentrated on the Jos plateau, and include columbite (niobium, tantalum, manganese, and iron ores), cassiterite (a tin ore), and lead, zinc, and silver ores. The Enugu coal basin, which was very important in the past and is now being revived, is located in the southern highlands; there are also iron deposits in the Lokoja area. Annual production of electrical energy, which is also supplied by two large hydroelectric plants including the Kainji dam on the Niger, is 9.935 billion kWh (1989).

Beginning from a solid foundation of local traditions—peanut pressing, palm oil refining, tobacco and sugar production, and cotton ginning and spinning—Nigeria has developed a modern industrial base that concentrates on these production sectors. The cotton-based textile industry is also of some significance, both in production areas (Kano) and in regions where textiles are consumed and sold (Lagos). Stimulated by increased demand for consumer goods on the domestic market, new industries have developed in the food-processing sector (beer and fruit juices), in construction (cement and furniture), and in mechanical engineering sectors that did not previously exist (household appliances, bicycles, automotive assembly). Heavy industry is also present, with a steel plant at Ajaokuta and an aluminum production complex at Port Harcourt. In general, industrial activities, which employ 16% of the work force (1989) and produce 25% of gross national product, focus primarily on the production of consumer goods. Four major industrial regions can be identified. The southwest comprises the most urbanized area of the country centered on the cities of Lagos and Ibadan, with food processing installations, cement plants, mechanical industries, textiles, and petroleum refining. A southeastern region, with the cities of Port Harcourt and Aba, is characterized by food-processing industries (palm oil plants), textiles, metallurgy, and petroleum. The other two regions are largely dominated by mining activities: one lies to the south with the Enugu coal basin and the iron deposits at Lokoja; the other is on the Jos and Bauchi plateaus, where tin, lead, and zinc ores are extracted. Industrial activities are also present in northern Nigeria, for example at Kano (peanut-oil processing, textile plants, vehicle assembly, and cement works), Sokoto (cement), and Maiduguri (vegetable oils).

Commerce and communications. Domestic commerce is traditionally dominated by the Hausa. Foreign trade, however, is overshadowed by petroleum, which monopolizes 95% of exports. This reflects the fragility of the Nigerian economy, which has made the nation entirely dependent on the whims of the world oil market. Trade relations have changed substantially; the United Kingdom, which in 1959 absorbed 50% of Nigeria's exports, now receives only 4%, while the United States is the destination for half the value of exported goods, followed by the EC countries. Imports, which comprise mechanical products (50%), textiles, chemicals, and foodstuffs, come from the EC, the United States, and Japan. The long run of years in which the balance of trade was positive ended in 1978, due to massive imports of capital goods and to other factors already mentioned. It later went back into the black thanks to a reduction in consumption imposed by a government austerity policy.

In addition to a good harbor system (besides Lagos and Port Harcourt, other important ports include Calabar, Warri, Sapele, and Bonny), Nigeria has a fairly dense road network (76,900 mi [124,000 km], of which 21,000 mi [34,000 km] were paved in 1985), especially in the southwest. The rail system is still inadequate for the country's needs (only 2173 mi [3505 km] in 1987). Lagos and Kano have international airports but tourism is still limited.

Historical and cultural profile. ***Ancient civilizations: the Yoruba.*** Some of the most important civilizations of black

Africa flourished in the basin of the lower Niger, following complex lines of historic development that until recently tended to interfere with the creation of a unified nation. In the 10th century, the southwestern region was occupied by a people who had come, in all probability, from the Nile valley and who knew the technology of bronze smelting: the Yoruba. Merging with the native population, they generated three prosperous and relatively stable kingdoms—the Yoruba itself, the Nupe, and the Benin—which in the 15th century grew rich by selling slaves to the Portuguese. These were divine monarchies, in which the person of the king had the supernatural function of acting as depository for the vital energy of his entire people. Whenever a sovereign died, a bronze bust was made for propitiatory purposes, transmitting his features to posterity. In the late 19th century these true sculptural masterpieces astounded the Europeans when some two thousand of them, together with equally lavish wood and ivory carvings, were brought to London as war booty following the conquest of the city of Benin.

The Hausa. The historical path taken by the northern regions of Nigeria was very different. Here groups of Berbers began to appear in the 8th century, first subjugating and then blending with the local black populations, and beginning a slow process of Islamization. This was the origin of the Seven States of the Hausa, a group that attained a high level of economic development through intensive agriculture, raising horses and cattle, and craft production and commerce (the main commodity once again being slaves). Their civilization is known to us both from the reports of Arab travelers and from original documents written in the Hausa language. We know from these sources that their political system was based on the city-state, ruled by elected monarchs with broad powers sustained by feudal lords of Muslim origin. The Hausa language was used not only for official records, but also for a rich written literature that supplanted classical Arabic (initially used by Muslim scholars) for the composition of religiously inspired poetry, writings of moral edification, and works exalting the teachings of the Prophet. This literature was especially important toward the end of the 18th century, when the production of poetic narratives, lyrics, homilies, and other religious compositions became part of the "holy war" declared by Emir Uthman Dan Fodio, designed to impose Arab culture on neighboring non-Muslim rulers. In the next century, it was again in Hausa that Alyiu Dan Sidi composed a poem on the dangers of European ingression.

European ingression. In the 19th century the British were indeed exploring the Niger delta and penetrating upstream, establishing trading companies and gradually subjugating the people they encountered. The first were the Ibo, who lived in the southeast corner of the region and had no government structures at all to defend them; this had made them frequent victims of the raids of neighboring slave-trader kingdoms. Shortly thereafter, crisis also overtook the Yoruba and Hausa kingdoms, weakened by the invasions of the Fulbe people from western Sudan. At the Conference of Berlin in 1885, Great Britain succeeded in obtaining formal recognition for its expansion in the Nigerian region, and proclaimed a protectorate over the area. Local rulers were allowed a minimum of autonomy in accordance with the British colonial theory of "indirect rule"; this structure allowed Britain to control an extremely large area, extending from its commercial bases at Lagos and in the Niger delta to embrace most of the

river's basin. At the beginning of the 20th century, the British possessions were subdivided into a protectorate in northern Nigeria, which included the Hausa lands, and a colony in southern Nigeria, comprising the rest of the country. Despite the fact that the two entities were formally combined into a federal state in 1914, there was no real commitment to national ethnic and cultural integration on the part of the British administration, which exploited tribal clashes to its own advantage. The independence movement itself was weakened by the habit of adhering to tribal structures, and not until the creation of the University of Ibadan in 1949 was there sufficient stimulus for some sort of national consciousness among the future ruling classes.

A difficult independence. In 1960 the Nigerian state achieved independence; it continued to maintain a federal structure, and grew in size a few years later by incorporating the former British Cameroon, an event that accentuated its composite nature. The young republic, led by Nmandi Azikiwe, found life difficult as a result of growing tensions among the various ethnic groups which flourished in an environment of severe economic and social difficulties. The precarious balance of the first years collapsed in 1966, with two successive military coups within a few months' time; at first, power was seized by the Ibo, who abolished all federal institutions; then the advantage was regained by the traditional northern castes of the Hausa, who restored the federation and installed General Yakubu Gowon as its leader. The ancient ethnic rivalries were now being complicated by the interests of the multinational corporations, and in 1967 the eastern region inhabited by the Ibo, with its abundant oil reserves, declared its independence as the "Republic of Biafra." This attempt at secession was quashed after three years of bloody warfare, which took a particularly heavy toll on civilians. Nigeria then began a phase of rapid economic development based on petroleum production, but festering ethnic and social tensions, rampant corruption, and foreign interference once again undermined this potential for development, and for the next few years the nation reeled from one military coup to the next. Gowon was succeeded in 1976 by Muhammad Murtala, followed by Lt. Gen. Olusegun Obasanjo. Shehu Shagari was elected president in 1979, marking the return of civilian rule and the beginning of an austerity program that included the expulsion of two million immigrants who had come from other African countries. The military returned to the scene with Muhammad Buhari (1984), then with Maj. Gen. Ibrahim Babangida, who committed himself to a program of transition to democracy by the end of 1992.

Literature. The British conquest introduced a lull in the figurative arts and literary expression of the various ethnic groups. Nigerian literature (mostly in Yoruba, Hausa, and English, but to a lesser extent in Ibo and other languages as well) began to develop again after World War II, as part of the independence movement and as a reflection of the complex social, ethnic, and cultural problems facing Nigeria. These problems have been addressed by a number of authors; among writers who use English, Ciprian Ekwenzi has offered a realistic portrait of his country in the period prior to decolonization, while Chinua Achebe has written a trilogy outlining the last century of Nigerian history. Historical and social themes have also been addressed in Yoruba, for example by F. Jeboda, who has convincingly described the squalor of the cities, while the poet and novelist A. Olabimtan has criticized the surrender of local culture to European models.

SENEGAL

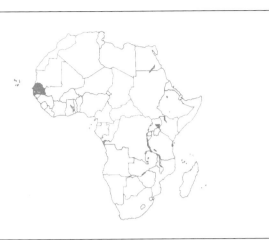

Geopolitical summary

Official name	République du Sénégal
Area	75,935 mi² [196,722 km²]
Population	6,892,720 (1988 census)
Form of government	Presidential republic; the head of government and National Assembly are elected by universal suffrage every 5 years.
Administrative structure	10 regions divided into 30 departments
Capital	Dakar (pop. 1,382,000, 1985 estimate)
International relations	Member of UN, OAU; associate member of EC. United with Gambia in the Federation of Senegambia
Official language	French; national language is Wolof (Sudanese)
Religion	Muslim 85%; approximately 300,000 Catholics
Currency	CFA franc

Natural environment. Senegal is bordered on the north by Mauritania, on the east by Mali, on the southeast by Guinea, on the southwest by Guinea-Bissau, and faces the Atlantic to the west opposite Cape Verde. It also shares a border with Gambia, which is entirely included within its territory except for its coastline.

Delimited to the north and east by the Senegal river and its left-bank tributary the Falémé, the terrain is fairly uniform, with a low average elevation (less than 160 ft [50 m]), and represents the surface of the sediments that, in the Tertiary epoch, filled a wide gulf of the ocean that was sealed off on the inland side by low hills of Paleozoic rock that at present reach elevations of less than 1600 ft [500 m] and lie between the Falémé and the region from which the Gambia and Casamance rivers flow toward the ocean, the latter very close to the nation's southern border. The

last element in this morphological picture is the complex peninsula on which the capital, Dakar, is located, consisting of the remains of Miocene and Quaternary basaltic lava flows. It should also be mentioned that the area which defines most of the course of the Gambia river belongs not to Senegal, but to Gambia. The Dakar peninsula, which extends into the Atlantic at Cape Verde, also divides the narrow northern coast, which is sandy and devoid of natural harbors, from the southern portion which is extremely indented and often swampy.

The great hydrographic node of western Africa, consisting of the Fouta Djallon Mountains (4969 ft [1515 m]) which are almost entirely contained within neighboring Guinea, is the source of the Senegal and Gambia rivers, which are navigable and have therefore always encouraged penetration into the region. Like the Casamance, they empty into the Atlantic, with an irregular flow that produces substantial flooding during the rainy season. The center of the country is more arid, and is crossed by seasonal watercourses such as the Ferlo, a left-bank tributary of the Senegal, and the Saloum, which is perennial only for about 30 mi [50 km] near its mouth. Enormous projects are now under way in order to exploit these rivers for irrigation, as sites for hydroelectric power plants, and to improve navigability.

In terms of climate, Senegal represents a transitional area, although in a definitely tropical context, between decidedly arid conditions (lower Senegal river area) and essentially humid ones. Precipitation increases quite steadily from about 15 in. [400 mm] annually in the north to over 60 in. [1500 mm] in the south (along the Casamance). This represents the transition from a steppe climate to a subequatorial climate, passing through all the climatic gradations of the savanna. The characteristic element is the alternation between a long, very dry season from October to June, and a brief rainy season from July to September. At Saint-Louis on the Senegal river, at the border with Mauritania, the rainiest month is August with 8 in. [200 mm] of precipitation, almost half the amount for the entire year (16.5 in. [423 mm]); this is also the hottest period of the year (81°F [27°C] in August and September). The dry season, on the other hand, is characterized by an average temperature of 68°F [20°C]. This climate has monsoon-like features, in which the place of the summer monsoon is taken by the southern-hemisphere trade wind which passes across the equator and changes direction to the northeast, while the function of the winter monsoon is performed by the *harmattan*, the dry Saharan wind that blows from the northeast. Farther south the contrast becomes less pronounced: the rainy season gets longer and annual temperature swings are smaller. In the extreme south, at Ziguinchor on the Casamance, annual rainfall is 60.6 in. [1555 mm], while January and July average temperatures are 74°F [23.2°C] and 79°F [26.2°C], respectively. The same contrasts are seen when moving from the coast toward the interior: temperature variations increase, and annual maximums reach

Climate data				
Location	**Altitude (ft asl)**	**Average temp. (°F) January**	**Average temp. (°F) July**	**Average annual precip. (in.)**
Dakar	10	72	82	21.1
Tambacounda	171	77	80	34.3
Conversion factors: 1 ft = 0.3 m; 1 in. = 25 mm; °C = (°F − 32) × 5/9				

decidedly torrid levels. At Tambacounda, at the fringes of the semidesert Ferlo region, annual average temperature is 82°F [27.9°C], and rainfall is 34.3 in. [879 mm] per year.

The vegetation reflects these changes in climate with latitude: about a third of the nation is covered with woodlands, which are present over the entire coastal band south of Dakar and the Casamance region, and also occupy much of Niokolo and Koba National Parks. Senegal's tropical forest is somewhat thin in structure, however, and does not include the most highly valued trees; the wooded savanna farther north is characterized by baobabs, acacias, and tall grasses. The interior is covered by the thorny savanna typical of a predesert environment, with xerophytic shrubs. Important natural areas under protection include the Djondj reserve in the north near the bend in the Senegal river, which provides a refuge for migratory birds; Saloum Delta National Park, which shelters marine fauna; Basse-Casamance Park, which protects terrestrial animals and amphibians typical of the savanna.

Population. The dominant ethnic group is Sudanese, including a number of smaller populations. The Wolof (44% of the population), sedentary farmers living in the northwest who achieved political dominance after independence, make up the administrative class within the government. The Serer (18%) live in this same area, while the northeastern districts are dominated by the Peul, Fulbe (12%), and Toucouleur, Arabicized nomadic herders. The Diola-Mandingo (9%), expert rice cultivators, live in Basse-Casamance. Senegal has not experienced the dramatic instability provoked by ethnic conflicts in other African states, not just because of French colonial policy—which regarded the colony as French territory, and helped create administrative ability among the population—but also because the nation, guided by the African poet Léopold Senghor, made the decision to construct a new culture to which both European and African elements contributed.

From a population of little more than 3 million at the time of independence in 1960, Senegal has now grown to about 7 million

(1988), posting an annual growth rate of 3%, consistent with the rest of the continent. Population density has risen from 44 per mi^2 [17 per km^2] to 91 per mi^2 [35 per km^2], although there is still a disparity between the west central area centered on Dakar, with a density twice the national average, and the northeastern region, with a small and widely scattered population. Basse-Casamance is a region of intermediate density, with interesting dynamics fostered by economic modernization. Senegal can therefore be divided into two regions: one urbanized, in the Cape Verde peninsula area that includes the cities of Dakar, Rufisque, Thiès, Diourbel, and Kaolack; and the other rural, comprising the rest of the country. More than 38% of Senegal's population is now urban, a legacy of the colonial era when Dakar was considered the spearhead for France's conquest of all of western Africa. The capital, located in an open area to the south of the Cape Verde peninsula, at first developed haphazardly around the harbor, then spread out into completely separate functional districts. Besides the installations at the port of Cajar and the M'Bao free industrial zone, and the airport at Yoff, the city comprises the modern business district, a European-style residential neighborhood, and an African quarter, La Medina. A modern university, numerous museums, an efficient hospital, and many hotels and skyscrapers all confirm its cultural importance. The rural landscape is dominated, especially in the north central area where the Wolof live, by small scattered villages consisting of clusters of circular mud huts with conical straw roofs, while the cities are characterized by European architecture, either modern (Dakar) or colonial (Saint-Louis).

Rapid population growth and a youthful population are characteristic of every developing country. In Senegal, 53% of the population is less than 20 years old, while only 6.2% are over the age of 60: this forces unemployed young people in rural areas to look for work in the cities and abroad. The urban areas, especially Dakar, have become focal points of unemployment and underemployment, with a large number of outlying shantytowns. This situation aggravates social conditions that are already precarious, characterized by high illiteracy (77.5%), a life expectancy of only 48 years, and an average per capita income of US$665.

Economic summary. The economy, based traditionally on peanuts, still exhibits the characteristic imbalances caused by colonialism. This legacy places the nation in a difficult position, squeezed between the limits imposed by monoculture (dependence on the international market), and a lack of resources that results in an inadequate food supply. The economic planners have attempted to "Senegalize" the economy with exports, by strengthening food self-sufficiency and building a modern agricultural base. But this requires colossal investments and a very long time frame: difficult climatic and soil conditions restrict the relatively fertile land to the area along the Senegal and in Basse-Casamance, and changes in eating habits have increased the consumption of rice, which Senegal does not produce in sufficient quantity. In addition, most capital (which comes largely from French and international sources) is poured into industry, which produces 26% of GNP, among the highest figures in Africa. The drought of the 1970s, which disrupted the entire Sahel ecosystem, thwarted many economic plans, and Senegal is now seeking investment in its promising tourist industry. The country is fairly evenly divided between farmland (26.6%), meadows and pastures (29%), and forests (30.2%), with only 14.2% of the territory

Administrative structure

Administrative unit (regions)	Area (mi^2)	Population (1988 census)
Dakar	212	1,490,450
Diourbel	1,683	619,680
Fatick	3,067	475,970
Kaolack	6,176	816,410
Kolda	8,106	606,790
Louga	11,267	490,400
Saint-Louis	17,033	680,220
Tambacounda	23,006	570,020
Thiès	2,548	948,100
Ziguinchor	2,837	394,680
SENEGAL	75,935	6,892,720

Capital: Dakar, pop. 1,382,000 (1985 estimate)
Major cities (1985 estimate): Thiès, pop. 156,200; Kaolack, pop. 132,400; Saint-Louis, pop. 91,500; Ziguinchor, pop. 79,464.

Conversion factor: 1 mi^2 = 2.59 km^2

Socioeconomic data

Income (per capita, US$)	650 (1988)
Population growth rate (% per year)	2.5 (1985–90)
Birth rate (annual, per 1,000 pop.)	46 (1989)
Mortality rate (annual, per 1,000 pop.)	15 (1989)
Life expectancy at birth (years)	45.8 (1985–90)
Urban population (% of total)	38 (1989)
Economically active population (% of total)	43.9 (1988)
Illiteracy (% of total pop.)	61.7 (1990)
Available nutrition (daily calories per capita)	2350 (1988)
Energy consumption (10^6 tons coal equivalent)	0.95 (1987)

unproductive; but the soil and climate are not conducive to modern agriculture. Peanut production is a significant factor in plans for agricultural restructuring, to liberate the country from this culture and expand the production of subsistence crops. Combined pressure from international peanut traders and domestic producers, urged on by Muslim associations who want to promote peanut cultivation, along with the profitability of the product, have all resulted in an extension of cultivation to include southern regions as well, although peanut growing dries out the land and contributes to the advance of the desert. Other plantation crops grown for the international market are cotton, oil and coconut palms, and bananas. In the area of subsistence crops, certain widely grown "low-grade" products such as millet and sorghum are being replaced by semi-industrial or truck-garden produce such as sweet potatoes, tomatoes, and other vegetables. Millet, corn, and manioc are the traditional products grown for immediate consumption; rice, introduced by the French, has become a favorite food that must be imported despite increased production in Basse-Casamance. Seventy percent of the labor force is engaged in agriculture, which retains an archaic structure except for substantial innovations in peanut, rice, and vegetable cultivation, and cooperative farming enterprises. Livestock is raised in the central and northern areas, and boasts more than 6 million head of cattle and sheep, although animal holdings were devastated by the drought. The fishing industry is steadily expanding, exploiting the rich waters around Cape Verde; in addition to providing food for the country, the catch supports a burgeoning canning industry, especially for tuna. The forest produces not only everyday materials for both fuel and construction, but also gum arabic, a latex extracted from *Acacia senegalensis.*

The most important mineral resources are phosphates (2.5 million t in 1989), extracted from deposits at Thiès, Taiba, and Pollo in western Senegal, where calcium and aluminum ores are also obtained. Titanium is of minor importance, while significant iron deposits discovered at Falémé in the eastern part of the country must await the construction of a dam at Manantali on the Senegal to produce electrical energy (along with 435 mi [700 km] of railroad to the port of Cajar at Dakar) before the product can be brought to market. There is a single salt complex at Wharf, producing 110,000 t of sea salt each year (1988). Minor reserves of natural gas are also present on the coast; production of electrical energy is still modest, and local wood is used for fuel.

The industrial structure is inadequate, although the large petroleum refinery at Mbour-Dakar produces raw materials for a nascent chemical industry. The most important sector is food processing, producing peanut oil, processed fish, leather, and hides for export, and internal consumer goods such as beer, cigarettes, sugar, canned goods, and canned meat. The textile industry is developing vigorously, and cement plants and automobile assembly operations are also important. Production of nitrogen-based fertilizers is becoming increasingly significant, with installations at Dakar and at Taiba, where phosphates are extracted. The capital is the focus of an industrial area consisting of the commercial port at Cajar, the free-trade zone of Mbour, the Cape Verde peninsula, the city of Rufisque (which has now become a bedroom suburb of greater Dakar), and the northern region comprising Thiès, Taiba, and the entire phosphate-mining region.

With 15% of the labor force, the services sector earns over half of GNP from export activities (in Dakar) and internal commerce, with lively produce markets at Kaolack. Tourism is developing strongly, and contributes 9% of GNP; 300,000 tourists visited the country in 1990, attracted by well-groomed beaches south of Dakar and in the extreme south, by nature parks in which savanna animals and plants can be observed, and by native and colonial cultural traditions. The balance of trade is heavily weighted toward imports, although the gap is now narrowing. Exports are based primarily on raw materials—peanuts, phosphates, fish—and on processed products such as vegetable oils, textiles, and petrochemicals. Imports include capital equipment, fuels, and food. The major trading partner is France, which accounts for almost 50% of both imports and exports, followed by Germany, the United States, Great Britain, and Italy for imports, and Ivory Coast, Mauritania, and Italy for exports. Communications with the outside world are well developed, with harbors at Cajar (Dakar) and Saint-Louis, and an international airport at Yoff (Dakar). Foreign trade passes through the ports of Kaolack and Ziguinchor on the deep estuaries of the Saloum and Casamanche rivers, while river traffic moves along the navigable portions of the Senegal and Saloum. The rail system (735 mi [1186 km] in 1984) consists of two main lines out of Dakar, one to Saint-Louis along the coast, and the other to Tambacounda in the east. There are 2521 mi [4066 km] of paved roads.

Historical profile. *The ancient kingdoms.* Stone tools and bone fragments from the Paleolithic period indicate the presence of ancient settlements near present-day Dakar. Megalithic monuments similar to Breton menhirs have been dated to the Neolithic, along with tumuli called "mbanar," evocative remains of a civilization of sun-worshippers. The first historical records, however, date to the first century of the Christian era, when the banks of the Senegal river were visited by Roman troops and later by Christianized Moorish tribes. By the 9th century the nation of Tekrur, along the lower course of the river, had become the primary center from which Islam radiated into western Africa. In succeeding centuries, with the assistance of the Almoravid brotherhood (a Berber group that had established itself in Morocco), Tekrur overthrew the empire of Ghana and founded a kingdom that linked its fate in some respects to that of Mali. The legacy of Tekrur and Mali was passed onto the Djolof empire, which in the 14th century extended over the regions of Walo, Kayor, Baol, Siné-Saloum, and Dimar, becoming the crucible in which the Wolof, the largest ethnic group in Senegal, forged their cultural unity.

The slave trade. In the 14th century, the first Europeans also arrived to explore the coast: Italians, Portuguese, and French were particularly attracted by the easy approach to the Cape Verde promontory and the island just offshore, which became a stopover for every caravelle on the routes to the Indies and Brazil. Bartolomeo Diaz passed through here, as did Vasco da Gama, the Jesuit missionary Francesco Saverio, and the Portuguese poet Luis de Camões. It was the starting point for explorations of the interior and for the first commercial exchanges—the slave trade. Over the next two centuries the Dutch and English arrived as well, and bitter battles were fought for control of the island which the Dutch called Gorée, the most important base in Africa for the lucrative traffic in slaves. The French ultimately gained the upper hand, securing not only Gorée but another strategic position on the coast, where they founded the city of Saint-Louis.

Colonization and independence. In the mid-19th century Saint-Louis was the jumping-off point for colonization of the interior. The effort was led by General Faidherbe, who pushed the Moors back to north of the Senegal river, subjugated the petty kingdoms that had emerged from the dissolution of the Djolof empire, and founded Dakar. Within fifty years the city had become capital of the new colony and the political and administrative center of all of French West Africa, signaling the decline of nearby Gorée. The French regarded Senegal as a sort of model colony: they introduced plantation crops, constructed roads, and built up the educational system so as to assimilate the population as much as possible to European models. But the spread of education produced results that were quite the opposite of those desired: in the early 20th century, there appeared on the scene a young intellectual elite that led the country toward independence, and became an ideological guiding light for the entire African anticolonial movement. Important roles were played in this process by philosophies such as Muridism, Tadjianism, and "layene," which oscillated between pacifism and armed struggle but were nonetheless committed defenders of the Islamic cultural heritage. Particularly decisive were personalities such as Lamine Gueye, Blaise Diagne, and Léopold Senghor. Senghor, a European-educated intellectual, had developed (together with Aimé Césaire of Martinique and the Guinean Léon Damas) the fundamental concept of "Negritude" which guided the nation after independence (1960), bringing it from a presidential republic to a multi-party state in 1978. Three years later Senghor resigned as president—an event unique in the history of contemporary Africa —to return to his literary work, turning over his office to his protégé Abdou Diouf.

Literature. The charismatic figure of Senghor dominates the panorama of Senegal's young and vibrant literature, which developed in French after World War II; traditional literary works are still passed on orally by "griots" in the languages of the various ethnic groups. Senghor was educated as a symbolist, and his poetry sings in passionate tones. Other authors include Birago Diop, who has developed local folk traditions in both prose and poetry, and David Diop, a poet with a strong sense of civic commitment. One noteworthy novelist is Cheikh Hamidou Kane: his *L'Aventure ambiguë* [The ambiguous adventure] addresses the theme of lost identity as Africans are cast adrift in the Western world. Senegalese historians and anthropologists, for their part, have investigated various aspects of local culture and traditional language; they include Omar Ba and Cheikh Anta Diop, after whom the university at Dakar is named.

SIERRA LEONE

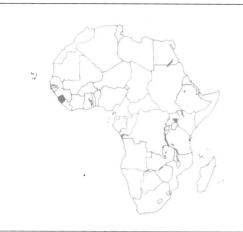

Geopolitical summary

Official name	Republic of Sierra Leone
Area	27,692 mi^2 [71,740 km^2]
Population	3,517,530 (1985 census); 4,140,000 (1990 estimate)
Form of government	Presidential republic. The present government is of a military nature.
Administrative structure	3 provinces divided into 13 districts, plus the capital territory (Western Area)
Capital	Freetown (pop. 469,776, 1985 census)
International relations	Member of UN, OAU, Commonwealth; associate member of EC
Official language	English; Krio and Sudanese languages widely spoken
Religion	Animist 51%; Muslim 39%; Protestant 6%; Catholic 2%
Currency	Leone

Natural environment. Sierra Leone is bordered on the north and east by Guinea, on the east and southeast by Liberia, and faces the Atlantic Ocean to the south. The territory consists of a large coastal plain some 60–90 mi [100–150 km] wide and more than 180 mi [300 km] long, which is flat and marshy near the Atlantic coast and dotted with lagoons. The terrain slopes gently upward toward the interior, forming rolling hills and plateaus in the ancient (Paleozoic) crystalline rocks of the continental basement; in the northeastern section at the border with Guinea, the topography rises to form well-defined highlands, which reach a maximum elevation of 6389 ft [1948 m] in the Loma Mountains. Scattered along the low coastline are numerous islands and several promontories, the largest of which is the peninsula on which the capital, Freetown, stands.

The hydrographic system consists of rivers that flow down from the slopes of the Fouta Djallon (Great Scarcies, Little

Climate data				
Location	Altitude (ft asl)	Average temp. (°F) January	Average temp. (°F) July	Average annual precip. (in.)
Freetown	33	80	78	136.6
Bo	276	81	77	120.7
Conversion factors: 1 ft = 0.3 m; 1 in. = 25 mm; °C = (°F − 32) × 5/9				

Scarcies), from the Guinean highlands (Rokel, Moa, and Mano-Morro), and from the Loma Mountains (Jong and Sewa) in a northeast–southwest direction, interrupted by frequent falls near the plain which make them unusable for navigation.

The climate, strongly governed by precipitation, is typically tropical (subequatorial): hot and wet, with two seasons. The summer rainy season lasts from April to November, with rainfall levels exceeding 120 in. [3000 mm] (even higher on the coast, where 160 in. [4000 mm] can fall each year), and ends with violent hurricanes caused by the southwest summer monsoon. The dry winter season is shorter, governed by the *harmattan*, the hot, dry wind that blows from the continent to the sea. Temperatures are very high and steady (averaging around 86°F [30°C], but exceeding 95°F [35°C] in summer), with a high humidity that makes living conditions very unpleasant and is mitigated only on the higher ground of the interior.

The evergreen rain forest, whose presence is a result of the hot, wet climate, abounds in plant and animal species; it has now been reduced to 4% of the country's land area due to intense exploitation and clearing for cultivation. The wooded savanna, with its typical baobab, karite (shea), acacia, and oil palm trees, is common in more inland areas, where the largest representatives of African wildlife live. The vegetation along rivers consists of fringing forests, while mangrove communities are common in coastal areas.

Population. Sierra Leone's population is predominantly indigenous, consisting of two principal ethnic groups: the Temne in the north, and the Mende in the south. Both are Sudanese groups, survivors of Paleonegritic peasant communities. The Peul, Muslim herders who have now become sedentary, also live in the north. The political and cultural transformations undergone by the country in the past were initiated by a minority of black Creoles from the United States, descendants of freed slaves who for the moment have lost their hold on political power. Sudanese dialects and the Krio language spoken by the Creoles are more common than English.

The population, with an average density of 150 per mi^2 [58 per km^2] and a 2.5% annual rate of increase, is rising too fast for the nation's difficult economic conditions. Settlement is predominantly rural, consisting of small isolated villages located either in clearings along the coast or in the interior, where black-magic rituals are still practiced in the context of a feudal social structure. The capital, Freetown, inhabited by the Creoles, betrays obvious Anglo-Saxon influences. Founded in 1792 as a destination for freed slaves (hence the name), in 1800 it became the capital of all British possessions in Guinea. Now it is an active economic center with one of the best harbors on the Gulf of Guinea, site of a university and many other institutions of learning. It was the venue for the 1980 conference of the OAU

(Organization of African Unity).

Roy Lewis has described the "creole" character of Freetown:

Freetown is essentially a creole metropolis.... though Freetown is neither Delhi nor Port Said as a meeting-place of races, there are classes, occupations, and twenty or more tribes to be distinguished–and above all, faces. Who's creole and who's not? There are African girls, slim and assured in print frocks that might have been bought in London's Oxford Street; African girls in native dress of every hue that Manchester can print....

Explosive development in diamond prospecting in the last 40 years has changed the structure of the country: more and more rural people have abandoned the countryside for the mining centers, attracted by the easy money to be made largely from clandestine extraction and the illegal gem trade with Liberia. This has reduced the number of persons employed in rural areas and increased the size of the disorganized urban proletariat. Average per capita income is still extremely low; despite the existence of educational facilities, 79.3% of the population is illiterate, and mortality is still high (2.3%). Living conditions in general are extremely backward, making Sierra Leone one of the poorest nations in Africa.

Economic summary. Sierra Leone's economic situation is very difficult, and the government must rely on international aid to survive. The economy is based largely on agriculture (occupying 62% of the working population), but the nation has not yet achieved self-sufficiency in food. The main product grown for internal consumption is rice, which is cultivated in the wet coastal regions. Production of other crops (corn, millet, sweet potatoes, sorghum, manioc) is falling further and further behind domestic demand. Commercial crops such as coffee, cocoa, ginger, and beeswax, which were particularly important during the colonial period, have collapsed due to mismanagement and other factors. In 1985, the government began a "green revolution" program to promote agricultural development, but it ultimately failed for lack of sufficient investment, and because of massive migration away from the countryside.

Although a third of the nation's land is used for meadows and pastures, animal husbandry is of marginal importance and is practiced only by the Peul in the northern regions.

The most profitable activities are associated with mineral extraction in the northern part of the country—especially dia-

Administrative structure		
Administrative unit (provinces)	Area (mi^2)	Population (1985 census)
Eastern	6,003	960,551
Northern	13,871	1,262,226
Southern	7,602	740,510
Western Area	215	554,243
SIERRA LEONE	27,691	3,517,530

Capital: Freetown, pop. 469,776 (1985 census)
Major cities (1985 census): Bo, pop. 39,000; Kenema, pop. 31,000; Makeni, pop. 26,000.

Conversion factor: 1 mi^2 = 2.59 km^2

Socioeconomic data

Income (per capita, US$)	160 (1990)
Population growth rate (% per year)	2.5 (1980–89)
Birth rate (annual, per 1,000 pop.)	48.2 (1989)
Mortality rate (annual, per 1,000 pop.)	23.4 (1989)
Life expectancy at birth (years)	39.4 (1989)
Urban population (% of total)	32.2 (1989)
Economically active population (% of total)	34.9 (1989)
Illiteracy (% of total pop.)	79.3 (1991)
Available nutrition (daily calories per capita)	1813 (1989)
Energy consumption (10^6 tons coal equivalent)	0.28 (1987)

monds, of which Sierra Leone is one of the world's leading producers—and are carried on for the most part by a joint-venture company. Other important mineral resources are rutile (from which titanium is extracted for aerospace applications), platinum, chromite, and bauxite. Iron ores were significant in the past, but production has dropped due to exhaustion of the deposits, and now covers only internal demand.

The industrial base consists of plants for processing food products, a petroleum refinery, and metallurgical installations for bauxite processing. The balance of payments is in the red: exports of diamonds (60% of total exports), cocoa, coffee, bauxite, and rutile are not enough to compensate for imports of machinery (30%), food products (19%), basic manufactured goods (17.8), and petroleum and derivatives (13.8%). Sierra Leone's principal trading partners are Great Britain, the United States, and the Netherlands for exports, and Great Britain, China, Japan, and Nigeria for imports.

The communications network includes only 4650 mi [7500 km] of roads, of which 600 mi [960 km] are paved (1989), concentrated in the western part of the country. The railroads consist of 370 mi [597 km] of line, built in the colonial period to haul iron ore and now largely unused (1985). The principal ports are at Freetown, Pepel, and Port Loko; air traffic passes through the international airport at Lungi near Freetown.

Historical profile. *The lure of freedom.* The first European explorers who set foot in the area in the 14th century gave the name "Pepper Coast" to the littoral of what is now Sierra Leone. The present name dates back to the 15th century and was given by the Portuguese, who are said to have recognized the outline of a lion in the shape of a certain mountain in the interior (or, according to another version, compared the thunder rumbling down from the mountains to the roar of a lion). In the two centuries that followed, the region became the area of Africa most frequently visited by English pirates and by slave traders. It was possibly for this reason that, in the late 17th century, this stretch of coast was selected by several English philanthropic societies as the destination for slaves freed at the end of the American Revolution. The first settlements, promoted by a private company in 1787, 1791, and 1792, had a difficult time, not least because of strained relations between the local tribes and the new arrivals, at that time called "Creoles."

The situation improved in the early 1800s when the British government converted the original settlements into a colony, which expanded as new land was acquired. The old nucleus of

Freetown continued to attract Africans freed from the yoke of slavery, and the influx grew after 1833, the year in which the slave trade was abolished: the occupants of every slave ship intercepted by the British navy were then brought to Sierra Leone. The territory expanded to its present borders with Liberia and French Guinea, and was divided into two sectors: a British colony around Freetown inhabited by Creoles, and a protectorate inland, subject to the authority of the tribal chiefs. Political and social progress was faster in the former than in the latter, and the local ethnic groups thus found themselves subordinated to the ruling class in Freetown.

Fundamental problems. The institutional and administrative distinction between these two sectors persisted until 1960, when unification became a prelude to the nation's independence which was proclaimed the following year. Power then passed from a pro-British party to a democratic group led by Siaka Stevens, which fought for government control over the iron and diamond mines, the country's principal resources. But government policies turned out to be incapable of responding adequately to Sierra Leone's fundamental problems; increasingly disastrous economic conditions and rising unemployment and crime resulted in serious disturbances, which spread throughout the country in 1984. Even in the years since then, the ruling party has been exposed to mass protest demonstrations and attempted military coups.

Culture. The most vigorous artistic manifestations are those of the ethnic groups living in the interior, who still follow very ancient paradigms (partly for ritual use, and sometimes merely for sale to tourists). One form already known to 16th-century Portuguese travelers, but undoubtedly dating back much further in time, is expressed by the anthropomorphic statuettes called "pomdo" or "pombo" by the Kissi, and "nomoli" or "momori" by the Mende: these little steatite figures, used in the ancestor cult, depict deceased nobles with realistic features and great sculptural force. Even today, the custom survives of burying a stone statuette, called "mormal," in the ground as a way of invoking divine favors. The Kissi also produce statuettes richly embellished with largely abstract surface decorations, called "pomtan," perpetuating a tradition that originated in the 13th century.

Ancient and modern, tradition and research are blended in these lines entitled "Up-Country," by the contemporary poet Abioseh Nicol:

> *Then I came back*
> *Sailing down the Guinea coast,*
> *Loving the sophistication*
> *Of your brave new cities:*
> *Dakar, Accra, Cotonou,*
> *Lagos, Bathurst, and Bissau,*
> *Freetown, Libreville.*
> *Freedom is really in the mind.*
>
> *Go up-country, they said,*
> *To see the real Africa.*
> *For whomsoever you may be,*
> *That is where you come from.*
> *Go for bush—inside the bush*
> *You will find your hidden heart,*
> *Your mute ancestral spirit.*
>
> *And so I went,*
> *Dancing on my way.*

TOGO

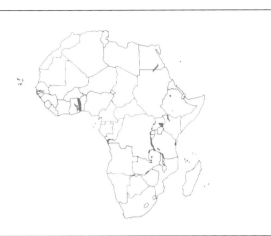

Geopolitical summary

Official name	République Togolaise
Area	21,919 mi^2 [56,785 km^2]
Population	2,700,982 (1981 census); 2,970,000 (1984 estimate)
Form of government	Presidential republic with a one-party National Assembly; the head of government serves a 7-year term, and the Assembly is elected every 5 years.
Administrative structure	5 regions, divided into 21 prefectures
Capital	Lomé (pop. 400,000, 1984 estimate)
International relations	Member of UN, OAU, *Conseil de l'entente*; associate member of EC
Official language	French; Sudanese languages commonly spoken
Religion	Predominantly animist; Catholic 21%, Muslim 17%
Currency	CFA franc

Natural environment. Togo consists of a narrow corridor extending approximately 375 mi [600 km] north–south and about 60 mi [100 km] wide, which faces the Atlantic Ocean (Gulf of Guinea) to the south along only 50 mi [80 km] of coastline; it is bordered on the west by Ghana, on the north by Burkina Faso, and on the east by Benin.

The coastal plain, dotted with lagoons and lakes (including Lake Togo), is covered by recent alluvial deposits that rest on a sedimentary layer (limestone) laid down during the Cretaceous and Eocene. Farther inland, the topography is modeled onto crystalline rocks (granite, gneiss, micaceous schist, quartzite), which form the framework of the hills crossing the country from north to south. These consist largely of an extension of the Togo Mountains (3444 ft [1050 m]), dropping down to a low plateau together with the Fazao Mountains in the central region, and the slopes of the Atakora at the border with northern Benin.

The hydrographic system consists of watercourses that descend mostly to the south, emptying directly into the ocean. The principal rivers are the Mono, the last 60 mi [100 km] of which marks the border with Benin to the southeast, and the Oti, a tributary of the Volta, which defines part of the border with Ghana to the northwest.

The climate is subequatorial on the coast, subject to the effect of the southwest monsoon (a moist summer wind from the ocean that brings rain), and the northeast wind, the winter *harmattan* that brings dry weather. Precipitation varies from moderate at Lomé (34 in. [875 mm]), to slightly higher readings in the foothills of the Togo Mountains (78 in. [2000 mm]), which are more exposed to moist ocean air masses.

The plant cover consists of equatorial forest on the coast and in the central mountainous areas, and wooded savanna, with baobab and acacia trees, on the northern lowlands. In the central region, wild animals are protected in the Malfakasso forest nature preserve and those of Fazao and Koné, which contain elephants, buffalo, antelopes, primates, and birds. There are also protected areas of savanna—the La Kéran reserve and the "lion pit" forest—and the Togado forest reserve, with abundant warthogs and monkeys, which extends along the banks of the Mono river and also contains crocodiles and hippopotamuses.

Population. Togo is inhabited by a mosaic of peoples of various ethnic groups, comprising some forty mutually hostile tribes with no national identity. In earlier times the area was subject to the Ashanti and the Abomey, then suffered from the slave trade practiced along the coast (the "Slave Coast") by various local chiefs on behalf of the Europeans. Predominant in the south are the Ewe, who live in agricultural villages, and have to some extent absorbed Western culture. More impoverished peoples live in the north: the Tem (animists), and the Peul, a Muslim group settled in villages of circular huts with conical roofs.

When it gained independence in 1960, Togo had fewer than 1.5 million people; with an annual average growth rate of 3%, its population has now more than doubled. The density at present is 135 per mi^2 [52 per km^2]; the most densely populated areas lie along the coast (where 38% of the population lives) and where the density can reach 520 per mi^2 [200 per km^2]. The only urban area is the capital, Lomé, which has some 400,000 inhabitants; other centers are much smaller (10,000–30,000 population). Lomé, situated near the ocean, embodies the entire modern life of Togo; it is frequently the site of international conferences—the conference of the Economic Community of West African States (ECOWAS) was held here in 1976—and has several international-class hotels and a modern administrative center. Farther east on the coast is historic Aného, the first capital of Togo during the colonial period. Other cities include Sokodé,

Climate data				
Location	**Altitude (ft asl)**	**Average temp. (°F) January**	**Average temp. (°F) July**	**Average annual precip. (in.)**
Lomé	16	81	76	34.1
Sokodé	1,371	78	76	54.6
Conversion factors: 1 ft = 0.3 m; 1 in. = 25 mm; °C = (°F − 32) × 5/9				

Administrative structure

Administrative unit (regions)	Area (mi²)	Population (1984 estimate)
Centrale	5,088	310,000
De La Kara	4,489	445,000
Maritime	2,469	1,148,000
Des Plateaux	6,552	708,000
Des Savanes	3,320	359,000
TOGO	21,919	2,970,000

Capital: Lomé, pop. 400,000 (1984 estimate)
Major cities (1981 census): Sokodé, pop. 48,098; Kpalimé, pop. 31,800; Atakpamé, pop. 27,100; Bassar, pop. 21,800.

Conversion factor: 1 mi² = 2.59 km²

chief town of the Centrale region, and Atakpamé in the Plateaux region. The population has an extremely high birth rate (4.9%), with a natural growth rate of 3.5%. The illiteracy rate is 60.9%. With a GNP of US$410 Togo is certainly not a rich country, but the policy of national unity and economic development instituted by the government has created social services (schools, hospitals, etc.) that guarantee a decent life for the people.

Economic summary. Togo is an agricultural nation, with about 70% of the working population employed in farming, which uses 85.2% of the territory, leaving 14.8% uncultivated and unproductive. Except during the 1970s, this sector provided enough food for national self-sufficiency. Export crops include coffee, cocoa, cotton, coconut and oil palms, peanuts, sugar cane, and fruits and vegetables; internal needs are met by corn, rice, manioc, and low-grade cereals. Cattle and sheep are raised for domestic consumption, with some cattle and fewer than a million sheep; the fishing industry is developing steadily.

The principal mineral resource is phosphates (3,685,000 t in 1989) at Hahotoé, which represent about half of the value of Togo's exports. Industrial activity is based on the conversion of agricultural products (cocoa, cotton, palm nuts); an industrial zone has developed around the port of Lomé, including a cement plant and a petroleum refinery.

The balance of trade is decidedly negative due to falling prices for phosphates; other important export products include cocoa (26.1%), coffee (9%), and diamonds (2%), while the most signifi-

Socioeconomic data

Income (per capita, US$)	390 (1989)
Population growth rate (% per year)	4.5 (1985–90)
Birth rate (annual, per 1,000 pop.)	50 (1988)
Mortality rate (annual, per 1,000 pop.)	14 (1988)
Life expectancy at birth (years)	53 (1985–90)
Urban population (% of total)	25 (1988)
Economically active population (% of total)	50 (1988)
Illiteracy (% of total pop.)	56.7 (1990)
Available nutrition (daily calories per capita)	2207 (1988)
Energy consumption (10^6 tons coal equivalent)	164 (1987)

cant imports are machinery, fuels, various industrial products, and foodstuffs. Plantation produce and local craft products are traded at markets organized by the Ewe.

The road network consisted in 1986 of more than 5000 mi [8000 km] (of which 930 mi [1500 km] are paved) and Togo has 326 mi [525 km] of railroad (1985); the main airport is at Tokoin near Lomé. Foreign tourists numbered 126,000 in 1989, with 50% coming from Europe.

Historical and cultural profile. Originally inhabited by Paleoafrican tribes practicing agriculture or nomadic herding, the region experienced some embryonic and ephemeral forms of state organization in the form of the principalities created in the north by the Mossi, aristocratic Sudanese horsemen who had arrived in the area from the east in the 11th century. The southern region was settled in the 14th century by the Ewe, who merged with the local tribes but never established a centralized government. The European arrival had a highly disruptive effect, plunging the region into anarchy; in the end, the resources offered by Togo were acquired by a number of European states that often were at odds with each other. The trade in slaves and palm oil was divided up between the Portuguese and the Danes. Then Germany made its move, in the person of the explorer Gustav Nachtigal. In 1882, an advantageous treaty was signed with the chief of the coastal tribes, in exchange for payment of a modest tribute. Two years later the Germans imposed a form of protectorate on the Ewe, but their penetration into the interior was blocked by the Dagomba, Kukuka, and Kabré tribes.

After World War I, during which Togo had been occupied by the French and British, the League of Nations divided the country into two lengthwise strips: the western part went to Great Britain and was annexed to Ghana, while the eastern strip was assigned to France, and was granted independence along with the other French possessions in 1960. The new republic did not prove stable and in January 1963 Togo gained the unenviable distinction of initiating the series of military coups that characterized newly independent Africa. In that year, a military takeover ousted president Sylvanus Olympio, and in subsequent elections Nicholas Grunitzsky was elevated to the office. In 1976 Colonel Gnassingbé Eyadema seized power, dissolved all existing political groups, and replaced them with a single party, the Rassemblement du Peuple Togolais (Togo People's Union). Re-elected president in 1979 and 1986, Eyadema promoted reprivatization of businesses, and followed a policy of rapprochement between French-speaking and English-speaking Africa. A movement to restore democracy began to gain strength in the country, however, and in 1991 Eyadema was deposed by a "High Council of the Republic," a transitional body created to lead Togo toward a multiparty system.

Literary works, transmitted either orally or in writing, have been created in the various languages of the Kwe and Volta groups. The first written text (on religious topics) in the Gemina language dates back to 1658, while the German colonial rulers collected and studied a number of works in Ewe, including those by the poet Duho and poetess Dzemavo. Very recently a Togolese literature in French has begun to develop, with no real links to the cultural movements of French-speaking western Africa. Togo's most famous writer is David Ananou, author of *Le fils du fétiche* [Son of the fetish] (1955).

WESTERN AFRICA

Images

1. *The statue in the monument erected at Lomé to commemorate the independence of Togo, which was proclaimed on April 27, 1960. Lomé was built quite recently, and owes its development to the construction of a harbor on the northern shore of the Gulf of Guinea, linked to the rail system and therefore capable of transshipping cargo from the interior. The city also enjoys considerable political prestige, having played host to conferences of the United Nations, the OAU, and the* Conseil de l'entente, *of which Togo is a member.*

2. *The great Fouta Djallon massif (Guinea), much of it situated at elevations above 3300 ft [1000 m], consists of a foundation of igneous and metamorphic rocks; in its western portion, this is covered by tablelands of Paleozoic sandstone and dolerite, elevated by tectonic forces and intersected by faults and magma intrusions. It is also an important hydrographic node, since the abundant rainfall on its Atlantic slopes spawns numerous rivers such as the Gambia and several spring-fed branches of the Senegal.*

3. *São Vicente island in the Cape Verde archipelago. Cape Verde, an island nation in the Atlantic Ocean some 300 mi [500 km] from the Cape Verde promontory in Senegal, con-sists of fifteen volcanic islands made of Quaternary basalt. São Vicente is one of the most active economically, containing the country's most important harbor, Porto Grande.*

4. *The Gambia river, 698 mi [1126 km] long, rises in the Fouta Djallon massif in Guinea, passes through Guinea and Senegal, then crosses the entire length of the nation called The Gambia, emptying into the Atlantic Ocean through a wide estuary. It is the only communication route between the interior and the coast, so all of The Gambia's commercial activities are concentrated along the river's alluvial valley, characterized by dense shrubby vegetation that becomes even thicker toward the interior.*

5. *The immense Niger delta area (Nigeria) is covered by a rich, dense rain forest that thrives on the region's plentiful rainfall. The Niger, which rises on the eastern slope of the Fouta Djallon in Guinea and runs northeast to the Atlantic Ocean, is subject to periodic floods twice a year; both are produced by the summer zenithal rains that fall simultaneously on the upper course of the river (at which time it is called the "great river") and on its lower course.*

6. *The southern Casamance region of Senegal, through which the Casamance river flows, is covered by typically equatorial vegetation that flourishes in a fairly humid climate with a rather long rainy season. Unlike surrounding areas, it contains basalt formations produced by eruptive activity associated with the formation of the river basin, which continued into the Quaternary period.*

7. *A Mandingo man, member of a group of Niger/Senegalese populations that speak a Mande language and live in a large area of western Africa including Guinea, Senegal, The Gambia, Mali, and Ivory Coast. Skilled farmers, herders, and metal-workers, the Mandingo are subdivided into several tribes, some of which are quite numerous and can look back on a proud history as the creators of great empires: Ghana (4th–8th centuries) and Mali (11th–17th centuries).*

8. *The Hausa ethnic group, settled in northwestern Nigeria and southern Niger, are a people of Sudanese origin with certain Berber characteristics, and number some 7 million in all. The most glorious period of their history began in the 10th century, when the Arab leader Abu Yaziad united several small states under his rule and converted them to Islam,* creating a powerful federation that ultimately, in the late 17th century, gave birth to a huge empire.*

9. *Among the Akan, an ancient African people whose traditions and customs constitute the cultural and artistic mainstream in Ghana, each house accommodates an extended family group as well as facilities for making pottery, the most widespread craft product. The complex consists of a large square or rectangular building, closed off on the outside by a perimeter wall with a single entrance.*

10. *A fortified village of the Mossi in Burkina Faso. The Mossi are one of the most important populations within the Volta group, in terms of both numbers and their well-established social and political structure, which is based on monarchical and patriarchal traditions and still survives almost unchanged. Their villages, called* yiri, *are groups of small round houses with conical straw roofs, along with typical egg-shaped granaries, made of wickerwork and held up with a few sticks pounded into the ground.*

11. *A nighttime view of the modern section of Abidjan. In the foreground is the Cathedral, an imposing building covering almost 50,000 ft^2*

[4500 m²], with twin spires 230 ft [70 m] tall. Capital of Ivory Coast since 1960, Abidjan is considered the second most important economic and cultural center (after Dakar) in French-speaking western Africa. A rapid increase in exports, and completion of a railroad system with Abidjan as its terminus, have led to the construction of a modern port, protected from silting by a deep underwater channel.

12. Freetown, a port city on the north side of the Sierra Leone peninsula, is also the capital of the Republic of Sierra Leone. One of the best harbors in western Africa, it is an important naval base and a commercial port for the export of agricultural products and valuable minerals. As the name recalls, Freetown was established in 1778 as a home for freed African slaves repatriated from America, and it has retained the appearance of a colonial American city.

13. The modern mosque at Kano (Nigeria), a Muslim city that in the 16th century was the capital of one of the seven Hausa states. Founded in the 9th century on the site of an older fortress, it is a market center that developed at the time when trans-Sahara trade flourished. Reminders of its origin include typical mud houses with inner courtyards, whose layout and external appearance recall the Islamic urban environment of Sahelian Africa.

14. The city of Monrovia, present-day capital of Liberia, stretches along the Mesurado peninsula on the banks of the river of the same name, and faces directly onto the Atlantic. It was founded in 1822 by the American missionary Jehudi Ashmun, who chose the name Monrovia to honor the American president James Monroe, a supporter of freedom for black African slaves.

15. The principal source of income for Senegal is peanuts, produced on shrubby plants that are cultivated especially in Siné Saloum near Kaolack. Initially held back by competition from India, this crop regained importance after the opening of the Suez canal. At the beginning of this century a railroad was constructed between Saint-Louis and Dakar, making the peanut crop easier to transport.

16. The cattle market at Dakar. Senegal has great potential in terms of livestock, with its extensive Sahel pastures that are used to graze cattle, goats, and camels.

17. Although Togo is a predominantly agricultural country, fishing is one of its major sources of subsistence despite primitive equipment. One factor impeding the development of this and other sectors of the national economy is an inadequate communications network, which hampers both internal trade and exports.

18. In addition to the agricultural products commonly found in many western African countries—coffee, cocoa, copra, cotton, pineapples, and cereals—the market at Abidjan (Ivory Coast) offers some unique local specialties: hardwoods prized by cabinet-makers, and many food plants grown in the southern part of the country, such as bananas, manioc, sweet potatoes, and yams.

19. The city of Kano (Nigeria) is the cradle of a very special activity, fabric dyeing, in which the dyes are prepared and heated in vats dug into the ground. The typical conical straw parasols are used to protect both the dyer and the dye-pots from the broiling sun. The quality of the product is so high that Tuareg horsemen come all the way from the Ahaggar Mountains in southern Algeria to buy the indigo-dyed fabric for their typical costume.

20. A sulfur extraction and processing plant near Dakar. The Senegalese capital is the largest industrial and commercial center in western Africa. Industry in general is slowly progressing, especially in the light-industrial sector and with the exploitation of agricultural products and mineral resources. One of the principal obstacles to rapid expansion is a lack of sufficient electric energy to meet the nation's needs.

21. Like other tropical countries, Guinea has substantial mineral resources. A particularly important ore is bauxite, of which this country is one of the world's leading producers. The only industrial sector of any significance is therefore metallurgy, including a large installation at Fria where bauxite is processed and smelted into aluminum.

22. A diamond prospector at work in the fields of Tortiya, site of a major deposit in Ivory Coast. The vigorous economic development experienced by this country has resulted not only from favorable natural conditions, but also from its political stability and the presence of massive foreign investment which supports the necessary industrial infrastructure.

23. A 17th-century bronze relief panel depicting a drummer, found at Benin (Nigeria) and now in the British Museum in London. Around the 15th century, the city of Benin became the capital of the Kingdom of Benin, as well as Nigeria's most important artistic center. Local art reflected the kingdom's social organization—centered on the person of the oba, the absolute monarch—and was exclusively a vehicle for royal glorification.

24. The polychrome bas-reliefs on the royal palace at Abomey (Benin) depict and exalt the warlike exploits of the king and the nobles who lived in Dahomey at the beginning of the 1600s, the century in which this kingdom attained its greatest prosperity. The reliefs are made of sun-baked clay and painted in vivid colors, telling of pivotal battles or illustrating proverbs and sayings, often associated with the figure of a courtier. The doorways, in particular, are decorated with carved wooden animal figures representing the power and strength of a sovereign, like the red buffalo that is the symbol of King Gezo.

25. One of the bas-reliefs from the royal palace at Abomey, depicting a lion. Artists and sculptors worked in clay, iron, wood, ivory, and precious metals under commission to the king or the nobility, whose deeds and greatness they were expected to glorify.

26. A Fulbe man in warrior dress. Nigeria is one of the western African countries that has retained a great deal of its traditional heritage, including periodic celebrations such as the durbar shown here, and the costumes worn by the many local ethnic groups.

27. A scene at the durbar in Ketsina (Nigeria). For the descendants of the ancient Fulbe, an ethnic group present in western Africa since the 15th century, a durbar is an occasion to celebrate the glorious battles of the past: in the early 19th century, several groups of Islamized warriors conquered the Hausa kingdom and established the Muslim empire of Sokoto, subdivided into many smaller states including Ketsina, which survived until the colonial takeover.

28. The social structure of Ivory Coast includes seven large ethnic groups distributed over four geographical areas, divided in turn into 59 tribes that speak 22 different languages. This intricate mosaic of peoples has many ceremonies, festivals, and folk practices, blending music, dance, and the use of masks and face paint. Celebrations are connected with family or village life, and are therefore linked to the seasons, the harvest, magic, and religious observances.

3

4

7

8

9

23

EQUATORIAL AFRICA

In the more than 2200 mi [3500 km] that separate the Atlantic coast at the mouth of the Ogooué from the Indian Ocean at the mouth of the Juba, the equator passes through an extraordinary range of environments, both ethnic and natural. To the west, the Atlantic washes a flat and marshy coast covered by dense evergreen forest, which dips into the ocean's water in the mangrove swamps. Then the topography changes, rising farther up the course of the Ogooué to the soft undulations of the crystalline plateau (a strip of the pre-Paleozoic continental basement) of Gabon, which culminates at an elevation of over 3300 ft [1000 m], still covered by the thick mantle of the rainforest. Here, again dominated by the forest, one finds the great depression housing the central section of the Congo basin: this is the true heart of black Africa, homeland of the Bantu peoples. Here, deep in the forest, in shelters made of sticks and leaves, live the Pygmies, feared by others and proud of the lethal poison of their blowguns. In this environment, where Stanley made his bold expeditions in search of Emin Pasha, rivers are still the primary communication route, rivaled only by the airplane, which constitutes the fastest and safest means of transport all over Africa.

The old Congo (now Zaire), encountered by the Portuguese in the 15th century, is only the second longest African river (2600 mi [4200 km]), but is the largest in terms of its drainage basin (1.43 million mi^2 [3.7 million km^2]), and is distinguished by its dense network of tributaries, on both the right bank (Sangha, Ubangi) and the left (Kasai, Busira, Lomami). Farther east, the Congo basin is bordered by the high crystalline and volcanic elevations (from Ruwenzori to Nyiragongo) that overlook the deepest of the eastern African tectonic depressions, partly filled by the great lakes (from Lake Albert to Lake Tanganyika) surrounding that veritable inland sea that is Lake Victoria (third largest in the world after the Caspian Sea and Lake Superior), with a surface area of some 27,000 mi^2 [70,000 km^2].

Between the two eastern African tectonic depressions—that of Lake Tanganyika and the Rift Valley—stretches the Africa of the high plateaus, with its savannas and park-like forests, where the excesses of the equatorial climate are tempered by altitude, and human settlement finds more favorable conditions. The contrast between this region of central Africa and the Congo basin is even more accentuated by the presence of the highest elevations on the continent, placed around the shores of Lake Victoria like majestic sentinels: Ruwenzori and Karisimbi to the west, Elgon, Kenya, Meru, and Kilimanjaro to the east. All are over 13,000–16,000 ft [4000–5000 m] high, and some, like Ruwenzori, Kenya, and Kilimanjaro, are even covered with snow and ice, another particularly striking contrast given the equatorial latitude. The great variations in elevations do indeed produce considerable changes in vegetation over short distances, particularly evident when climbing the slopes of these titans: from the arid steppes and grasslands of the Masai plateau, one passes first through a luxuriant tropical forest and alpine meadows, and then into a genuine tundra that precedes the snow-covered environment at the peak.

There are enormous contrasts not only in the physical context, but also in human terms. Although the entire area of the Congo basin is inhabited predominantly by Bantu tribes (like the Ba-Luba and BaLolo), peoples of Sudanese origin (like the Dinka) and Nilotic races (like the Masai) appear in the more easterly regions. Lastly, Hamitic infiltrations are also present especially in areas near Somalia and Ethiopia, while a fairly substantial Asian presence is evident on the shores of the Indian Ocean.

As everywhere else in Africa, the present-day geopolitical structure is a legacy of the subdivisions imposed during the colonial era, which has also left a linguistic imprint: a distinction can still be made between francophone nations (Zaire, Congo, Cameroon, Central African Republic, Rwanda, and Burundi), and English-speaking countries (Uganda, Kenya, Tanzania, and Zambia). Even the economy remains closely linked to the dispensations of the former colonial power, meaning plantation agriculture and mineral exploitation, with a high proportion of raw materials exports to western European countries. Industrialization is still sparse, despite possibilities for local energy production (especially hydroelectric). Transportation infrastructure also still seems to depend on the logistical needs of the main areas of economic exploitation, like the old railroad linking the mining districts of Katanga and Kasai to the Angolan port of Benguela. The total population living in this area of some 2.3 million mi^2 [6 million km^2] (one fifth of the continent) is barely 125 million people, with a density of 52 per mi^2 [20 per km^2].

BURUNDI

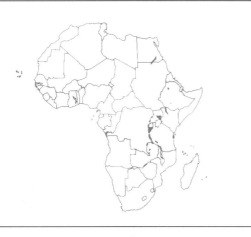

Geopolitical summary

Official name	République du Burundi / Republika y'Uburundi
Area	10,744 mi² [27,834 km²]
Population	4,782,406 (1986 census); 1990 estimate: 5,382,459
Form of government	Presidential republic governed by the military; the National Assembly is elected at the direction of the single party.
Administrative structure	15 provinces divided into 114 districts
Capital	Bujumbura (pop. 272,600, 1986 census)
International relations	Member of UN and OAU; associate member of EC
Official language	French and Kirundi (Bantu); Swahili is used in commerce
Religion	One-third animist, two-thirds Christian (mostly Catholic); 1% Muslim
Currency	Burundi franc

Natural environment. Burundi is bordered on the north by Rwanda, on the east and south by Tanzania, and on the west by Zaire.

Like neighboring Rwanda, Burundi is distinguished by its mountainous terrain and high average elevation. The country comprises at least three identifiable morphological regions: a section of the Rift Valley depression, through which the Ruzizi river runs and in which Lake Tanganyika is located; a narrow, elongated mountain range running north–south, constituting the eastern pillar of the Rift Valley and rising to about 6000 ft [1800 m]; and a series of high plateaus stepping down toward Lake Victoria. Every drop of water that comes down from the western slopes of the mountains flows via the Ruzizi (either directly or through the Malagarasi river) into Lake Tanganyika,

and from there via the Lukuga river into the Congo. From the eastern slope, the watercourses flow into Lake Victoria and thence into the Nile via the Ruvubu river, one of the spring-fed branches of the Kagera.

The search for the sources of the Nile, one of the most fascinating chapters in the history of modern exploration, was brought to a conclusion by Grant, Speke, and Baker in 1863:

> *[July 18th] ... as it appeared all-important to communicate quickly with Petherick, and as Grant's leg was considered too weak for traveling fast, we took counsel together and altered our plans. I arranged that Grant should go to Kamrasi's direct with the property, cattle, and women, taking my letters and a map for immediate dispatch to Petherick at Gani, while I should go up the river to its source or exit from the lake, and come down again navigating as far as practicable....*
>
> *We started all together on our respective journeys; but, after the third mile, Grant turned west, to join the high road to Kamrasi's, while I went east for Urondogani, crossing the Luajerri, a huge rush-drain three miles broad, fordable nearly to the right bank, where we had to ferry in boats, and the cows to be swum over with men holding on to their tails. It was larger than the Katonga, and more tedious to cross, for it took no less than four hours, musquitoes in myriads biting our bare backs and legs all the while. The Luajerri is said to rise in the lake and fall into the Nile due south of our crossing-point. On the right bank wild buffalo are described to be as numerous as cows, but we did not see any, though the country is covered with a most inviting jungle for sport, with intermediate lays of fine grazing grass. Such is the nature of the country all the way to Urondogani, except in some favored spots, kept as tidily as in any part of Uganda, where plantains grow in the utmost luxuriance....*
>
> *[July 21st—Urondogani.] Here at last I stood on the brink of the Nile. Most beautiful was the scene; nothing could surpass it! It was the very perfection of the kind of effect aimed at in a highly-kept park; with a magnificent stream from 600 to 700 yards wide, dotted with islets and rocks, the former occupied by fishermen's huts, the latter by sterns and crocodiles basking in the sun, flowing between fine high grassy banks, with rich trees and plantains in the background, where herds of the n'sunnu and hartebeest could be seen grazing, while the hippopotami were snorting in the water, and florikan and Guinea-fowl rising at our feet. Unfortunately, the chief district officer, Mlondo, was from home, but we took possession of his huts—clean, extensive, and tidily kept—facing the river, and felt as if a residence here would do one good....*
>
> *We were now confronting Usoga, a country which may be said to be the very counterpart of Uganda in its richness and beauty. Here the people use such huge iron-headed spears with short handles, that, on seeing one to-day, my people remarked that they were better fitted for digging potatoes than piercing men. Elephants ... were very numerous in this neighborhood.... Lions were also described as very numerous and destructive to human life. Antelopes were common in the jungle, and the hippopotami, though frequenters of the plantain garden and constantly heard, were seldom seen on land in consequence of their unsteady habits.*

The equatorial climate is mitigated both by altitude and by the presence of the great lakes. Precipitation is distributed into two seasons—March to May and October to December—and while it is less than 40 in. [1000 mm] on Lake Tanganyika and in the Ruzizi valley, it exceeds 55 in. [1400 mm] on the high plateaus.

Climate data				
Location	**Altitude (ft asl)**	**Average temp. (°F) January**	**Average temp. (°F) July**	**Average annual precip. (in.)**
Bujumbura	2575	73	73	32.5

Conversion factors: 1 ft = 0.3 m; 1 in. = 25 mm; °C = (°F – 32) × 5/9

Average monthly temperatures are mild considering the latitude, ranging between 73° and 77°F [23–25°C]; on the high plateau and on higher elevations, they do not exceed 68°F [20°C]. The dominant plant community consists of a semi-arid wooded savanna: only on mountain slopes does evergreen forest appear, with a thick understory that gradually changes into mountain meadows at higher elevations. The lake basins are fringed with papyrus and other aquatic plants. The typical fauna, comprising the large herbivores and carnivores of the savanna (antelope, zebra, giraffe, elephant, rhinoceros, lion, and tiger) does not receive any particular protection.

Population. The ethnic groups making up the population are the BaHutu (85%), agriculturists of Bantu stock; the BaTutsi (14%), predominantly herders; and the BaTwa (1%).

Total population according to a 1990 estimate is about 5.5 million, with an annual growth rate of 2.9%. Average population density is 536 per mi^2 [207 per km^2], rather high compared to other African nations. The most densely populated areas are the north-central high plateaus, where the climate is more temperate.

The population is 95% rural; only 5% of Burundians live in the urban centers of Bujumbura and Gitega. Bujumbura, the capital, located on Lake Tanganyika at an elevation of 2560 ft [780 m], is a commercial port active primarily in the export of coffee and cotton; the city has a university founded by the Jesuits in 1964. Gitega, located on the central high plateau, is an important crossroads and a trading center for hides, coffee, and agricultural produce.

With an annual per capita income of US$210, Burundi is considered one of the twenty poorest countries in the world: this situation is maintained by its very high population density and a growth rate that outstrips the nation's food resources. In some areas hunger and diseases caused by malnutrition claim many victims.

Economic summary. Burundi, classified as one of the world's least developed countries, receives economic aid from Belgium, France, and Germany, and from the EC Development Fund and the African Development Fund.

Agriculture, which employs 56% of the work force and supplies 55% of GNP, centers on plantation products, especially coffee (41,800 t in 1991), cotton, and tea. Minor crops include tobacco, rice, and sugar cane. Subsistence agriculture, which is insufficient for domestic needs, produces manioc, sweet potatoes, bananas, and low-grade cereals. Stockraising is very widely practiced (450,000 head of cattle, and 1,268,000 sheep and goats in 1990), but yields poor-quality animals. Lake Tanganyika is a very rich source of fish, with about 12,870 t caught each year (1989).

Mineral production is almost nonexistent; limited quantities of kaolin are extracted, along with tiny amounts of gold and cassiterite. Industrial activity, which employs only 15% of the work force and contributes 15% of GNP, is concentrated around Bujumbura, with plants for processing coffee, cotton, tea, and oil-bearing seeds. Other industries include textile mills, shoe factories, and a few cement plants.

Commercial activities are impeded by inadequate communications: the road network, covering some 3658 mi [5900 km] (1988), is only partly paved. Lake Tanganyika is used as a navigation route to transport goods to the terminus of Zaire's railroad. The capital has a harbor and an international airport. The balance of trade is negative: exports of coffee, cotton, and tea are not sufficient to cover even half the cost of imported goods (machinery, petroleum products, and food).

Historical outline. The kingdom of Burundi arose along the banks of Lake Tanganyika between the 15th and 16th centuries, after the original peasant population of the area, the BaHutu, had been deprived of their land and reduced to slavery by the BaTutsi warrior aristocracy, who had migrated earlier from the region of the Upper Nile. This feudal structure helped support the colonial domination that began at the end of the 19th century, when Burundi (then called Urundi) was incorporated, along with Uganda and Tanganyika, into German East Africa.

After the end of World War I, the territory of Ruanda-Urundi was transferred as a mandate to Belgium, which governed it through a general resident associated with the Congolese colonial system; nevertheless, no formal restrictions were placed on the traditional institution of monarchy, centered on the *mwami* of the BaTutsi caste. After World War II, the United Nations left the colony under the trusteeship of Belgium, and Burundi maintained its monarchical form of government even

Administrative structure		
Administrative unit (provinces)	**Area (mi^2)**	**Population (1990 estimate)**
Bubanza	422	224.652
Bujumbura	515	647,173
Bururi	971	406,803
Cankuzo	749	140,787
Cibitoke	633	263,855
Gitega	768	608.398
Karuzi	563	283,419
Kayanza	474	479,084
Kirundo	660	395,481
Makamba	761	174,501
Muramvya	591	468,366
Muyinga	704	344,309
Ngozi	567	519,142
Rutana	733	199,239
Ruyigi	913	227,250
BURUNDI	10,023(*)	5,382,459

(*) Excluding 721 mi^2 of inland waters
Capital: Bujumbura, pop. 272,600 (1986 census)
Major cities (1986 census): Gitega, pop. 95,300; Muyinga, pop. 18,458; Bubanza, pop. 12,394.
Conversion factor: 1 mi^2 = 2.59 km^2

Socioeconomic data

Income (per capita, US$)	210 (1990)
Population growth rate (% per year)	2.9 (1980-89)
Birth rate (annual, per 1,000 pop.)	48 (1989)
Mortality rate (annual, per 1,000 pop.)	15 (1989)
Life expectancy at birth (years)	49 (1989)
Urban population (% of total)	5 (1989)
Economically active population (% of total)	52.3 (1988)
Illiteracy (% of total pop.)	50 (1991)
Available nutrition (daily calories per capita)	2253 (1989)
Energy consumption (10^6 tons coal equivalent)	0.09 (1987)

after independence (1962). This was followed by years of disputes and border squabbling with neighboring Rwanda, which had become a republic, and especially by a period of ethnic conflicts, with periodic rebellions by the BaHutu population that were then brutally repressed by the BaTutsi. It has been calculated that more than 100,000 people were slain in 1972, in just one such internecine incident. Also murdered in the course of this massacre were many representatives of an intellectual elite that was beginning to take shape in Burundi, establishing itself as a meeting ground between African and Western cultures, and between traditional values and new ones. One of the most important of these Burundian intellectuals, another victim of the massacres at Gritenga in 1972, was Michel Kayoya, author of a history of Burundi titled *Sur les traces de mon père* [In my father's footsteps], a book of sorrowful reflections on the fate of his people, from which the following passage is taken:

> *There is no time to lose*
> *But let us not forget that,*
> *Like everything that grows, people need limits*
> *The wild fruit tree is one that has neither garden nor*
>   *gardener;*
> *Its fruits finally degenerate,*
> *Become even wilder.*
> *Thus our work must always be to remain within our limits.*
>
> *After one colonization perhaps we are about to suffer another*
> *Another, more terrible one,*
> *A colonization made of the baseness that lurks within each*
>   *heart*
> *Of sloth and pride,*
> *Burdens that weigh on our souls and stop us from growing*
> *The struggle for liberation becomes a struggle between*
>   *brothers who devour each other.*

Changes have since taken place on the political scene (changeover to a republic following the coup d'état in 1966, another coup in 1976, and promulgation of a second constitution in 1981), but they have not eroded the power of the BaTutsi, especially the Hima group who still hold key positions in the army and in the single political party (Unité pour le Progrès National [UPRONA]). The country was ravaged by new outbreaks of violence in 1988.

CAMEROON

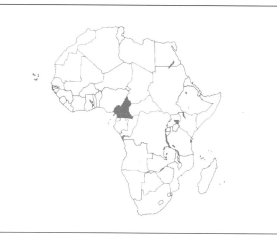

Geopolitical summary

Official name	République du Cameroun / Republic of Cameroon
Area	183,521 mi² [475,442 km²]
Population	7,131,833 (1976 census); 1990 estimate: 11,540,000
Form of government	Presidential republic; the head of government and the National Assembly are elected by universal suffrage for 5-year terms.
Administrative structure	10 provinces, subdivided into 49 departments
Capital	Yaoundé (pop. 653,700, 1986 estimate)
International relations	Member of UN and OAU; associate member of EC
Official language	French and English; Bantu and Sudanese languages are widely spoken
Religion	Animist 40%, Muslim 22%, Catholic 21%, Protestant 15%
Currency	CFA franc

Natural environment. Cameroon is easily identifiable among the nations of tropical Africa by the extremely irregular outline of its borders and by its great length, often constricted down to a narrow corridor, stretching between the banks of Lake Chad and the Gulf of Guinea (which it faces along a low and marshy coastline). As a result, the country exhibits an enormous diversity of environments, in terms of geomorphology, biology, and climate. Cameroon is bordered on the north by Nigeria and Chad, on the east by the Central African Republic, on the south by Congo, Gabon, and Equatorial Guinea, and on the west by the Gulf of Guinea.

Geological structure and relief. The south central region of the country, physically the largest, contains most of the Sanaga river basin as well the crystalline massif of Adamawa (with elevations of about 6500 ft [2000 m]), which defines the Sanaga basin to the north, separating it from the Niger basin and Lake Chad. Toward the southwest, the Adamawa uplands lead into a major volcanic ridge that includes Mt. Oku, 9866 ft [3008 m], while toward the south, the tributaries of the Sanaga basin cut into a series of low plateaus that mark the watershed with the Congo and Ogooué basins.

Geologically, Cameroon's territory consists of outcrops of Precambrian crystalline basement rock, variously dislocated and crisscrossed by deep fractures that have allowed the emergence of basaltic magmas and the formation of volcanic structures of significant size. Among these, on the shores of the Gulf of Guinea, rises the isolated mass of Mt. Cameroon, a 13,353-ft [4070-m] gigantic volcano that first became active in the Cretaceous period. Activity continued, with long quiet intervals, into the lower Tertiary and the Miocene, as well as the entire Quaternary. The last great eruptions occurred in 1909 and 1922.

North of the Adamawa massif, the country includes some of the upper basin of the Bénoué, the major left-bank tributary of the Niger, and further north, beyond the small and arid volcanic massif of the Mandara Mountains (4730 ft [1442 m]) stretches the western part of the alluvial plain of the Chari, ending at the marshy shores of Lake Chad.

Hydrography. Because of its specific geographic situation and the fact that it is traversed by watersheds of continental significance, Cameroon possesses a rather disarticulated hydrographic system, in which the Adamawa massif and the volcanic ridge constitute reference points. The only large waterway that flows entirely within the nation's borders is the Sanaga, some 322 mi [520 km] long, which flows into the Atlantic (Bight of Biafra) through a broad delta. The stretch of coastline north of Mt. Cameroon represents the extreme western limit of the huge delta formed by the Niger. In addition to Lake Chad, the largest body of water in Sahelian Africa (although now greatly reduced in area), the southern portion of which belongs to Cameroon, the country also contains numerous crater lakes and several artificial reservoirs of modest size, located in the high valleys of the Bénoué (Lagdo dam) and the Sanaga (M'Bakaou dam), and on the volcanic ridge.

Climate. Climatic conditions are quite varied, but a general distinction can be made between an arid Cameroon north of the Adamawa, and a wet Cameroon south of the massif. It is on the Adamawa itself, and on the volcanic ridge and Mt. Cameroon, that the southeast trade winds drop their rain, raising annual average precipitation to 60–80 in. [1500–2000 mm] and more (the figure for Douala is almost 160 in. [4000 mm]). This same high ground blocks the effects of the *harmattan*—the hot, dry wind from the Sahara—leaving it to rage through the upper Bénoué valley and the Chad depression, where annual precipita-

tion of only 23–31 in. [600–800 mm] is recorded. Temperatures are generally high, with considerable swings in the northern regions; on the volcanic ridges and other elevations, the effects of altitude bring a considerable mitigation in temperature extremes, giving these areas an extremely pleasant climate.

Flora and fauna. Cameroon's vegetation is extremely varied, thanks in large part to the varied climatic conditions. Along the coastal strip the climate is equatorial (hot and humid), favoring the growth of thick mangrove swamps. The regions of the southern high plains, drained by the Sanaga and other brief watercourses (Ntem, Nyong, etc.), are covered by a dense evergreen forest with numerous gigantic trees (acajou mahogany, limba, azobé, etc.), which are also home to an extremely varied fauna, including many primate species, especially chimpanzee and gorilla. Other mammals are also present, including the bongo (an extremely rare antelope), elephant, and panther. North of the Sanaga, the forest thins out and gives way to savanna, although it retains the typical tunnel configuration along the waterways. The savanna, studded with trees and rich in both grasses and shrubs, also covers the Adamawa massif and the volcanic ridge, alternating with cultivated areas and patches of forest. Wild animal populations in these regions—densely settled because of the favorable conditions created by the high-altitude climate—have been considerably reduced. North of the Adamawa in the Bénoué basin, the savanna is still liberally studded with trees, and provides a habitat for an extremely diverse wild fauna characterized by more typical species, for whose protection several national parks and preserves—Faro, Bénoué, Bouba-Ndjidah, etc.—have been established. Farther north, all the way to the shores of Lake Chad where climatic conditions become more arid, is the realm of the dry, treeless savanna and the steppe, home to the large wild animals that are protected in Waza National Park on the banks of the Logone.

Population. Because of its great north–south length, Cameroon is inhabited by a wide variety of peoples. The northern savannas are populated by people of Sudanese stock while near Lake Chad and in the Logone valley live the Choa, seminomadic herders of Arab origin. Also here are the Kotoko, descendants of the ancient Sao, who practice subsistence agriculture, and the Massa and Mousgoum, fishermen and rice farmers with animist traditions who are gradually converting to Islam. The lowlands of the Bénoué basin are inhabited by Fulbe populations, resident here since the early 19th century; fervent supporters of Islam, they are both farmers and stockraisers. Some groups within the Fulbe, like the Bororo, still follow a nomadic way of life. Animist populations, generally referred to as the Kirdi, live in mountainous areas such as the Mandara range, and include a great variety of subgroups; all are characterized, however, by a patriarchal type of family organization and by an intensive style of agriculture that uses all available land, including densely terraced hillsides. Their villages consist of houses made of stone or dried mud bricks. On the central plateaus of the Adamawa Mountains and in the remainder of the country live people who speak Bantu and semi-Bantu languages; some are autochthonous, while others arrived in a succession of migratory waves. Among the latter are the BaMileke, Tikar, and BaMum. The BaMileke, in particular, organized into powerful chiefdoms, represent the most important ethnic group in Cameroon. The south-

Climate data				
Location	Altitude (ft asl)	Average temp. (°F) January	Average temp. (°F) July	Average annual precip. (in.)
Yaoundé	2493	76	73	61.2
Douala	43	79	75	158.5
Conversion factors: 1 ft = 0.3 m; 1 in. = 25 mm; °C = (°F − 32) × 5/9				

Administrative structure

Administrative unit (provinces)	Area (mi²)	Population (1976 census)
Adamaoua	24,241	336,150
Centre	27,059	1,098,680
Est	42,923	342,850
Extrême-Nord	–	448,296
Littoral	8,029	841,456
Nord (Bénoué)	–	
Nord-Ouest	6,875	914,912
Ouest	5,211	968,856
Sud	18,926	294,928
Sud-Ouest	9,446	580,360
CAMEROON	181,497(*)	7,131,833

(*) Excluding 2,023 mi² of inland waters
Capital: Yaoundé, pop. 653,700 (1986 estimate)
Major cities (1981 estimate): Maroua, pop. 81,861; Garoua, pop. 77,856; Bafoussam, pop. 75,832; Douala, pop. 1,029,731 (1986 estimate).

Conversion factor: 1 mi² = 2.59 km²

western regions, dominated by the rainforest, are inhabited by Bantu groups like the Bassa and BaKoko, Djem, and Pahouin; the last are characterized by the persistence of powerful tribal allegiances, and by their linear villages. Several thousand Pygmies still live in the forests of southeastern Cameroon.

The linguistic landscape is also extremely diverse: Sudanese and Bantu languages are often mixed with either Arabic or European languages like French and English (and, earlier than either, with German). As far as religion is concerned, although Islam and Christianity (both Protestant and Catholic) are widespread and officially recognized by the government, animism remains the basis of everyday religious thinking.

The most densely populated areas are in the southwest, comprising the Adamawa Mountain plateaus and the valleys which descend to the coastal plain of the Gulf of Guinea, where the most populous towns such as Bamenda, Foumban, Bafoussam, Kumba, and Nkongsamba are located. Besides the capital, Yaoundé, a city of modern appearance on the high plateau between the Sanaga and Nyong, the other important urban center in Cameroon is Douala, the country's principal commercial port.

In an effort to improve its citizens' social conditions, the government has intensified initiatives in the area of public education; as a result, Cameroon has one of the highest school attendance rates in Africa. A university is located at Yaoundé.

Economic summary. Excellent availability of natural resources, along with a low level of demographic pressure, are among the factors that have allowed balanced development of Cameroon's economy. The government's economic policy has also been based on a planned laissez-faire approach that leaves a great deal of latitude for private enterprise, and attempts to restrict imports while exploiting domestic resources.

The economic foundation of Cameroon—which in recent years has had to confront serious difficulties resulting from excess labor force and fluctuations in the price of certain raw materials such as coffee, cocoa, and petroleum—is agriculture,

which provides the majority of exports. The crops for these exports are grown on an industrial scale: cocoa in the southern regions, coffee in the uplands of the west and northwest, sugar cane northeast of Yaoundé, and cotton in the northern savannas. Rubber trees and oil palms are grown in the coastal and forest areas. Other crops being developed include tea, peanuts, tobacco, pepper, and bananas. The predominant crops in the drier northern regions are cereals such as millet, manioc, corn, and rice, most of which are destined for domestic consumption.

Stockraising, still practiced traditionally (total livestock holdings in 1990 were more than 13 million head of large animals), especially in the Adamawa, is constantly impoverished by epizootic diseases, while fishing, which supplies the basis of the local diet in the regions around Lake Chad and the Logone and Chari rivers, is also carried out by traditional methods.

The latest governmental five-year plans included incentives for development of infrastructures and energy sources. Energy is obtained from a number of hydroelectric plants as well as from petroleum, extensive reserves of which are located off the Atlantic coast (Kole-Marine and Kribi fields). Substantial deposits of bauxite have also been identified in the Adamawa Mountains, but are not yet being systematically exploited; at present, the aluminum processed in the plants at Edéa comes from Guinea. A large paper mill has recently been built at Edéa, and a petroleum refinery is operating at Limbé. Other active industries include cement plants, shoe factories, and breweries.

The deficit in the balance of trade has become more severe in the last few years, particularly because of massive imports of capital goods. Major trading partners include France, the Netherlands, the United States, and Germany.

The communications network, not yet completely efficient, includes more than 39,680 mi [64,000 km] of road (1986), and a railroad that connects the port of Douala with the capital and with the high plateaus of the Adamawa (Ngaoundéré). Air transport is much better developed, with major airfields at Douala, Yaoundé, and Garoua, which is also a river port on the Bénoué. The merchant fleet is still small. The most important seaports, besides Douala, are Limbé, Kribi, and Bonabéri.

Historical and cultural profile. *The Bantu kingdoms.* The name "Cameroon" derives from the Portuguese word *camarão* (shrimp), and the Portuguese were indeed the area's first explorers; they included Fernando Póo, who near the end of the 15th century reached the stretch of the Guinean coast around the Sanaga river estuary. Here they established trading posts, which

Socioeconomic data

Income (per capita, US$)	1090 (1990)
Population growth rate (% per year)	3.2 (1980–89)
Birth rate (annual, per 1,000 pop.)	44 (1989)
Mortality rate (annual, per 1,000 pop.)	12 (1989)
Life expectancy at birth (years)	57 (1989)
Urban population (% of total)	40 (1989)
Economically active population (% of total)	39.2 (1988)
Illiteracy (% of total pop.)	43 (1991)
Available nutrition (daily calories per capita)	2161 (1989)
Energy consumption (10^6 tons coal equivalent)	2.9 (1987)

for three centuries were used by the European powers as bases for trade (especially in slaves and ivory) with the populations of the interior. The area that is now Cameroon was at that time a fairly complex system of tribally based states. The coastal region was occupied by the Bantu potentates of the Yaoundé and Douala tribes, who grew rich from their commercial exchanges with whites. The Adamawa massif was the land of the Fulbe, who later would establish a powerful emirate there. Between the coast and the central highlands arose the city of Foumban, capital of the Bantu kingdom of the BaMum, one of the most culturally interesting groups. There is considerable evidence of their artistic tradition, including a series of terra cotta statues dating back to the 15th century, along with richly decorated objects for both ritual and everyday use. In the 19th century, under the guidance of King Njoya, a process of unification began between the tribes of the highlands and those of the Sanaga basin, and a common ideographic/phonetic writing system was even developed. This process continued alongside European penetration, which over the course of the century finally prevailed, disrupting local cultures.

The colonial period. Those who ventured into the interior were mostly British traders and missionaries, but in 1860 German colonists also appeared. Among them was the explorer Gustav Nachtigal, the first to succeed in imposing a European protectorate on Douala. This colonization, completed in 1902 with the conquest of the Adamawa, brought considerable political prestige to Germany, which gained additional reinforcement from its treaty with France in 1911 extending the colony's borders to the Congo and the Ubangi. During World War I the German protectorate was occupied by the Allies, and at the end of the conflict it was subdivided into two regions, assigned by international mandate to France (south central area) and Great Britain (northern area), which governed them differently under their respective colonial systems. Reunification emerged as the principal goal of Cameroonian nationalism; it was not completely achieved, however, since part of British Cameroon would later be definitively ceded to Nigeria.

Toward independence. During World War II, the French colony shifted its support from the collaborationist Vichy regime to the Free French (1940), and in 1946 was entrusted to France by the UN as a trusteeship. In the politically heated atmosphere of the 1950s, calls for independence and national unity grew louder. Finally, a true revolutionary movement was established, led by the Marxist Union des Populations Camérouniennes (UPC) whose leader, Um Nyobé, was killed in 1958. But once independence was achieved in 1960, the UPC shifted toward reformist positions, and Cameroon aligned itself ideologically with the moderate nations of French-speaking Africa. The following year the British mandate was also terminated, and the area controlled by Great Britain was combined (at least in part) in a federal system with the Republic of Cameroon that had recently been created. Under the leadership of President Ahmadou Ahidjo, the country has worked since that time to achieve political, economic, and social integration between the two regions. In 1972 a large majority approved a new constitution, which established a unified nation with a single governmental system, a monocameral National Assembly, and a centralized administrative apparatus.

On the economic level, Cameroon embarked on a "planned laissez-faire" policy of development, based on supplies of foreign capital and on the entrepreneurial spirit of the nation's middle class, which provided a promising start. Nevertheless, Ahidjo's successor, Paul Biya, found himself faced in 1981 with serious internal problems (corruption, ethnic and social strife) which continue to impede the country's modernization efforts.

Literature. In the last fifty years an interesting body of literature in French has appeared in Cameroon, closely linked to the struggle for independence and national unity. Poetry has drawn on the local tradition of popular songs (for example in the collection *Kamerun! Kamerun!* [Cameroon! Cameroon!] by E. E. Jondo), or from recent historical events, as in the poem "Why everyone lives free" by Hadgi al Mukran, written as a eulogy for Um Nyobé, Cameroon's national hero.

Prose works began to appear in the mid-1950s. In their novels and short stories, writers like Alessandro Biyidi (who writes under the pseudonyms Era Boto and Mongo Beti), F. Oyono, and B. Matip address the social and spiritual problems of Cameroon's people. Mongo Beti's novels—*Ville cruelle* [Cruel city], *Le Roi miraculé* [The healed king], *Mission terminée* [Mission accomplished]—are particularly important. Conflicts between the local population and the French colonizers are the most common themes of Cameroonian prose, which focuses principally on illuminating Cameroon's recent past. Acute conflicts of this kind are at the heart of Oyono's novels *Une Vie de boy* [Houseboy], *Le Vieux Nègre et la médaille* [The Old Man and the Medal], and *Chemin d'Europe* [Road to Europe]. Matip, who has written essays on African economy and sociology, speaks of village youths and their struggles and hopes in his brief novel *Afrique, nous t'ignorons* [Africa, we do not know you].

After independence was achieved in 1960, the work of organizing a variety of cultural and artistic groups began. One younger author worth noting is Francis Bebey, whose poem, "One day you will know," appears below. Dedicated to his son, it says much about the condition of black Africans:

> *One day you will know*
> *That you have black skin and white teeth*
> *And hands with white palms*
> *And a pink tongue*
> *And hair as curly*
> *As the vines of the virgin forest.*
> *It means nothing.*
> *But if you ever learn*
> *That you have red blood in your veins,*
> *Then burst out laughing,*
> *Clap your hands together,*
> *Pretend to go mad with joy*
> *At this unexpected news.*
> *And after this moment of apparent joy,*
> *Take on a serious air*
> *And ask those around you:*
> *The red blood in my veins,*
> *Is it enough to make you believe*
> *That I am human?*
> *My father's flesh,*
> *It too, has red blood in its veins.*

Other important cultural figures include R. Philombé, storyteller and author of *Un sorcier blanc à Zangali* [A white magician at Zangali], and the satirical humorist G. Oyôno-Mbia, who draws inspiration from everyday tribal life.

CENTRAL AFRICAN REPUBLIC

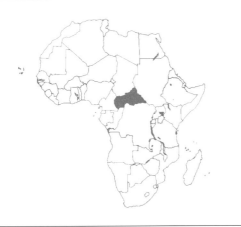

Geopolitical summary

Official name	République Centrafricaine
Area	240,260 mi² [622,436 km²]
Population	2,054,610 (1975 census); 2,878,253 (1989 estimate)
Form of government	Presidential republic governed by the military; the head of government is elected every six years by universal suffrage, and the single-party National Assembly every five years.
Administrative structure	14 prefectures, 2 economic prefectures, and 1 autonomous municipality (the capital)
Capital	Bangui (pop. 596,776, 1988 estimate)
International relations	Member of UN, OAU; associate member of EC
Official language	French; national language is Sangho (Sudanese)
Religion	Animist 57%, Muslim 8%, Protestant 15%, Catholic 20%
Currency	CFA franc

Natural environment. The Central African Republic is bordered on the west by Cameroon, on the north by Chad, on the east by Sudan, and on the south by Zaire and Congo.

The country, which extends more than 800 mi [1300 km] from east to west, comprises a large portion of the watershed separating the contiguous basins of Lake Chad, the Congo river, and the Nile. The predominant landform consists of a series of high plateaus cut out of the pre-Paleozoic crystalline basement rocks, which rise to an average elevation of more than 1650 ft [500 m] and are partly covered by Mesozoic sediments. The topography rises at the two ends of the country, with the Bongos Mountains (4487 ft [1368 m]) to the east constituting the watershed with the White Nile, while to the west, the Yadé Mountains (4658 ft [1420 m]) act as the boundary with the basin containing Lake Chad and the Sanaga river (Cameroon).

The highly complex hydrography can be divided into two major basins: Lake Chad to the north (40% of the nation's land area), fed by the Chari and Logone rivers; and the Congo to the south (60% of the area), fed by its most important right-bank tributary, the Ubangi. For a large part of its course, the Ubangi follows the border with Zaire, and constitutes the primary communication route between the Congo drainage area and the upper Nile valley. Despite numerous rapids, it is navigable in any season.

The climate varies considerably with latitude: in the southern regions it is tropical, with two seasons, mean annual temperatures around 75°F [24°C], extreme daily temperature swings, and precipitation in excess of 80 in. [2000 mm] annually during the rainy season. To the north, the climate becomes more Sahelian, with rainfall decreasing to less than 40 in. [1000 mm], concentrated in a brief, summer rainy season.

Reflecting these different climatic conditions is a variety of vegetation: the northern regions are dominated by an arid savanna characterized by scrub and thorny acacia, while to the south, vegetation becomes more dense until it finally merges into true rain forest in the southwest part of the country. The fauna comprises more wild animals than anywhere else in Africa: elephants are common, along with lions, leopards, antelopes, gazelles, and many species of monkey. Three large game preserves, along with the André Felix, Bamingui Bangoran, and Saint Floris national parks, help preserve these natural treasures.

Population. The original population consisted of Bantu Negroid peoples, joined at later periods by several groups that have combined to produce a composite population consisting of Banda (31% of the total), Baya (29%), Mandjia (8%), and Sara (8%). The Banda, of Sudanese origin, settled in the east central region of the country and are divided into numerous tribes; they live by farming, hunting, and fishing, and have kept their ritual traditions alive. The Baya and Mandjia, dwelling in poor villages scattered through the western regions far from population centers, are hunters who also grow meager cereals. The Pygmies, hunter-gatherers now reduced to only a few thousand individuals, still survive in the forests.

The Central African Republic has never had a large population, even though it has a high birth rate (4.2%) and 2.7% annual population growth. Average population density is about 13 per mi² [5 per km²]. Bangui, the capital, is located at the confluence of the Mpoko and Ubangi rivers, on the border with Zaire. Surrounded by dense tropical vegetation, the city is the Republic's main river port and its only urban and economic center. There are other fairly large towns—Bambari, Bouar, Berbérati, and Bossangoa—but none has more than

Climate data

Location	Altitude (ft asl)	Average temp. (°F) January	Average temp. (°F) July	Average annual precip. (in.)
Bangui	1266	80	77	60.0
Ndélé	1673	83	77	55.7

Conversion factors: 1 ft = 0.3 m; 1 in. = 25 mm; °C = (°F − 32) × 5/9

Administrative structure

Administrative unit (prefectures)	Area (mi²)	Population (1989 estimate)
Bamingui-Bangoran	22,465	35,760
Basse-Kotto	6,795	223,306
Haute-Kotto	33,447	58,307
Haute-M'Bomou	21,435	45,516
Haute-Sangha	11,658	267,589
Kemo-Gribingui	6,641	90,495
Lobaye	7,425	186,121
M'Bomou	23,604	153,948
Nana-Mambéré	10,268	229,927
Ombella-M'poko	12,288	150,558
Ouaka	19,261	248,090
Ouham	19,396	311,081
Ouham-Pendé	12,391	289,741
Vakaga	17,949	29,487
Ibingui (economic prefecture)	7,718	–
Sangha (economic prefecture)	7,493	72,459
Bangui (autonomous municipality)	26	385,757
CENTRAL AFRICAN REPUBLIC	240,260	2,878,253

Capital: Bangui, pop. 596,776 (1988 estimate)
Major cities (1988 estimate): Bambari, pop. 52,092; Bouar, pop. 49,166; Berbérati, pop. 45,432; Bangassou, pop. 36,254.

Conversion factor: 1 mi² = 2.59 km²

50,000 inhabitants; all are located along the principal communication routes.

The Central African Republic is one of the poorest nations in Africa, with annual per capita income of US$440 and a persistently high mortality rate (1.5%) due to malnutrition and endemic tropical diseases. Illiteracy is still widespread (62.3% of the population) as a result of inadequate infrastructure, although the educational system provides primary, secondary, and vocational schooling as well as a university at Bangui.

Economic summary. Despite the high percentage of the population engaged in agriculture (66%), only 3.2% of the entire country is arable land. The most fertile areas are used to raise commercial produce, primarily cotton, coffee, tobacco, peanuts, sisal, and palms. The most important subsistence crop is manioc (616,000 t in 1990), followed by corn and minor cereals. Livestock holdings, limited to some 2.6 million cattle (1990) and slightly more than 1 million sheep and goats, are low due to the presence of trypanosomiasis (sleeping sickness) transmitted by the tsetse fly. The equatorial forest which covers the southwestern region of the country (forests and woods cover a total of 57.5% of the country) is rich in tropical woods such as mahogany and ebony, which constitute some 12% of exports.

Mineral resources consist principally of diamonds in the western Bangui region (annual production approximately 340,000 carats in 1989), and uranium in the Bakouma area to the east. Limited quantities of iron ore, gold, and graphite are also extracted. The hydroelectric plant at Boali, powered by the falls of the M'Bali river, produces 90% of the nation's energy requirements.

The industrial sector concentrated around the capital is extremely modest, employing only 12% of the labor force and based primarily on the processing of agricultural products. Cotton ginning plants, peanut-oil processing factories, textile industries, and facilities for producing cigarettes and beer are of particular importance.

Service activities employ 22% of the work force, and contribute 40% of gross national product. The balance of trade is negative and is supported by foreign aid, especially from France, the country which receives the Republic's exports of cotton, coffee, lumber, diamonds, and uranium, and which provides most imports, especially industrial products. Since the Republic has no railroads, the major transportation routes are the Ubangi and Sangha rivers; the roads (total length of some 12,500 mi [20,000 km] in 1988) become unusable during the rainy season.

The country offers fairly good accommodations for tourists; the Boali falls (60 mi [90 km] from Bangui) are one of the most popular attractions.

Historical profile. Originally inhabited by Banziri populations, the Ubangi-Chari territory was first colonized by the French beginning in the 1880s, when Bangui was founded and protectorate treaties were signed with a series of small local kingdoms. French interest in this impenetrable inland region of equatorial Africa was not directly related to potential economic exploitation, as was the case for the regions on the Atlantic coast; France simply needed to link its territories in the basin of the lower Congo river with those in Chad and the Nile valley. This interest did not diminish even after the loss of France's possessions on the Nile following the Fashoda incident; the Ubangi-Chari region was still felt to play an essential role as a connection between French possessions in Algeria, western Africa, and the Congo. To conquer this strategically important region, the French did not hesitate to confront the powerful forces of the slave-trader Rabah, who had established his capital in the Chari valley.

In 1910, after the conquest was complete, Ubangi-Chari was declared a colony within French Equatorial Africa. A desperately poor, backward, sparsely populated country with no outlet to the sea, it was always overlooked in colonial development plans. The lack of a road system and railroads was particularly keenly felt; once roads were finally introduced in 1920, they proved useful mostly for military operations during World War II, offering Allied troops a route to northern Africa. Nationalist sentiment began to grow after the war, giving rise to a new political class whose leading personalities were Antoine Darlan and

Socioeconomic data

Income (per capita, US$)	440 (1990)
Population growth rate (% per year)	2.7 (1980–89)
Birth rate (annual, per 1,000 pop.)	42 (1989)
Mortality rate (annual, per 1,000 pop.)	15 (1989)
Life expectancy at birth (years)	51 (1989)
Urban population (% of total)	46 (1989)
Economically active population (% of total)	48.4 (1988)
Illiteracy (% of total pop.)	62.3 (1991)
Available nutrition (daily calories per capita)	1980 (1989)
Energy consumption (10⁶ tons coal equivalent)	0.13 (1987)

Barthélémy Boganda: both fought against French colonial rule, although they took different approaches. The historian Endre Sík has described Boganda as follows:

> In 1952 Boganda founded a political party, the Mouvement pour l'Evolution Sociale de l'Afrique Noire (M.E.S.A.N.). At the elections held in March 1952 his party obtained the majority of African seats in the Territorial Assembly (17 of 20).... He [adopted] a less sharp tone than Darlan, and demanded independence for the country, stressing every time that after independence he wanted friendly relations with France, not separation from her. He had partly this, partly his repeated anti-Communist pronouncements to thank for the fact that the French colonial authorities raised no obstacles to his political activity.... "Decolonization" was his constant slogan, but in none of his speeches did he fail to assure France of his loyalty. His radical diatribe roused the indignation of some French colonial officials, but the Council members stood up for him as one man: the Council at its meeting of November 16, 1957, unanimously voted him total confidence and requested the Governor to punish the officials who refused to comply with the spirit of the loi-cadre.
>
> After De Gaulle came to power in France, Boganda continued his double-faced policy: he demanded independence for his country but did not miss a single opportunity to express his own pro-French feelings. At the time of De Gaulle's visit to Brazzaville on September 1, 1958, Boganda cordially thanked him 'for having kept his promises,' but at the same time he handed him a petition requesting the French government to recognize in the new Constitution the right to independence of the African colonies.
>
> In the September 1958 referendum 98.8 per cent of the electorate, following Boganda's advice, voted 'yes,' that is, chose autonomous status within the French Community instead of real independence.

Boganda died in an airplane crash in 1959, a year before his country achieved independence. Ubangi-Chari then became the Central African Republic within the French Union, but the republican form of government proved not to be very durable. In 1966 General Jean-Bedel Bokassa overthrew David Dacko, the president, abolished Parliament and the constitution, and instituted a reign of terror; a decade later he transformed the country into the "Central African Empire," and assumed the title of Emperor. Although at the time he assumed power he had voiced support for a reform program intended to emancipate the rural masses, Bokassa's personal dictatorship was upheld by a class of bureaucrats and plantation owners who never permitted serious modernization. Nonetheless, the proceeds from the country's diamond mines contributed to a perceptible improvement in general living conditions. Bokassa's regime began to totter in 1979, when a street demonstration was suppressed with particular brutality: public opinion in both Africa and the West condemned the dictator, and even France, which until that time had never refused assistance to Bokassa, supported the coup d'état which restored the republic and brought David Dacko back to power (1981). He was succeed by General André Kolingba, who in 1987 agreed to the election of a new parliament and in 1991 to a multi-party system.

CONGO

Geopolitical summary

Official name	République du Congo
Area	132,012 mi^2 [342,000 km^2]
Population	1,909,248 (1984 census); 2,264,300 (1990 estimate)
Form of government	Partly presidential republic with a National Assembly elected every five years and a Senate every six; the head of government is also the chief executive.
Administrative structure	9 regions divided into 46 districts, plus the capital district
Capital	Brazzaville (pop. 760,300, 1990 estimate)
International relations	Member of UN, OAU; associate member of EC
Official language	French; Bantu languages widely spoken
Religion	Animist 47%, Protestant 9.6%, Muslim 3%, Catholic 40%
Currency	CFA franc

Natural environment. Congo is bordered on the west by Gabon, on the north by Cameroon and the Central African Republic, on the east and south by Zaire, on the south by Angola (Cabinda), and on the southwest by the Atlantic Ocean.

The country extends for about 450 mi [700 km] along the Congo river and its tributary the Ubangi. Three main topographical regions can be distinguished from the Atlantic to the interior: the first consists of the coastal strip, some 90 mi [150 km] long, low and sandy as far as Pointe-Noire where it becomes rocky and steep. The central part of the country consists of a wide ridge, deeply incised on the southern slope, consisting of a continuation of the Gabon massif of Chaillu together with the Bateké uplands. The third region covers 60% of the territory, with low inland plateaus (1000–1300 ft [300–400 m]) descending toward the center of the Congo basin, which is covered by substantial Tertiary sedimentary deposits and Quaternary river and lake alluvium. From a hydrographic point of view, the country lies almost entirely within the Congo river basin, into which water flows either directly or via tributaries, chiefly the Sangha and the Ubangi. The northern coastal portion contains the basin of the Niari-Kouilou river, which carries water draining off the southern slopes of the Chaillu massif.

The climate is equatorial throughout almost the entire country, characterized by high average temperatures all year and annual temperature swings of only a few degrees. Annual rainfall exceeds 78 in. [2000 mm] only near the Gabon highlands, while in the northeastern region it is less than 70 in. [1800 mm]. Precipitation varies according to the sun's position at the zenith, with two seasons: a dry season in summer when the sun is north of the equator, and a winter rainy season when it is to the south. Near the coast and as far inland as Brazzaville, the cold Benguela current depresses rainfall to less than 62 in. [1600 mm], leading to a drier climate (with wooded savanna). The winter rains are heaviest in November and April, with precipitation of 12–16 in. [300–400 mm] per month, and average temperatures of 78.3°F [25.7°C] in January and 71.1°F [21.7°C] in July.

The typical rain-forest vegetation (covering 62% of the land area) contains many valuable woods such as okoumé, ebony, and mahogany. Equatorial flora and fauna are protected in Odzala National Park and reserve, while the wooded savanna is preserved in the Léfini reserve in the upland plain region.

Population. Congo's population has resulted from a number of mixtures at various times in the past, all involving Bantu-speaking peoples who replaced the Pygmies in those areas most suitable for cultivation. The most numerous groups are the BaKongo (52% of the population) who live in the southern regions, and the BaTeke (47%), settled farther inland as far as the Alima river. The Mboshi (12%) live in the lower river valley, while the Kota (5%) live to the north. The Babinga Pygmies today number only a few thousand, and live in the forest or in settlements close to Bantu villages.

A low population density (18 per mi^2 [7 per km^2]) is characteristic of the country and of the entire Congo basin. As late as 1901 there were only 400,000 inhabitants in what is now Congo; this figure doubled during the first fifty years of this century, and has now exceeded 2 million (1990 estimate). About two thirds of the population live in rural areas. The most heavily populated region is the coastal strip, with a density greater than 26 per mi^2 [10 per km^2]; the central and northern areas are practically unoccupied. Villages are sheltered by trees and contain 100–200 inhabitants; huts are protected against the damp ground by clay-paved platforms. The principal cities are Brazzaville, the capital, a flourishing industrial center, major commercial nexus, and site

Administrative structure

Administrative unit (regions)	Area (mi^2)	Population (1984 census)
Bouenza	4,734	187,143
Cuvette	28,892	135,744
Kouilou	5,286	369,073
Lékoumou	8,087	68,287
Likouala	25,493	49,505
Niari	10,014	173,606
Plateaux	14,822	109,663
Pool	13,120	184,263
Sangha	21,539	46,152
Brazzaville (Federal District)	25	585,812
CONGO	132,012	1,909,248

Capital: Brazzaville, pop. 760,300 (1990 estimate)
Major cities (1990 estimate): Pointe-Noire, pop. 387,774; Loubomo, pop. 62,073.
Conversion factor: 1 mi^2 = 2.59 km^2

of the university; Pointe-Noire, a modern city with an efficient commercial harbor; and Loubomo, a commercial center.

Thanks to low mortality (1.5%) and infant mortality (11.5%) rates, population growth has remained high (3.4%), although only average for Africa. Low values for life expectancy at birth (54 years) and available calories per capita per day (2512—close to the subsistence limit) are characteristic of a country that, although it is developing with a great deal of attention to social problems (only 43.4% illiteracy), still has much to do before achieving real security.

Economic summary. After independence, the nation seemed unable to select an economic model, swinging from a permissive laissez-faire attitude to strict planning, and from worker management of enterprises to a mixed economy. Its economic potential is high: Congo is geographically well-placed within the overall transport system of equatorial Africa; it has fertile land, extensive forest with valuable hardwoods, enormous hydroelectric resources, and substantial mineral deposits.

The country's agricultural activities (10% of GNP and 34% of the work force) have developed only near the coast and along the railroads and rivers: archaic methods for producing subsistence crops (manioc, sweet potato, banana, corn) are being replaced by modern cooperative production, and new crops are being grown both for internal consumption and for export (sugar cane, oil palm, rubber, coffee, cocoa). The forests (covering 62% of the country) are one of Congo's principal resources, producing 116 million ft^3 [3.3 million m^3] of hardwoods in 1989 (okoumé, ebony, mahogany), which can easily be transported by river.

Oil wells produce much of the nation's wealth. Potassium salts (potassium chlorate and carnabite), extracted at Hollé near Pointe-Noire, constitute the third most important export item, while copper, lead, tin, and zinc ores are of very modest importance. Hydroelectric energy production is enormous (397 million kWh in 1989), coming almost exclusively from the Djoué plants near Brazzaville. Industry is significant and well developed: it produces 38% of GNP and employs 25% of the work force.

Climate data

Location	Altitude (ft asl)	Average temp. (°F) January	Average temp. (°F) July	Average annual precip. (in.)
Brazzaville	1014	78	73	57.0

Conversion factors: 1 ft = 0.3 m; 1 in. = 25 mm; °C = (°F − 32) × 5/9

Socioeconomic data

Income (per capita, US$)	1010 (1991)
Population growth rate (% per year)	3.4 (1980–89)
Birth rate (annual, per 1,000 pop.)	48 (1989)
Mortality rate (annual, per 1,000 pop.)	15 (1989)
Life expectancy at birth (years)	54 (1989)
Urban population (% of total)	40 (1989)
Economically active population (% of total)	39.7 (1988)
Illiteracy (% of total pop.)	43.4 (1991)
Available nutrition (daily calories per capita)	2512 (1989)
Energy consumption (10^6 tons coal equivalent)	0.76 (1987)

Principal specialties include agricultural processing (with oil plants and breweries at Brazzaville and Pointe-Noire), cement factories at Loutété, sugar plants at Moutela, cotton processing at Kisoundi, and tobacco manufacturing facilities at several locations. Wood is processed at over a dozen centers that produce plywood and semi-finished products; crude oil is refined at Pointe-Noire.

The balance of trade is largely positive: oil, lumber, potassium, and sugar are exported, while imports include machinery, foodstuffs, and textiles. Communication routes are limited, consisting of the colonial railroad directly to Gabon (495 mi [798 km] in 1985), 5394 mi [8700 km] of roads, and especially the river corridors of the Congo and Ubangi. The main airports are Brazzaville/Maya-Maya and Pointe-Noire.

Historical and cultural profile. In the pre-colonial era, the region was the site of Bantu states ruled by the BaVili and BaTeke; by the 15th century, Europeans were already aware of the Congolese kingdom of Loango. But the Bantu in turn had supplanted the first true possessors of the land, the Pygmies, as related by Folco Quilici:

> In earlier times, so Nanshen told me, the Pygmies were the absolute rulers of equatorial Africa. The Bantu blacks who now occupy these lands arrived long after them, coming from the East only two or three thousand years ago, and when they began to penetrate into the great forests, they conquered and subjugated the Pygmies. In the eyes of the little people of the forest, the invaders were gigantic beings, who dominated them not only by their physical strength and their superior military and social organization, but mostly by the force of their religious beliefs and their superstitions, which had made them masters of terrifying magic and secret liturgies. The "strong races" thus came to occupy the Pygmies' hunting grounds, built large villages, and took possession of their herds and grazing grounds. Then all that was left for the Pygmies was to take refuge in distant valleys, in the deepest and most inaccessible parts of the forest. To placate the fierce newcomers, the Pygmies took up the habit of going in the dead of night to the fringes of the "invader" villages, and leaving fruit and game that they had collected. As time passed, these forms of subjugation became a voluntary submission to the other black races of equatorial Africa, and the practice is still common today.

The first European explorer to venture into this region was the Frenchman Pierre Savorgnan de Brazza, who in 1880 signed a "treaty of friendship" with a local chief named Makoko that is regarded by some as the first document of political annexation in western Africa. Brazza then founded, on the right bank of the Congo river at Ntamo, a town that would be the nucleus of modern Brazzaville. The boundaries of France's possessions were defined in long negotiations among the European powers; enlarged in 1888 with the annexation of Gabon, in 1910 the colony was called Middle Congo, and became part of French Equatorial Africa. A broad-based nationalist movement developed between the two World Wars, and after the colony declared its allegiance to Free France (1940), it was at Brazzaville that the conference was held establishing the French Union, intended as a way station on the road to self-government for the African nations. After independence (1960) and three years of personal dictatorship under President Fulbert Youlou, a popular uprising brought to power a revolutionary government led by Alphonse Massamba-Débat. The new government opened the way for socialism, nonalignment, and detachment from the neocolonial model, a trend that became more pronounced in the late 1960s, when a group of extreme leftist military officers took power. General Marien Ngouabi transformed Congo into a people's republic governed by a single party, the Congolese Labor Party. In the years that followed, this approach led to nationalization of several production sectors (including petroleum) and, in foreign policy, to stronger links with the socialist countries. It was a choice that would face serious difficulties, plagued by conflicts and violence within the governing elite; the political debate was still occasionally obscured by outbreaks of tribally-based disputes.

With a rich and ancient tradition of oral literature, in this century Congo has also developed an original corpus of poetry and prose literature in French. The most important novelist is J. Malonga, author of *La Légende de Mfumu Ma Mazono* (1955). Significant poets include G. F. Cikaya U Tamsi, who addressed (in *Épitome*) the theme of the "tragedy of the African continent," and Martial Sinda. Sinda has made effective use of the rhythms of popular African verse to express anticolonialist sentiments. Here is an excerpt from *Tam-Tam* entitled "Tam-Tam-toi":

> Silence.
> Always silence.
> We no longer speak.
> We no longer dance.
> We no longer shout.
> Because we are not free.
> Because we are no longer among our own.
> O Africa of long ago!
> O Africa subdued!
> O Africa, ohoé! Our Africa....
>
> Africa, land of sadness!
> Africa, land of misunderstanding!
> Africa, land without joy, without dancing, without songs!
> Africa, land of wailing and laments ...
>
> O sweet jazz trumpet!
> O melodious xylophone!
> O n'tsambi of the Congo!
> And you griots from my beloved Dakar!
> And you dancer Zannie Amaya from Bangui!
> Cherish always what you are, cherish it always
> until you have created a new Africa,
> but an Africa still Black.

GABON

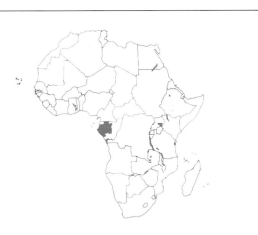

Geopolitical summary

Official name	République Gabonaise
Area	103,319 mi^2 [267,667 km^2]
Population	950,007 (1970 census); 1,299,000 (1978 estimate)
Form of government	Presidential republic; the head of government is elected for seven years by universal suffrage. Legislative power is exercised by a National Assembly.
Administrative structure	9 provinces divided into 37 departments
Capital	Libreville (pop. 352,000, 1987 estimate)
International relations	Member of UN, OAU; associate member of EC
Official language	French; Bantu languages also spoken
Religion	Christian 84%, the remainder animist
Currency	CFA franc

Natural environment. Gabon is bordered to the north by Equatorial Guinea and Cameroon, to the south and east by Congo, and faces the Atlantic Ocean to the west.

The country sits astride the Equator, and consists predominantly of pre-Paleozoic crystalline basement rock. Three distinct morphological areas are evident: the coastal strip of alluvial origin, with Cape Lopez jutting out to the north, fringed with lagoons and sandbanks south of the mouth of the Ogooué; the central section, comprising a few high mountain ranges arranged symmetrically around the Ogooué which passes through them—the Tembo Mountains (3950 ft [1200 m]) and Cristallo Mountains (2920 ft [890 m]) to the north, and the Chaillu massif (5150 ft [1575 m]) to the south—and lastly, to the east, the interior plateau, dissected by numerous tributaries of the Ogooué river. The drainage basin of this river, which has several spring-fed branches rising in neighboring Congo, covers almost the entire country. Other rivers include the Gabon, which rises in the Cristallo Mountains, and the Nyanga, draining the southern slopes of the Chaillu massif.

The climate is markedly equatorial—hot and humid. The area of greatest precipitation is the coastal strip near the border with Equatorial Guinea, where rainfall is always more than 100 in. [2500 mm] per year, and can exceed 150 in. [4000 mm]. Average monthly temperatures range from 75 to 81°F [24–27°C], with little variation.

Huge stretches of dense equatorial forest cover almost all of Gabon, although the south central plateaus are covered by savanna vegetation, and the coastal regions are fringed with impenetrable belts of mangrove swamp. The large African animals, which range mostly in the savanna, are protected in Wonga-Wongué, Okanda, and Petit Loango national parks.

Population. Almost the entire population is of Bantu stock. The largest ethnic group is the Fang, which constitutes one-third of the population, followed by the Eshira in the Ogooué delta region, the Mbédé, the BaKota, and the Omyené. Small groups of Babinga Pygmies, considered the first inhabitants of the country, still live in the forest.

Gabon is a sparsely populated country. According to a UN estimate its population was about 1.3 million in 1978, with an annual average growth rate of 3.7% and an average population density of just over 13 per mi^2 [5 per km^2], distributed unevenly: the most densely populated regions are the agricultural areas to the north and the southern Ngounié basin, as well as the recently settled mining regions; the forest is practically uninhabited. The principal urban center is Libreville, the capital; located on the north bank of the Gabon river, it has an active commercial harbor and an international airport. The most important industrial city is Port Gentil, on the Cape Lopez peninsula, which is the nation's petroleum port. Other smaller centers include Lambaréné (seat of the hospital founded by Dr. Albert Schweitzer) and Franceville.

Despite its natural wealth, Gabon exhibits enormous internal imbalances among regions. Its illiteracy rate was 60.7% in 1991, one of the highest in Africa. Higher education is provided at the university in Libreville. The quality of social infrastructures, especially hospitals, is high.

Economic summary. Gabon's economy has developed rapidly thanks to large reserves of crude oil, uranium, and manganese, and intensive exploitation of the rain forest.

The agricultural sector uses only 1.7% of the country's land area, and is quite weak. Subsistence crops, meeting only about 15% of domestic needs, consist of manioc, corn, sweet potato, and yams. Commercial crops are also limited, and include cocoa, coffee, oil palms, peanuts, and bananas. Forests and woods cover

Climate data

Location	Altitude (ft asl)	Average temp. (°F) January	Average temp. (°F) July	Average annual precip. (in.)
Libreville	13	80	75	98.8
Franceville	1246	77	73	73.2
Conversion factors: 1 ft = 0.3 m; 1 in. = 25 mm; °C = (°F − 32) × 5/9				

Administrative structure

Administrative unit (provinces)	Area (mi²)	Population (1978 census)
Estuaire	8,006	359,000
Haut-Ogooué	14,107	213,000
Moyen-Ogooué	7,155	49,000
Ngounié	14,572	118,000
Nyanga	8,216	98,000
Ogooué-Ivindo	17,785	53,000
Ogooué-Lolo	9,797	49,000
Ogooué-Maritime	8,836	194,000
Woleu-Ntem	14,847	166,000
GABON	103,319	1,299,000

Capital: Libreville, pop. 352,000 (1987 estimate)

Major cities (1983 estimate): Port Gentil, pop. 123,300; Franceville, pop. 38,030; Lambaréné, pop. 26,257 (1978 estimate).

Conversion factor: $1 \ mi^2 = 2.59 \ km^2$

Socioeconomic data

Income (per capita, US$)	3450 (1990)
Population growth rate (% per year)	3.7 (1980–89)
Birth rate (annual, per 1,000 pop.)	42 (1989)
Mortality rate (annual, per 1,000 pop.)	15 (1989)
Life expectancy at birth (years)	53 (1989)
Urban population (% of total)	45 (1989)
Economically active population (% of total)	44.3 (1988)
Illiteracy (% of total pop.)	39.3 (1991)
Available nutrition (daily calories per capita)	2396 (1989)
Energy consumption (10^6 tons coal equivalent)	1.24 (1987)

74.7% of the country. Herding and fishing are limited.

Crude oil, production of which reached 85 million bbl [13.5 million metric t] in 1990, is extracted along the entire coastal strip. The situation in the Gabonese petroleum industry was summarized as follows by A. Foglio, in a report on OPEC markets published in 1982:

> Interest in petroleum exploration dates back to 1930. For a while, however, prospecting was unsuccessful, until efforts were focused on the western part of the sedimentary basin, where the first strike was made at Ozouri, followed immediately by another at Pointe Clarette; these discoveries were followed by others at Port Gentil in 1956. Deep-water exploration began in 1959, leading to the first offshore strike in 1962. The large Gamba field was discovered in 1963, followed by other interesting discoveries between 1968 and 1973 at Grondir, Barbier, Torpille, and Mandaros.
>
> Actual petroleum production began in 1957, and developed gradually until 1976 when it reached a stable level of about 70 million bbl [11 million metric t] (equivalent to about 220,000 bbl per day). Significant production increases were achieved between 1971 and 1973, leading to a doubling in production. At present, three oil companies account for 94.2% of production: Elf-Gabon with 68.4%; Shell-Gabon with 8.1%; and Snea with 17.7%. The remainder is distributed among Odeco, Ocean, Gulf, Mobil, Murphy, and Ensearch.
>
> ... Gabon has always shown a certain interest in research and prospecting, both direct and indirect, by cooperating with foreign oil companies. This interest was demonstrated in 1977, when four production co-participation contracts were signed between Gabon and British Petroleum Wintershall, Ina Naftaplin Wintershall, Acron, and Snea.

Production of manganese from the Moanda region has reached 1.2 million t in 1990. Uranium ore, mined near Franceville, yields some 770 t of metal per year. Iron ores are present in the northeastern part of the country. Another rich resource is lumber, including hardwoods such as okoumé (or Gabon mahogany), of which Gabon is the leading producer (some 40 billion ft³ [1,137,000 m³] in 1990).

The nation's industrial base is small; the most common industries are food processing, wood working, and cement plants.

The balance of trade has been comfortably positive for years, thanks to increases in crude oil prices and massive exports of uranium, manganese, and lumber. Internal communications are still inadequate (the country had only 4672 mi [7535 km] of road in 1986), and improvement prospects are presently limited to construction of the TransGabon Railway. International traffic moves through the harbor at Port Gentil and the airports at Port Gentil and Libreville.

Historical and cultural profile. Before colonization, the territory of Gabon was a mosaic of small, ethnically-based states, each jealous of its autonomy and seldom in contact with its neighbors. In 1484 the coastal strip was explored by the Portuguese navigator Diego Cão, and it was the Portuguese who gave the name gabão (hooded cloak) to the river which then gave its name to the entire country. Later centuries saw the arrival of the Dutch, English, and French, all eager, like the Portuguese, for any opportunity to exploit the region's potential for the slave trade. Of all these European powers, France was the first to sense the advantages that it could derive beyond this commerce, and between 1839 and 1849 used the pretext of an effort to combat slavery to establish Libreville and settle in Gabonese territory. The interior was conquered later, particularly under Pierre Savorgnan de Brazza, who reached the source of the Ogooué river in 1878. Annexed to the French Congo (1888) and then to French Equatorial Africa (1910), Gabon followed the same historical path as the other colonies in the region, obtaining autonomy in 1958 and full independence two years later. The presidential republic instituted in 1960 was led for the first few years by Léon M'Ba, and after his death (1967) by Albert Bongo. The policy defined for their country by these two political figures was based on gradual economic development and exploitation of local resources (especially minerals and forest products), making extensive use of French financial and technical assistance (still a pivotal economic factor today).

Among Gabon's ethnic groups are peoples such as the BaKota, Fang, and BaKwele who retain traditional forms of social organization and artistic expression associated with secret societies. Typical features of this art, in addition to masks (which resemble Oriental ones in the case of the Mpongive), are reliquaries in which the bones of ancestors are preserved, supported or surmounted by wooden statues that are sometimes covered with sheets of metal.

EQUATORIAL GUINEA

Geopolitical summary

Official name	República de Guinea Ecuatorial
Area	10,828 mi^2 [28,051 km^2]
Population	300,000 (1983 estimate); 417,000 (1990 estimate)
Form of government	Presidential republic; the head of government is elected by universal suffrage for 7 years, the People's Representative Assembly every 5.
Administrative structure	7 provinces, divided into 3 regions
Capital	Malabo (pop. 30,710, 1983 census)
International relations	Member of UN, OAU; associate member of EC
Official language	Spanish; Bantu dialects and Creole are widely spoken
Religion	Majority Catholic (80%)
Currency	CFA franc

Natural environment. Equatorial Guinea is bordered on the north by Cameroon and on the east and south by Gabon; it faces the Atlantic Ocean to the west.

The mainland portion of the country is predominantly mountainous, forming part of the highlands that define the western rim of the Congo basin. These uplands consist of Precambrian crystalline rocks, deeply incised by river valleys and descending to a narrow, undulating coast fringed by islands in the Baie de Corisco (Corisco, Elobey Grande, and Elobey Chico) offshore from the Mitémélé estuary. The insular portion consists of the large island of Bioko (rising to 9866 ft [3008 m] on Santa Isabel Mountain) closest to the shore of the Gulf of Guinea, and the small island of Pagalu (Annobón) 370 mi [600 km] southeast of Bioko. The islands are volcanic in origin, and are closely related to the great fault that led to the formation of Mt. Cameroon and the islands of São Tomé and Príncipe.

The hydrographic system consists essentially of the Río Muni,

Climate data

Location	Altitude (ft asl)	Average temp. (°F) January	Average temp. (°F) July	Average annual precip. (in.)
Malabo	16	77	73	78.0

Conversion factors: 1 ft = 0.3 m; 1 in. = 25 mm; °C = (°F – 32) × 5/9

which rises in the Tembo Mountains of Gabon and crosses the mainland portion from east to west. It is fed by several tributaries, and reaches the sea through a deep estuary that is navigable as far as Niefang. Running almost parallel to the Muni is the Mitémélé river, which opens into the broad Baie de Corisco and also, for a short distance, marks the border with Gabon.

The climate is typically equatorial. Rainfall, which can reach annual levels of between 80 and 120 in. [2000–3000 mm], is particularly frequent in March and October. Temperatures do not vary much, remaining at about 77–79°F [25–26°C]. At Malabo, average annual temperature is 77.7°F [25.4°C], and average annual rainfall is 77 in. [1985 mm].

Vegetation is characterized by the presence of rain forest with many tropical hardwoods, while the coastal swamps are rife with mangroves. The typical fauna is represented by smaller animals: monkeys, birds, reptiles, and insects, including the tsetse fly which prevents the raising of livestock. Numerous crocodiles live in the swamps.

Population. Ethnic groups are different in the two geographical regions. On the island of Bioko, the dominant group consists of the Bubi tribe of Bantu (at present numbering only 15,000); on the continent, the population consists predominantly of the Fang, the same group present in Gabon (also speaking a Bantu language), along with residual bands of Babinga Pygmies in the forest.

According to a UN estimate, the population of Equatorial Guinea was 417,000 in 1990, and is growing at an annual rate of 2.3%. Approximately three-quarters of the people live in the Mbini region on the mainland, where density is very low (18 per mi^2 [7 per km^2] in 1983) and settlements are rural; on Bioko island, 50% of the population is urban, and the density was

Administrative structure

Administrative unit (provinces)	Area (mi^2)	Population (1983 census)
Annobón	7	2,006
Bioko Norte	300	46,221
Bioko Sur	479	10,969
Centro Sur	3,833	52,393
Kié-Ntem	1,522	70,202
Litoral	2,573	66,370
Wele-Nzas	2,115	51,839
EQUATORIAL GUINEA	10,828	300,000

Capital: Malabo, pop. 30,710 (1983 census)
Major cities (1983 census): Bata, pop. 32,734.

Conversion factor: 1 mi^2 = 2.59 km^2

Socioeconomic data

Income (per capita, US$)	310 (1990)
Population growth rate (% per year)	2.3 (1988–89)
Birth rate (annual, per 1,000 pop.)	43.8 (1989)
Mortality rate (annual, per 1,000 pop.)	19.2 (1989)
Life expectancy at birth (years)	44.4 (1989)
Urban population (% of total)	30.2 (1989)
Economically active population (% of total)	41.9 (1988)
Illiteracy (% of total pop.)	68.5 (1991)
Available nutrition (daily calories per capita)	–
Energy consumption (10^6 tons coal equivalent)	0.03 (1987)

88 per mi^2 [34 per km^2] in 1983. The most important urban center is the capital, Malabo (formerly Santa Isabel), located on the north coast of Bioko island; it is an active commercial port and an administrative, economic, and cultural center. Bata on the mainland is a lively port town associated with plantation activities.

The two regions are also characterized by a considerable social imbalance: while the population in Mbini lives in rural villages and practices low-level subsistence agriculture, per capita income on the densely populated island of Bioko is almost three times as high. Illiteracy afflicts 68.5% of the overall population. On the whole, Equatorial Guinea is a poor country that survives on international aid.

Economic summary. The economic structure of Equatorial Guinea is particularly backward, not only because it has very few resources, but also due to a lack of incentives for development which characterized both the colonial era and the dictatorial period that followed it. Per capita income is not too low in terms of the average for the African continent, but it conceals great inequalities between the more developed insular portion and the more backward mainland. Having maintained close relations with the former mother country, Equatorial Guinea has been able to count on significant Spanish aid for its economy, although even this is not enough to make up for the country's insufficiencies, which are particularly acute in the services sector.

The economy is based primarily on plantation agriculture; the most important export crop is cocoa, of which the island of Bioko is one of the leading producers in Africa (annual average production for the three years 1987–1989 was approximately 9000 t), while in the Mbini region wealth is created only by exploiting the forest (covering 82% of the territory), which yields tropical hardwoods such as ebony, rosewood, and okoumé. Subsistence crops—manioc, sweet potatoes, coconut and oil palms—are not sufficient to meet the nation's food requirements.

Fishing, which produces 4400 t of product each year (1989) from the rich surrounding waters, could be developed further. Mineral and energy production is nonexistent: exploratory oil wells are being drilled by French companies, and an electrical power plant is under construction. What little industry exists is based on processing of agricultural products, especially cocoa (for which large processing plants have been built at Bioko) and lumber, which is processed on the coast from which it is then exported.

The balance of trade stays slightly positive thanks to cocoa, which represents 50% of exports, and to limited imports of products required for domestic use. Internal communication routes are inadequate: the road network consists primarily of hard-packed dirt tracks, and there are no railroads. Most transport occurs by sea. Two international airports are in operation, at Malabo and Bata.

Historical and cultural profile. The nucleus of present-day Equatorial Guinea was the island discovered in 1471 by the Portuguese navigator Fernando Póo; initially called Formoso, this island then assumed the name of its European discoverer. Occupied by the Portuguese for trading purposes, in 1778 it was ceded by them to the Spanish, who abandoned it a few years later, decimated by yellow fever. In 1827 it was England's turn to settle on Fernando Póo, having obtained from Spain the right to use two naval bases on the island in order to suppress the slave trade, which had been declared illegal at the beginning of the century. But they, too, left no trace of their presence.

The true colonial scenario began to unfold later, in 1844, when the Spaniards not only returned to the island, but united it administratively with the mainland territory of Río Muni, acquired at that same period on the basis of agreements with local tribal chiefs; the borders of this possession were not defined until later, in the last years of the 19th century.

Fernando Póo and Río Muni thus constituted a colony of the Bourbon crown called Spanish Guinea, and retained that status for more than a century until administrative autonomy was granted in 1964 and the territory adopted its present name. This was the prelude to complete independence, which followed in 1968. Equatorial Guinea became a presidential republic, governed by Francisco Macias Nguema with an increasingly overt style of capricious violence. In 1979 he was deposed by the leaders of a military coup d'état, who released all political prisoners and allowed 100,000 exiles to return to the country. Macias was captured and put to death for "crimes against humanity"; the number of victims of his despotic rule has been estimated at over 30,000.

Some interesting artistic forms are present in the mainland portion of the country, expressed particularly by the Fang, a group also widespread in Gabon. Their highly-prized wooden sculptures symbolize the guardian spirit of the family, and are placed atop reliquaries in which the skulls of ancestors are preserved. Also of interest, in addition to wooden masks, are the tokens used in a game called "abia," with designs carved into fruit stones. There are also dance and music traditions; François Reumax describes some of the latter in the following passage:

"Mvett," a Fang word, describes a musical instrument (a kind of zither), a type of music, and a particular form of sung story. When an itinerant musician arrives in a village or a neighborhood, he invites his audience to select from an extensive repertory of tales drawn from the oral tradition: the most famous of these tells the story of Oveng Obame, chief of the Flames tribe, who succeeded in establishing and maintaining peace by eliminating all the metals from which weapons were made. The little crowd gathered around the storyteller chooses a story by acclamation, and the spectacle begins: it is a complete performance, presented outdoors by the vagabond–musician–dancer–actor–singer and enlivened by the audience, whose questions and comments reinforce and articulate the tale. An authentic Gabonese epic, the "mvett" can last until dawn; it is sustained by a plot that is never the same twice, full of wicked sorcerers, philosophical animals, and mysterious forests.

KENYA

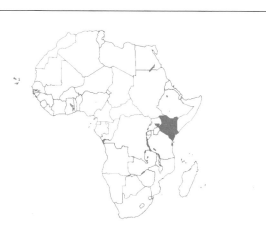

Geopolitical summary

Official name	Jamhuri ya Kenya
Area	224,901 mi^2 [582,646 km^2]
Population	15,327,061 (1979 census); 23,882,000 (1989 estimate)
Form of government	Presidential republic, with a Chamber of Representatives occupied by a single party
Administrative structure	7 provinces plus the capital region
Capital	Nairobi (pop. 1,429,000, 1989 estimate)
International relations	Member of UN, OAU, Commonwealth; associate member of EC
Official language	Swahili; Bantu-Nilotic languages and English also spoken
Religion	Majority animist; Protestants approximately 12%, Catholics 18%, approximately 300,000 Muslims
Currency	Kenyan shilling

Natural environment. Kenya is bordered on the north by Ethiopia and Sudan, on the west by Uganda, on the south by Tanzania, and on the east by Somalia; it faces the Indian Ocean to the southeast.

Geological structure and relief. Kenya's territory, with elevations ranging from sea level to more than 17,000 ft [5000 m] on Mt. Kenya, is characterized by extremely diverse environmental conditions, in terms of both morphological characteristics and precipitation and climate. The coastline, which extends for more than 260 mi [420 km] along the Indian Ocean, is highly indented, with numerous bays and small peninsulas and islands, often connected by coral reefs; traces of ancient terraces are also visible. Moving inland, the relief rises gently; a large erosion surface of Tertiary origin, cut into the Precambrian endogenous basement and presenting a flat, monotonous appearance rarely

punctuated by *Inselberge* (residual rock formations), extends as far as the central volcanic highlands. To the north, and especially in the region extending between the west shore of Lake Turkana and the Somali border, the lowland Kenyan plain takes on the characteristics of an arid semi-desert expanse, partly covered by recent sediments similar to the Kalahari series. The region of the west central plateaus, predominantly igneous and metamorphic, has an average elevation of some 5000 ft [1500 m]; it reaches much greater heights at certain major complexes of volcanic origin (especially Mt. Kenya at 17,053 ft [5199 m] and Mt. Elgon, 15,180 ft [4628 m], located on the border with Uganda), which formed near the great tectonic trench of the Rift Valley, the eastern branch of which cuts through the region from north to south.

In the passage below, Ernest Hemingway describes not only a safari, but the verdant countryside of the Rift Valley:

> That had been a fine hunt. The afternoon of the day we came into the country we walked about four miles from camp along a deep rhino trail that graded through the grassy hills with their abandoned orchard-looking trees, as smoothly and evenly as though an engineer had planned it. The trail was a foot deep in the ground and smoothly worn and we left it where it slanted down through a divide in the hills like a dry irrigation ditch and climbed, sweating, the small, steep hill on the right to sit there with our backs against the hilltop and glass the country. It was a green, pleasant country, with hills below the forest that grew thick on the side of a mountain, and it was cut by the valleys of several watercourses that came down out of the thick timber on the mountain. Fingers of the forest came down onto the heads of some of the slopes and it was there, at the forest edge, that we watched for rhino to come out. If you looked away from the forest and the mountain side you could follow the watercourses and the hilly slope of the land down until the land flattened and the grass was brown and burned and, away, across a long sweep of country, was the brown Rift Valley and the shine of Lake Manyara.
>
> We all lay there on the hillside and watched the country carefully for rhino. Droopy was on the other side of the hilltop, squatted on his heels, looking, and M'Cola sat below us. There was a cool breeze from the east and it blew the grass in waves on the hillsides. There were many large white clouds and the tall trees of the forest on the mountain side grew so closely and were so foliaged that it looked as though you could walk on their tops. Behind this mountain there was a gap and then another mountain and the far mountain was dark blue with forest in the distance.
>
> Until five o'clock we did not see anything. Then, without the glasses, I saw something moving over the shoulder of one of the valleys towards a strip of the timber. In the glasses it was a rhino, showing very clear and minute at the distance, red-colored in the sun, moving with a quick waterbug-like motion across the hill. Then there were three more of them that came out of the forest, dark in the shadow, and two that fought, tinily, in the glasses, pushing head-on, fighting in front of a clump of bushes while we watched them and the light failed. It was too dark to get down the hill, across the valley and up the narrow slope of mountain side to them in time for a shot. So we went back to the camp ...
>
> We were excited that night because we had seen the three rhino ...

In the central region, small volcanic cones of recent origin are scattered within this colossal fracture, along with numerous

Climate data

Location	Altitude (ft asl)	Average temp. (°F) January	Average temp. (°F) July	Average annual precip. (in.)
Nairobi	5448	65	60	37.8
Kisumu	3723	73	71	69.5
Mombasa	52	81	76	47.2

Conversion factors: 1 ft = 0.3 m; 1 in. = 25 mm; °C = (°F − 32) × 5/9

lakes (including Turkana, Baringo, Nakuru, and Naivasha). Along its edges rises a majestic system of mountains, with the Aberdare range (13,100 ft [3994 m]) rising on one side, and the Mau Escarpment (9870 ft [3009 m]) on the other. To the southwest, the complex of high plateaus descends gently toward the wide basin of Lake Victoria.

Hydrography. Kenya's hydrographic structure is extremely simple: all the major watercourses radiate outward from the central highlands, or from the southern formations of the Ethiopian plateau, and flow either toward the Indian Ocean (Tana, 439 mi [708 km]); Galana (339 mi [547 km]) or toward Lake Victoria (Mara, 180 mi [290 km]). Although these waterways are extremely numerous, very few of them are perennial. Because of evaporation and the endorheic drainage, many bodies of water within the Rift Valley are very saline.

Climate. Its equatorial latitude, topography, and outlet to the Indian Ocean, along with the continental location of some of the farthest inland regions, represent just some of the many factors capable of influencing climatic conditions in Kenya, which are therefore rather diverse. Because they are subject to the northeast monsoon—a dry, dust-laden wind that blows with particular intensity from November to May—the northern and central regions are predominantly characterized by a semi-desert climate (less than 20 in. [500 mm] of rain per year); this merges into true desert in the area surrounding Lake Turkana (less than 10 in. [250 mm]). The western edge of the country, however, receives much more frequent precipitation, which can reach 60 in. [1500 mm] per year in the southwestern corner of the country; this region enjoys a sub-equatorial climate thanks to the favorable influence exerted on both temperature and precipitation by the huge mass of water in Lake Victoria. In the highlands, temperature is moderated by altitude, reducing seasonal averages considerably and promoting more frequent precipitation (especially on the eastern slopes of high ground, better exposed to the prevailing winds, which receive an average of 70 in. [1800 mm] of rain every year). The Indian Ocean coast, exposed to the southeast tradewinds loaded with moisture (which blow throughout the period from March to June), is characterized by frequent precipitation and a hot, humid equatorial climate. Temperatures along the coast do not exhibit very significant seasonal variations (the maximum is 88°F [31°C] in the hottest months, and 81°F [27°C] in the coolest); temperature swings in the highlands, however, are slightly greater; daily maximums range between 79°F [26°C] in February and 70°F [21°C] in July and August.

Flora and fauna. The distribution of natural vegetation corresponds approximately to the major climatic regions: the humid zone facing the Indian Ocean, often fringed with a thick band of mangroves, typically has fairly dense vegetation, with some residues of equatorial forest (also present in the region around Lake Victoria) on the hilltops. The characteristic plant cover of the inland regions, however, consists of a variety of associations of species, most of them xerophytic and capable of surviving in the arid environment (savanna or parkland). Only on higher ground can the vegetation become more lush; then as elevation increases, the hillside forests become progressively thinner, blending into savanna and finally alpine meadows. Despite the indiscriminate hunting to which it has been subjected for decades, and progressive loss of habitat due to the expansion of human activities, even today the typical fauna of the African savanna survives quite remarkably well in Kenya. In fact, measures were already being taken in the colonial period to set aside a small portion of the country's territory under strict protection. Besides the national parks (Amboseli, Tsavo, Nakuru, etc.) and the game reserves (Masai Mara, Marsabit, Samburu, etc.), which at present cover 5% of Kenya's land area, several small underwater parks have also been created (Malindi, Watamu, and others), aimed at preserving the most beautiful areas of the barrier reef.

Population. Before the arrival of the Europeans, Kenya's population (predominantly of Bantu stock) lived permanently only in the wet southwestern regions and along the fertile coastal plain, where commercial contacts with the Arabs, dating back to the 10th century, had stimulated the growth of centers such as Mombasa, Lamu, and Malindi. The flat lands of the interior, still practically uninhabited like the highland regions, were home to nomadic tribes of hunters and herders of Nilotic and Hamitic origin (such as the Masai and Turkana), who had begun to supplant the local populations beginning in the 16th century. With the progressive expansion of agricultural activities, practiced mostly by the Bantu peoples and in particular by the large Kikuyu group, an increasingly sharp contrast emerged between settled farmers and nomadic herders, creating a clear separation line between the two groups. This division was further strengthened with the coming of the Europeans, who first constructed a rail-

Administrative structure

Administrative unit (provinces)	Area (mi²)	Population (1989 estimate)
Central	5,086	3,550,000
Coast	32,271	2,065,000
Eastern	61,718	4,193,000
North-Eastern	48,984	611,000
Nyanza	6,239	4,174,000
Rift Valley	67,113	5,128,000
Western	3,227	2,732,000
Nairobi (district)	264	1,429,000
KENYA	224,901(*)	23,882,000

(*) Including 4,335 mi² of inland waters
Capital: Nairobi, pop. 1,429,000 (1989 estimate)
Major cities (1984 estimate): Mombasa, pop. 425,600; Kisumu, pop.167,100; Nakuru, pop. 101,700.
Conversion factor: 1 mi² = 2.59 km²

road designed to connect the port of Mombasa with the interior, and then settled in the highland regions (which would later, under the 1939 law which prohibited the local populations from purchasing land, come to be called the "White Highlands"). Colonization proceeded with the assistance of the docile Kikuyu, while the prouder and more independent Masai were pushed into the more arid regions of the interior.

Karen Blixen (writing as Isak Dinesen), who lived for many years among the Masai, described their pride and elegance:

> *A Masai warrior is a fine sight. Those young men have, to the utmost extent, that particular form of intelligence which we call chic;—daring, and wildly fantastical as they seem, they are still unswervingly true to their own nature, and to an immanent ideal. Their style is not an assumed manner, nor an imitation of a foreign perfection; it has grown from the inside, and is an expression of the race and its history, and their weapons and finery are as much a part of their being as are a stag's antlers....*
>
> *The great contrast, or harmony, between these swollen smooth faces, full necks and broad rounded shoulders, and the surprising narrowness of their waist and hips, the leanness and spareness of the thigh and knee and the long, straight, sinewy leg give them the look of creatures trained through hard discipline to the height of rapaciousness, greed, and gluttony.*
>
> *The Masai walk stiffly, placing one slim foot straight in front of the other, but their movements of arm, wrist and hand are very supple. When a young Masai shoots with a bow and arrow, and lets go the bow-string, you seem to hear the sinews of his long wrist singing in the air with the arrow.*

Even today, these populations of nomads and seminomads, who continue to follow an archaic pastoral way of life, are almost completely excluded from Kenya's economic life. The arrival of the Europeans, and in particular certain needs associated with the building of the railroad, also brought the Indians, who were imported in large numbers by the British as laborers and later proved capable, along with the Arabs, of managing the country's entire commercial network. But after independence was granted, the previously very large Indian population experienced a sharp drop (like the white community, which at present numbers only 46,000 individuals), and today consists of only 180,000 individuals. Still active in developing commercial activities and in the service sector in general, the Indians, like the whites, are concentrated in the urban centers, particularly in Nairobi (the capital created at the beginning of this century at the center of the highlands region) and in the port city of Mombasa. The sedentary native populations, on the other hand, whose livelihood is largely agricultural, are predominantly rural and live in scattered villages and smaller towns. Since rainfall—and therefore the agricultural potential of Kenya's cultivated land—is unevenly distributed, population density is extremely inhomogeneous. Only a fifth of the country can be readily cultivated, and approximately 80% of the population lives on that portion. Thus while certain southwestern areas are characterized by extremely high densities, the great expanses of the interior are sparsely populated even today.

The severe population pressure now being exerted on the cultivated areas will certainly become even worse, since Kenya has experienced very high population growth rates for several decades. In 1931 (the year of the first census), the population was approximately 3 million; in 1989 it was 24 million, and it is conceivable that it could reach 37 million within the next decade. A steady decrease in mortality (which has dropped from 2.5% in 1948 to 1.1% in 1988, and will no doubt decrease further, especially for younger age groups) and a persistently high birth rate (4.6% annually) together give Kenya an annual average growth rate (3.9%) that is among the world's highest. The age structure of the population is therefore decidedly lopsided, with more than half the population less than 14 years old. This enormous demographic pressure imposes high costs on the country in terms of both education and health (there are only 65 hospitals in Kenya today, and 40.8% of the population is still illiterate), while per capita income (US$370) continues to rank among the lowest on the continent.

Economic summary. Although Kenya possesses practically no mineral resources, it is characterized by an economy in which ample leeway is given to private enterprise and private capital, even from foreign sources, placing it in a much better situation than is typical of many neighboring countries. This is the result partly of the development of cash crops for export, and partly of a modest expansion in the industrial sector; also critically important are the proceeds from Kenya's fortunate status as an international tourist destination.

In contrast to the course followed by many other African states, the newly independent Kenyan nation, although striving for an autonomous organization of its local economy, did not indiscriminately expropriate property; those white farmers and Indian merchants who wished to pursue their livelihoods were also permitted to continue to do so. This allowed the country to retain a fairly efficient economic structure, although it was affected by a series of problems of colonial origin that still persist today (including the excessive influence of foreign finance within the industrial sector, and the presence of a land ownership structure that, despite efforts at reform, is still extremely unequal).

Agriculture and livestock. Although agriculture occupies only a small fraction of the country's area, it represents the most important sector of Kenya's economy; this activity not only employs the overwhelming majority (70%) of the work force, but also provides a large proportion of GNP and constitutes the principal source of export income. This is largely the result of the success of the plantation-style agriculture initiated by the European colonizers in the highlands. Commercial agriculture, which is enormously important in terms of the balance of pay-

Socioeconomic data	
Income (per capita, US$)	370 (1990)
Population growth rate (% per year)	3.9 (1980–89)
Birth rate (annual, per 1,000 pop.)	46 (1989)
Mortality rate (annual, per 1,000 pop.)	10 (1989)
Life expectancy at birth (years)	59 (1989)
Urban population (% of total)	23 (1989)
Economically active population (% of total)	40.2 (1988)
Illiteracy (% of total pop.)	40.8 (1991)
Available nutrition (daily calories per capita)	1973 (1989)
Energy consumption (10^6 tons coal equivalent)	2.27 (1987)

ments (tea and coffee alone account for almost half the value of Kenya's exports), is now practiced not only on the large privately owned or government-run estates, but also, as promoted by a specific system of ordinances, by most small, personally owned farms. As a result, although the government pursues a policy of self-sufficiency in food, less attention is devoted to crops for domestic consumption (corn, millet, sorghum, wheat, sweet potatoes, etc.), which are generally grown using mostly outdated methods and on marginal and less-productive land. Animal husbandry, which is an important activity especially in the Rift Valley, supplies good quantities of dairy products, meat, wool, and hides.

Energy resources and industry. Kenya's territory, although it exhibits traces of a considerable variety of minerals, seems to lack any substantial deposits; one exception is diatomite, large quantities of which are found at the Pleistocene base of the Rift Valley. Energy needs are met largely with imported crude oil; besides wood, which provides almost 90% of the energy consumed for household use, the only local resource is hydroelectric power, which is produced in modest quantities (over 2.469 billion kWh in 1989) by a dozen large installations.

By comparison with the situation in many other African countries, Kenya's industry can be considered a relatively prosperous sector. After receiving a substantial boost in the 1960s from a customs union (now dissolved) with Tanzania and Uganda—over which Kenya was already enjoying a certain advantage—conversion activities are now capable of attracting a considerable flow of foreign investment, thanks to a policy of facilitation and tax relief promoted by the government. Processing activity is located for the most part around Nairobi and the port of Mombasa, centering predominantly on petroleum refining and processing of agricultural products. The most important elements in overseas trade, perennially characterized by a severe deficit, are petroleum (imported as crude oil and partially reexported and refined), machinery, and agricultural products.

Commerce and communications. The excess of imports over exports caused by an internal lack of finished goods, energy-related materials, and food products, is partly mitigated by a strong influx of foreign currency brought in by tourists. Visitors from abroad (676,900 in 1989), attracted by Kenya's sparkling white-sand beaches and internationally famous national parks, enter the country through the two international airports of Nairobi and Mombasa, and can make use of numerous other small airfields (including Malindi and Kisumu) for domestic travel. Commercial traffic proceeds not only by air, but also in substantial volumes through the harbor at Mombasa, whose hinterland also includes landlocked countries such as Uganda. Transportation by road and rail is of minor importance, since local infrastructures are generally poorly developed. The railroads, in particular, rely on a network barely 1201 mi [1937 km] in length (1985), with its principal axis being the main line between Mombasa and Nairobi. The highway network is better developed, comprising some 33,581 mi [54,163 km] of roads; however, only a small portion of this (4500 mi [7000 km]) is paved and therefore usable year-round.

Historical and cultural profile. *Arabs and Portuguese on the coast.* Little is known of the early inhabitants of the region of Kenya; they were of Hamitic and Bantu stock, and their rock carvings and burial sites, of fairly recent date, have been discovered in the country's interior. The coastal strip, where Arab colonies were established between the 10th and 12th centuries, is much richer in archeological material. Here the cities of Lamu, Mombasa, and Malindi were founded, focal points in a commercial network which spread out to every country on the shores of the Indian Ocean. Significant structures dating to the 12th century remain at Gedi as evidence of this prosperous age. The arrival of the Portuguese (1498) began a long struggle for commercial hegemony, which ended two hundred years later when they were forced to evacuate Mombasa, which they had conquered and equipped with the imposing defensive structures of Fort Jesus. Once again the Arabs took the place of the Portuguese, and between the 16th and 17th centuries, this 125 mi [200 km] stretch of Kenya's coastline was a possession of the imams of Oman. This period was important not only for commercial development, but also in cultural terms, since it saw the development of a literature, written in an alphabet adapted from Arabic, in the Swahili language. Between the 17th and 19th centuries this Swahili literature, based on both local tradition and Arab influences, produced a large number of works in prose and poetry; of particular importance are the "*utendi*" poems, epic works with heroic and religious subjects that also touch on everyday life.

Protectorate and colony. While this historical development was unfolding along the Indian Ocean coastline, changes were also occurring in the interior, with major movements of peoples taking place during this same time period. The expanses of what would later be called the Rift Valley were occupied first by nomadic tribes of Nilotic origin (Dorobo, Luo), then by Kikuyu farmers of Bantu stock (17th century), and later still by the Masai, who even today retain ties to their ancient traditions.

In the last decades of the 19th century, both the coastal strip and the interior became targets for European colonial ambitions. Both Germany and Great Britain were interested in exploiting the region, but the latter gained the upper hand; the British were granted a lease on the coast from the sultan of Zanzibar (who in the meantime had broken away from the Omani imams), and in 1895 made it into a protectorate of their own. From here they governed the interior, which was then declared a colony; in 1901 this region was opened up to the European advance by the construction of the rail line between Mombasa and Uganda. By reserving the best land for whites, the British colonization of Kenya created a dichotomy (destined to become more acute in more recent times) between the highly developed British economic sector, and the indigenous sector mired in its archaic methods of subsistence agriculture. This contradiction had obvious repercussions on African society and traditional cultures; some groups, like the Masai, became detribalized, ending up confined to reservations or forced to migrate to the urban centers, where they were reduced to a proletarian mob.

Karen Blixen's description of her first arrival at Mombasa is also a perceptive historical picture of the different ethnic groups that had washed into Kenya and the city in successive waves:

> *When you first come to the country, landing at Mombasa, you will see, amongst the old light-grey Baobab-trees,—which look not like any earthly kind of vegetation but like porous fossilizations, gigantic belemnites,—grey stone ruins of houses, minarets*

and wells. The same sort of ruins are to be found all the way up the coast, at Takaunga, Kalifi and Lamu. They are the remnants of the towns of the ancient Arab traders in ivory and slaves.

The dhows of the traders knew all the African fairways, and trod the blue paths to the central market-place of Zanzibar. They were familiar with it at the time when Aladdin sent to the Sultan four hundred black slaves loaded with jewels, and when the Sultana feasted with her Negro lover while her husband was hunting, and was put to death for it.

Probably, as these great merchants grew rich, they brought their harems with them to Mombasa and Kalifi, and themselves remained in their villas, by the long white breakers of the Ocean, and the flowering flaming trees, while they sent their expeditions up into the highlands.

For from the wild hard country there, the scorched dry plains, and unknown waterless stretches, from the land of the broad thorn-trees along the rivers, and the diminutive, strong-smelling wild flowers of the black soil, came their wealth. Here, upon the roof of Africa, wandered the heavy, wise, majestic bearer of the ivory. He was deep in his own thoughts and wanted to be left to himself. But he was followed, and shot with poisoned arrows by the little dark Wanderobos, and with long, muzzle-loaded, silver-inlaid guns by the Arabs; he was trapped and thrown into pits all for the sake of his long smooth light-brown tusks, that they sat and waited for it at Zanzibar.

Here, also, little bits of forest-soil were cleared, burned, and planted with sweet potatoes and maize, by a peace-loving shy nation, which was not much good at fighting, or at inventing anything, but wished to be left to themselves, and which, with the ivory, was in great demand on the market.

The greater and lesser birds of prey gathered up here....

The cold sensual Arabs came, contemptuous of death, with their minds, out of business times, on astronomy, algebra, and their harems. With them came their young illegitimate half-brothers the Somali,—impetuous, quarrelsome, abstinent and greedy, who made up for their lack of birth by being zealous Mohammedans, and more faithful to the commandments of the prophet than the children got in wedlock. The Swaheli went along with them, slaves themselves and slave-hearted, cruel, obscene, thievish, full of good sense and jests, running to fat with age.

Up country they were met by the Native bird of prey of the highlands. The Masai came, silent, like tall narrow black shadows, with spears and heavy shields, distrustful of strangers, red-handed, to sell their brothers.

In such a difficult situation, it is obvious that a strong movement for independence would arise, and in the 1920s, several Kikuyu intellectual circles founded associations such as the Young Kikuyu Association and the Kikuyu Central Association, which put forward political and constitutional claims. This movement became radicalized after World War II, when these associations were absorbed into the nationalist party called the Kenya African National Union (KANU), headed by Jomo Kenyatta. In 1952 a violent peasant-based anti-foreigner rebellion—the Mau-Mau—broke out; it was repressed with particular severity by the British authorities. With 40,000 dead and 100,000 deported (including Kenyatta himself), the country's population was stripped of its already slender constitutional guarantees, and Kenya remained under a state of emergency until 1960.

In 1953 Jomo Kenyatta was put on trial as an instigator, an event he described to Folco Quilici as follows:

I had something to do with the independence movements ... but not with the ones called Mau-Mau, nor was I the head of that organization. If the British found me guilty, it was because they had also found Gandhi, Nehru, Nkrumah, and many other freedom fighters guilty: they were imprisoned because they did not want to continue living under foreign domination. That is why they found me guilty: because I struggled to liberate Africans from a colonial regime.

Independence and neocolonialism. None of this was sufficient to turn aside the liberation movement, which in 1963 brought Kenya to independence. But the new republic created the following year, with Kenyatta as president, very soon moved away from the radical positions of earlier years, adopting a moderate, pro-Western policy in terms of international relations, and pursuing capitalism as an approach to internal development. This permitted Kenya to achieve a level of economic prosperity superior to that of the other nations of eastern Africa, although at the price of severe social inequality: agrarian reforms were not implemented, and the pre-eminence of the former white colonists was reaffirmed, with their land holdings remaining essentially untouched. In 1978 Kenyatta died, and the vice-president, Daniel arap Moi, was proclaimed his successor with no public confirmation. Moi began a program to combat widespread corruption in government offices, but otherwise continued along previously defined political lines.

Literature. Alongside traditional Swahili poetry, which still retains its vitality, 20th-century literary trends included nationalist works often written in English. The years of struggle for independence are expressed, for example, in the novels of James Ngugi, who also addressed the themes of alienation and rootlessness. In *Weep Not, Child*, Ngugi has written:

"Then came the war. It was the first big war. I was then young, a mere boy, although circumcised. All of us were taken by force. We made roads and cleared the forest to make it possible for the warring white man to move more quickly. The war ended. We were all tired. We came home worn out but very ready for whatever the British might give us as a reward. But, more than this, we wanted to go back to the soil and court it to yield, to create, not to destroy. But Ng'o! The land was gone. My father and many others had been moved from our ancestral lands. He died lonely, a poor man waiting for the white man to go. Mugo had said this would come to be. The white man did not go and he died a Muhoi on this very land. It then belonged to Chahira before he sold it to Jacobo. I grew up here, but working ... (here Ngotho looked all around the silent faces and then continued) ... working on the land that belonged to our ancestors—"

In anthropology, an important contribution was made by Kenyatta himself, who wrote a fundamental study of Kikuyu civilization entitled *Facing Mount Kenya*, published in 1938.

RWANDA

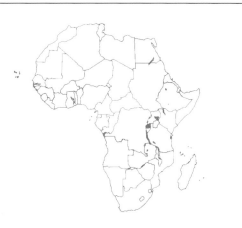

Geopolitical summary

Official name	République Rwandaise / Republika y'u Rwanda
Area	10,166 mi² [26,338 km²]
Population	7,100,000 (1991 census); 5,662,000 (1983 estimate)
Form of government	Presidential republic with a single-party legislative assembly (National Development Council)
Administrative structure	10 prefectures
Capital	Kigali (pop. 234,472, 1991 census)
International relations	Member of UN, OAU; associate member of EC
Official language	French and KinyaRwanda (Bantu)
Religion	Predominantly Catholic (56%); animist 11%, Protestant 13%, Muslim 9%
Currency	Rwanda franc

Natural environment. Rwanda is bordered on the north by Uganda, on the west by Zaire, on the south by Burundi, and on the east by Tanzania.

The country can be subdivided into three natural regions running north–south: a portion of the Rift Valley to the west, comprising part of Lake Kivu (elevation 4790 ft [1460 m]); a narrow band of mountains rising abruptly to 8200–8900 ft [2500–2700 m] near the lakeshore and containing two volcanic structures, Karisimbi (14,783 ft [4507 m]) and Muhavura (13,537 ft [4127 m]), which belong to the Virunga mountain group; lastly, to the east, the topography descends gently into a series of plateaus at between 5000 and 6000 ft [1500–1800 m], studded with volcanic crater lakes.

Rwanda's hydrography is complex. Part of the territory

Climate data

Location	Altitude (ft asl)	Average temp. (°F) January	Average temp. (°F) July	Average annual precip. (in.)
Kigali	4890	67	70	39.4
Conversion factors: 1 ft = 0.3 m; 1 in. = 25 mm; °C = (°F − 32) × 5/9				

belongs to the Nile basin, via the Kagera river which flows into Lake Victoria and drains a dense network of minor streams. To the south, water flows via the Ruzizi river into Lake Tanganyika, and thence via its effluent the Lukuga, into the Congo. There are also endorheic areas.

The climate is equatorial but greatly modified by elevation, and varies from region to region; the central portion enjoys a temperate climate, with annual rainfall of between 40 and 60 in. [1000–1500 mm] distributed in two seasons. Precipitation is highest over the Virunga Mountains. Average temperature varies, depending on elevation, from about 68° to 77°F [20–25°C]. The area around Lake Kivu has a pleasant climate that has resulted in dense human settlement. Vegetation is very lush and varied, although virgin forest now survives only on the slopes of the Virunga Mountains. The eastern uplands support a rich vegetation of grasslands and eucalyptus forests, while papyrus flourishes along the rivers. The typical African fauna, once extremely abundant, is now protected within the vast Kagera National Park which extends to the Tanzanian border, and in the Virunga Mountain park between Lakes Kivu and Bulera.

Population. The dominant group is the BaHutu (90%), agriculturalists of Bantu stock who over the course of time have supplanted the BaTwa (1%), hunter-gatherers related to the Pygmies, who were the first inhabitants of the region. The BaTutsi, a Nilo-Hamitic group from Ethiopia, represent a minority with warlike traditions who live on the plateaus and raise cattle. They have been now been almost decimated by migrations and massacres following the violent overthrow of their supremacy over the BaHutu, which ended in 1962.

Administrative structure

Administrative unit (prefectures)	Area (mi²)	Population (1983 estimate)
Butare	712	683,000
Byumba	1,922	624,000
Cyangugu	864	344,000
Gikongoro	844	402,000
Gisenyi	925	566,000
Gitarama	864	706,000
Kibungo	1,596	420,000
Kibuye	508	501,000
Kigali	1,250	835,000
Ruhengeri	681	581,000
RWANDA	10,166	5,662,000

Capital: Kigali, pop. 234,472 (1991 census).

Conversion factor: 1 mi² = 2.59 km²

With a population density of 699 per mi² [270 per km²], Rwanda is one of the most densely populated countries in Africa. Population growth rate is 3.2% with a total population of almost 7 million (according to the 1991 census). The most densely populated areas are those around Lake Kivu and on the upland plains, where the climate is healthier. The country's only urban center, the capital city of Kigali, is located in the central region at an elevation of almost 5000 ft [1500 m]; it is the major marketplace for agricultural produce and livestock, and performs administrative functions. Other centers include Nyanza, Butare, Gisenyi, and Gitarama.

Most of Rwanda's population lives at the edge of survival: the high growth rate and high density are creating severe overpopulation problems. The educational system provides free, mandatory primary instruction, along with secondary schooling and higher education at the university at Butare. The government's most pressing task is to promote literacy, since the illiteracy rate is high (50.6%).

Economic summary. The economy does not present a very encouraging picture. The dominant activity, employing 82% of the work force and accounting for 37% of GNP, is agriculture, primarily on plantations. Production includes 47,300 t of coffee (1991), 14,300 t of tea, 4400 t of tobacco, and 19,800 t of peanuts. In the food sector, garden produce is important along with cereals (corn, millet, and sorghum).

Local conditions are favorable for livestock: holdings in 1990 were 630,000 head of cattle, and more than a million sheep and goats. The forests (21.1% of the country's area) produced over 205 million ft³ [5.8 million m³] of timber. Fishing in Lake Kivu yielded 1619 t of salable merchandise (1989).

Industrial development is very limited, since the authorities, after trying to promote it with an incentive program, have decided to focus their support activities on agriculture in order to make that sector more productive and efficient. With 7% of the labor force, industry is mainly based on the processing of agricultural products, with textile mills, soap factories, and breweries. A tin processing plant is operating at Karuruma. Limited quantities of cassiterite (a tin ore), tungsten, gold, and columbite are mined. Natural gas reserves exist beneath Lake Kivu.

The balance of payments is decidedly negative: exports cover less than a third of the value of imports. The principal exports are coffee, tea, hides and leather, cassiterite, tungsten, and pyrethrum, while machinery, textile products, and fuel are imported. The major trading partners are the Benelux countries, Germany, China, Canada, and Japan. Rwanda's geographical position complicates international trade links, except for navigation on Lake Kivu (with Zaire). In 1987 the internal transportation system consists exclusively of 7483 mi [12,070 km] of roads (including 1350 mi [2200 km] of major highways). Two international airports at Kigali and Kamembe provide air transportation.

Historical and cultural profile. The BaTwa tribe of hunters and gatherers was succeeded in the region of Rwanda, at a period that is difficult to define, by BaHutu populations, who cleared the forests and lived by farming. This group was conquered in the 16th century by the BaTutsi, a people of Ethiopian origin and warlike traditions who unified the country and incorporated it into a feudal structure that relegated the BaHutu to the status of serfs. The implacable hostility between these two peoples never ceased to smolder; in more recent times the colonial occupation certainly did nothing to resolve it, since colonial rule was in fact built on those same feudal structures. At the end of the 19th century, Rwanda became part of German West Africa, administratively combined with Burundi, the region bordering it to the south, to constitute the territory of Ruanda-Urundi.

At the end of World War I this area was assigned to Belgium as a League of Nations mandate, which in 1946 was converted into a UN trusteeship. These were not, however, the decisive events in the history of Rwanda: the turning point came in 1959 when the BaHutu revolted against the BaTutsi elite. This uprising led to Rwanda's separation from Burundi, and to the forced departure of its last king, or *nwami*, Kigeri V. The end of the monarchy was confirmed in 1962 by an institutional referendum in which a republican form of government was selected, and by political elections held under the auspices of the UN, which brought the Republican Party (Parmehetu) to power. Independence from Belgium (on July 1, 1962) did nothing to improve living conditions in the country; violent clashes with racial overtones continued for years, fomented by neighboring Burundi, which until 1966 continued to be ruled by a monarch. An attempt at rapprochement between the two nations took place in 1969, when Grégoire Kaybanda, president of the republic and prime minister since independence, participated in a conference at Gisenyi, with the aim of creating an economic cooperation organism for the countries of southern Africa. But the situation worsened again in the years that followed: in 1973 there was an actual shooting war between the two states, whose relations began to improve only after the proclamation in Rwanda of a second republic headed by General Juvénal Habyarimana, who found himself faced with the difficult task of reconstructing a nation exhausted by both ethnic strife and an economic crisis.

In 1975 the military allowed civilians to hold positions both in the government and in the dominant party, the Revolutionary Movement for Development, and this action helped calm the situation and begin building a consensus around the figure of Habyarimana, who was returned to office in the presidential elections of both 1978 and 1983 (in both of which he was the only candidate). For a few years Rwanda enjoyed relative ethnic peace, until it was invaded in 1990 by armed groups of BaTutsi who had taken refuge in Uganda. The advance of this force toward the capital, Kigali, was halted in bitter fighting, in which French and Belgian paratroopers and Zairian troops participated alongside Rwanda's own army.

Socioeconomic data	
Income (per capita, US$)	310 (1990)
Population growth rate (% per year)	3.2 (1980–89)
Birth rate (annual, per 1,000 pop.)	52 (1989)
Mortality rate (annual, per 1,000 pop.)	17 (1989)
Life expectancy at birth (years)	49 (1989)
Urban population (% of total)	7 (1989)
Economically active population (% of total)	49.1 (1988)
Illiteracy (% of total pop.)	50.6 (1991)
Available nutrition (daily calories per capita)	1786 (1989)
Energy consumption (10^6 tons coal equivalent)	0.2 (1987)

SÃO TOMÉ E PRÍNCIPE

Geopolitical summary

Official name	República Democrática da São Tomé e Príncipe
Area	386 mi^2 [1,001 km^2]
Population	96,611 (1981 census); 118,000 (1989 estimate)
Form of government	Presidential republic, with a one-party People's Assembly
Administrative structure	2 provinces
Capital	São Tomé (pop. 34, 997, 1984 estimate)
International relations	Member of UN, OAU; associate member of EC
Official language	Portuguese; Creole also spoken
Religion	Predominantly Catholic
Currency	Dobra

Natural environment. The territory of this country—one of the smallest in Africa, along with the Seychelles—consists of two volcanic islands with outcrops of trachitic and basaltic rocks: São Tomé and Príncipe, which rise from the waters of the Gulf of Guinea near the Equator. They are also aligned on the great tectonic axis (submarine Cameroon ridge) that has led to the formation of the neighboring islands of Annobón (Pagalu) to the southwest and Fernando Póo (Bioko) to the northeast, as well as Mt. Cameroon itself.

The larger island is São Tomé (332 mi^2 [859 km^2]); the equator passes just off its southern tip. The island consists of a mas-

Climate data

Location	Altitude (ft asl)	Average temp. (°F) January	Average temp. (°F) July	Average annual precip. (in.)
São Tomé	16	79	75	37.4

Conversion factors: 1 ft = 0.3 m; 1 in. = 25 mm; °C = (°F − 32) × 5/9

Administrative structure

Administrative unit (provinces)	Area (mi^2)	Population (1981 census)
Príncipe	55	5,255
São Tomé	332	91,356
SÃO TOMÉ E PRÍNCIPE	386	96,611

Capital: São Tomé, pop. 34,997 (1984 estimate)

Conversion factor: 1 mi^2 = 2.59 km^2

sive volcanic cone (Pico São Tomé) rising to 6639 ft [2024 m] above sea level, surrounded by a forbidding coastline. The island of Príncipe, which is both smaller (55 mi^2 [142 km^2]) and lower (3109 ft [948 m]), is located 90 mi [150 km] to the northeast.

The climate of these two islands is distinctly equatorial, with precipitation throughout the year: annual rainfall at São Tomé is about 37 in. [950 mm], and average monthly temperatures range between 79°F [26°C] in January and 75°F [24°C] in July. A few substantial stretches of the lush tropical forests that once covered the islands still remain in the deepest interior; the rest is occupied by plantations and numerous farming villages.

Population and economy. Following the departure of the former Portuguese colonists after independence (1975), the population consists of black *"angolares"* and mixed-race mulattoes. The principal city is the capital, São Tomé, with a population of about 35,000 (one third of the entire nation). Most of the other inhabitants live in villages scattered among the plantations.

The economy, almost completely nationalized, is based essentially on tropical agriculture, yielding cocoa, coconuts, and sugar cane. Coffee production is declining, however. Other crops such as bananas and oil palms are now being developed. Exports are directed toward Europe (Portugal, Netherlands, Germany) and a few African countries (including Angola). The balance of trade is consistently negative, since there are no substantial local production facilities. Communications are provided by some 248 mi [400 km] of roads in 1988, the harbor at Ana Chaves, and the airport at São Tomé.

Historical profile. Because of their strategic location in the Gulf of Guinea, the islands of São Tomé and Príncipe were among the first territories reached by Europeans as they explored around the African continent: the Portuguese landed there in

Socioeconomic data

Income (per capita, US$)	380 (1990)
Population growth rate (% per year)	2.8 (1985–90)
Birth rate (annual, per 1,000 pop.)	38 (1989)
Mortality rate (annual, per 1,000 pop.)	7 (1989)
Life expectancy at birth (years)	62 (1985–90)
Urban population (% of total)	41.3 (1989)
Economically active population (% of total)	27.7 (1985)
Illiteracy (% of total pop.)	38 (1987)
Available nutrition (daily calories per capita)	2529 (1989)
Energy consumption (10^6 tons coal equivalent)	0.02 (1987)

1471 and found the islands uninhabited.

This was the report set down by the Italian navigator Alvise da Ca' da Mosto, sailing in the service of the Portuguese king:

X. – Description of the Island of São Tomè, of the island called Príncipe, of Anobon island, and of the city of Povoasan.

The Island of São Tomè, which was discovered now some eighty or more years ago by the captains of our king, having been unknown to the Ancients, is circular in shape: in its diameter it is sixty Italian miles wide, or one degree; and it is located below the line of the Equinox, with its horizontal passing through the two Arctic and Antarctic Poles: its day is always equal to its night, nor is even the slightest difference seen whether the Sun be in Cancer or in Capricorn. The Arctic Pole star is invisible: but the Pointers may be seen to rotate a little, and the stars called the Cross are visible very high up. This island has off its eastern shore a little islet called Príncipe, one hundred twenty miles away, which is at present inhabited and cultivated: and the income derived from the sugar belongs to the elder son of our king; which is why it is called "the Prince's island." Towards the southwest there is another uninhabited islet, called Anobon, which is entirely rocky: there is good fishing there; and those who live in São Tomè habitually go there to fish: it is forty leagues distant, two degrees below the line towards the Antarctic pole: an abundance of crocodiles is there, and venomous snakes. This Island of São Tomè, when it was discovered, was everywhere a very dense forest, with straight, green trees that reached to the sky, of many kinds, but sterile. They do not have branches like those we know, some spreading out crosswise and some growing straight; but these all proceed quite straight upward. Having some years ago deforested a large portion, they have built a principal city there, which they call Povoasan, which has a good harbor; it looks to the northeast: the houses are all made of wood, covered with planks. There is a bishop, who at present is of the Villa di Condi, ordained by the supreme Pontiff at the request of our king, with the Corregidor who administers justice. And there are between six hundred and seven hundred households. Many Portuguese, Castilian, French, and Genovese merchants live there; and those of any nation who wish to come and live there are gladly accepted: everyone has a wife and children. And those who are born on this island are white like us: but on occasion it happens that the merchants' white wives will die, and they will take a black one: and no great difficulty is found with this, the Negro inhabitants there being rich and of great intelligence, raising their children in our manner of living and dressing; and those born of these black women are dark in color ...

Subsequent colonization was intense: the first sugar cane plantations were cultivated by deportees and political exiles, who were then replaced, on the coffee and cocoa plantations, by black slaves from nearby Angola.

The first nationalist stirrings, led by the MLSTP (Movement for the Liberation of São Tomé and Príncipe), began in the 1950s, and brought this Portuguese "overseas province" to full independence in 1975. The first president of the republic was Manuel Pinto da Costa, a reformer; four years later he enlarged his own powers by eliminating the post of prime minister, and assuming those duties himself.

SEYCHELLES

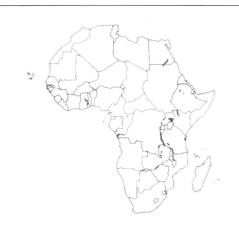

Geopolitical summary

Official name	Republic of Seychelles / République des Seychelles / Repiblik Sesel
Area	175 mi^2 [453 km^2]
Population	61,898 (1977 census); 68,598 (1987 census)
Form of government	Presidential republic with socialist orientation; the People's Assembly represents a single party.
Administrative structure	Central group and minor islands
Capital	Victoria (pop. 23,334, 1977 census)
International relations	Member of UN, OAU, Commonwealth; associate member of EC
Official language	French Creole; English and French are also spoken
Religion	Catholic 90%, Anglican 8%
Currency	Rupee

Natural environment. The Seychelles archipelago, located between the Equator and the island of Madagascar in the Indian Ocean, is another of the smallest African states, although its numerous islands (approximately 90), most of them coral formations, emerge from an expanse of ocean extending over more than 230,000 mi^2 [600,000 km^2]. This area includes part of an underwater ridge that divides the Somali Basin to the north from the Mascarene Basin to the south.

Climate data

Location	Altitude (ft asl)	Average temp. (°F) January	Average temp. (°F) July	Average annual precip. (in.)
Victoria	7	80	78	93.5

Conversion factors: 1 ft = 0.3 m; 1 in. = 25 mm; °C = (°F – 32) × 5/9

Administrative structure

Administrative unit	Area (mi²)	Population (1984 estimate)
Central group (granite)	92	64,300
Outer islands (coral)	83	400
SEYCHELLES	175	64,700

Capital: Victoria, pop. 23,334 (1977 census)

Conversion factor: 1 mi² = 2.59 km²

Geographically, the territory of the Seychelles consists of discrete island groups separated from one another by deep ocean. The actual Seychelles group emerges from a huge semi-submerged granite platform, at the center of which rises the island of Mahé (59 mi² [153 km²]) with an elevation of 2968 ft [905 m] above sea level. Adjacent to it, to the southwest, stretches the Amirante island group, which gives its name to the underwater basin from which the island of Coëtivy emerges. Farther to the southwest, the Amirante trench (more than 13,000 ft [4000 m] deep) separates all these islands from the Farquhar group, east of which lies the Aldabra island group; both groups consist of coral atolls that have developed around submerged volcanic formations, the bases of which are more than 13,000 ft [4000 m] deep.

The climatic conditions of the archipelago are essentially equatorial, with precipitation during every month of the year (more than 78 in. [2000 mm] of rain falls every year at Victoria on Mahé island), and consistently high average monthly temperatures (79–81°F [26–27°C]); the rains are more intense during the period between October and April, when the northeast monsoon is blowing.

The luxuriant tropical vegetation that once covered these islands has been substantially reduced, and is best preserved on the uninhabited islets which, between May and October, host a wide variety of migratory birds (terns, frigate birds, gannets, etc.). Sea turtles, however, are extremely rare, having been intensely hunted in the past. An extremely common representative of the local flora is the sea coconut (*Ladoicea sechellarum*), a unique palm characterized by its double nut.

Population and economy. Most of the population of the archipelago, consisting of Creoles of French origin as well as Asian, black, and Malagasy minorities, is concentrated on the island of Mahé. The only major city is the capital, Victoria, a

Socioeconomic data

Income (per capita, US$)	4670 (1990)
Population growth rate (% per year)	0.7 (1980–89)
Birth rate (annual, per 1,000 pop.)	24 (1989)
Mortality rate (annual, per 1,000 pop.)	8.1 (1989)
Life expectancy at birth (years)	67.3 (1989)
Urban population (% of total)	47.2 (1989)
Economically active population (% of total)	28 (1985)
Illiteracy (% of total pop.)	42.7 (1991)
Available nutrition (daily calories per capita)	2117 (1989)
Energy consumption (10⁶ tons coal equivalent)	0.77 (1987)

popular tourist destination with an international airport and a lively commercial harbor; it is also the nation's political and administrative center.

The archipelago's economy, traditionally based on fishing and agriculture (coconut palms, tea, cinnamon, vanilla) as well as exports of sea salt and guano (which yields phosphates for fertilizer), has been further invigorated by the advent of tourism, transforming this nation into an extremely popular international destination and bringing development of accommodations and logistical structures. The international airport at Mahé is very busy. Nonetheless there is a substantial deficit in the Seychelles' balance of trade: exports cover less than 20% of the value of imports, which include primarily machinery and transportation equipment, oil and petroleum products, and foodstuffs.

Historical profile. Long known to Arab merchants and Portuguese explorers, the Seychelles were first visited by an English expedition in 1609, then explored and conquered by the French in 1742. At the end of the 18th century they were transferred to the British, but a significant French presence remained. In the period between the 18th and 19th centuries, it was in fact a French noble, J.-B. Quéau de Quinnsy (or de Quincy) who played the decisive role in the islands' history. Clarisse Desiles describes him as follows:

This Parisian aristocrat, a strong, courageous, capable man of acute intelligence, was called the "Talleyrand of the Indian Ocean." He was given the thankless task of defending this French possession against the cupidity of England. The very same year he arrived, a British naval squadron showed up under the command of Commodore Newcome, who sent a message asking permission to collect water and food for his crew. Quinssy replied: "Honor prevents me from giving aid to the enemies of my country." Newcome demanded that the French garrison surrender. Quinssy, with only 46 soldiers and eight cannon, was forced to yield and to draw up the act of capitulation in his own hand. But the English sailed off again without leaving any garrison of their own behind, so Quinssy remained as the French governor of these now officially English islands.

When a French ship arrived, he hoisted the Tricolor; when an English patrol was sighted, he avoided hostilities by raising a flag with the words "Seychelles Capitulation."... The English respected him, and he remained practically the ruler of the islands until his death in 1827 (in 1810, following the conquest of Mauritius, the Seychelles had also definitively become an English possession). The significance of his work lies perhaps not so much in his repeated attempts to establish the cultivation of spices—all of which were more or less successful—but in fact that he introduced elements of French culture that can still be discerned in the islands' inhabitants.

For a century the British governed the Seychelles as a dependency of Mauritius, after which it was converted to a Crown colony. Although the islands were intensively cultivated during the age of slavery, after abolition the transition was made to extensive agriculture, favoring the plantation method which required little manpower. The slow and gradual road to independence culminated in 1976. Initially a republic with a multi-party government, the Seychelles ended up a year later being governed by the single party of France—Albert René, called the People's Progressive Front. In 1981 he was overthrown in a coup d'état supported by the Republic of South Africa.

TANZANIA

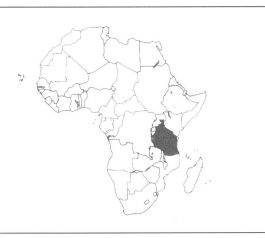

Geopolitical summary

Official name	Jamhuri ya Muungano wa Tanzania
Area	362,635 mi² [939,470 km²]
Population	23,174,336 (1988 census); 25,635,000 (1990 estimate)
Form of government	Presidential republic; the head of government is elected by universal suffrage every 5 years; legislative power is exercised by the National Assembly
Administrative structure	22 regions; the territory of Zanzibar retains special autonomy
Capital	Dodoma (pop. 203,833, 1988 estimate)
International relations	Member of UN, OAU, Commonwealth; associate member of EC
Official language	English; national language is Swahili
Religion	Christian approximately 40% (including over 3 million Catholics); Muslim 33%, animist 23%, Hindu 4%
Currency	Shilling

Natural environment. Tanzania is bordered on the north by Kenya and Uganda, on the northwest by Rwanda and Burundi, on the west by Zaire, on the south by Zambia, Malawi, and Mozambique, and faces the Indian Ocean to the east.

Geological structure and relief. The country occupies a large portion of the central African high plateau, located along the great tectonic trenches partially filled by lake basins. It also faces the Indian Ocean along a rather uneven coastline, fringed with extensive outcroppings of coral reefs and by several large islands (Pemba, Zanzibar, Mafia) which extend along the coast. These islands consist of fossil-bearing, pebbly sediments of fairly recent date (Upper Tertiary and Quaternary), which are also found along the coast itself and form the majority of the outcrops

of the plain furrowed by the lower course of the Great Ruaha river, extending then for about 60 mi [100 km] inland to the banks of the Ruvuma river and continuing into Mozambique. These outcroppings rest on sediments of marine origin deposited during the Jurassic and Cretaceous periods along the entire eastern margin of the African continent. Between the border with Kenya and the Ruaha valley, all these materials are cut off at a great fracture line at which they contact the crystalline rocks of the Precambrian continental basement. The fractures then continue toward the southwest as far as Lake Nyasa (Lake Malawi), constituting, along the upper courses of the Kilombero and Ruhuhu rivers (which flow into Lake Nyasa), a tectonic trench which interrupts the continuity of the basement rock and is filled with materials of the Karroo series (with sandstones, conglomerates, clays, limestones, schists, and coal- and iron-bearing strata). The remainder of the Tanzanian plateau (which lies at an average elevation of more than 5000 ft [1500 m]) consists of basement rocks, covered only in places by more recent sediments, like the Paleozoic ridge that extends from Lake Tanganyika to the shores of Lake Victoria (the Bukoba system, with dolomitic limestones, variegated sandstones, argillaceous schists, and basaltic intrusions), or by Tertiary and Quaternary deposits, as in the Dodoma depression (upper basin of the Ruaha river) or around the shores of Lake Victoria. Lastly, the entire mountainous northeastern area, including the massif containing Kilimanjaro (at 19,336 ft [5895 m], the highest peak in Africa), neighboring Mt. Meru (14,976 ft [4566 m]), and the saline basins of the Rift Valley (Lake Natron, Lake Eyasi, etc.), consists of thick volcanic deposits that originated in the first half of the Tertiary era, resulting from the ascent of trachitic magma along the huge fractures that had formed in the continental substrate. Later tectonic phases, with upwellings of basaltic and trachitic magma, occurred during the Miocene and Pliocene, resulting in another 3000 ft [1000 m] of lava and tuff deposits.

Hydrography. The structure of the hydrographic system in Tanzania is irregular, due to the presence of numerous endorheic basins and the continental watershed. The exorheic basins flow into the Indian Ocean, the Mediterranean (Nile), and the Atlantic (Congo river). The Indian Ocean is the destination of the many watercourses that descend from the plateau, the most important of which are the Pangani which flows down the slopes of Kilimanjaro; the Great Ruaha which drains the Rungwe massif (10,414 ft [3175 m]) and the Kipengere range north of Lake Nyasa; and the Ruvuma which marks the border with Mozambique. The Atlantic receives the waters of Lake Tanganyika and its principal tributary, the Malagarasi, which runs down from the central plateau between the Mahari Mountains (7783 ft [2373 m]) and the Bukoba ridge.

The European discovery of Lake Tanganyika, which occurred on February 18, 1874, later allowed the English explorer V. L. Cameron to establish that it communicated with the Congo river, and did not in fact belong to the hydrographic basin of the Nile:

> *At first I could barely realize it. Lying at the bottom of a steep descent was a bright-blue patch about a mile long, then some trees, and beyond them a great gray expanse, having the appearance of sky with floating clouds. "That the lake?" said I in disdain, looking at the small blue patch below me. "Nonsense!" "It is the lake, master," persisted my men.*

It then dawned on me that the vast gray expanse was the Tanganyika, and that which I had supposed to be clouds were the distant mountains of Ugoma, while the blue patch was only an inlet lighted up by a passing ray of sun.

[After having reached Ujiji, and recovered and sent to the coast Livingstone's maps, Cameron decided to circumnavigate Lake Tanganyika in order to discover its outlet, which Livingstone considered to be the headstream of the Nile.–*Ed.*]

The beauty of the scenery along the shores of the lake requires to be seen to be believed. The vivid greens of various shades among the foliage of the trees, the bright-red sandstone cliffs and blue water, formed a combination of color seeming gaudy in description, but which was in reality harmonious in the extreme.

Birds of various species—white gulls with gray backs and red legs and beaks, long-necked black darters, divers, gray and white kingfishers, and chocolate-colored fish-hawks with white heads and necks, were most numerous; while the occasional snort of a hippopotamus, the sight of the long back of a crocodile, looking like a half-tide rock, and the jumping of fish, reminded one that the water as well as the air was thickly populated.

[After the first few days of sailing they had rounded the treacherous Cape Kabogo, and the little vessel, which Cameron had christened "Betsy," entered a tranquil bay. This was where Livingstone and Stanley had put ashore at the end of their voyage through the northern part of the lake, striking out from there toward Unyanyembe.–*Ed.*]

Lastly, water flows into the Mediterranean from the upper basin of the Nile, with Lake Victoria (the southern half of which is on Tanzanian territory) and its numerous tributaries, including the Kagera, the spring-fed headstream of the Nile.

Climate. In terms of climate, the area of Tanzania exhibits distinctly tropical conditions, although it is greatly influenced by altitude and by the prevailing winds (like the trade winds and monsoons from the Indian Ocean, the former active in the winter months, the latter in the Southern Hemisphere's spring), and by the particular exposure of individual mountain slopes (the eastern being wetter), all of which produce considerable contrasts at different elevations and in different environments. Annual average temperatures are fairly high—between 68° and 77°F [20–25°C], with only minor swings. Average values for January range between 72° and 82°F [22–28°C], and between 70° and 73°F [21–23°C] in July. Precipitation occurs seasonally, and is concentrated in the period from November to May, with annual average totals from 33 in. [850 mm] (Tabora) to 59 in. [1500 mm] (Zanzibar). On higher ground, precipitation normally falls as snow, like that which feeds the glaciers of Kilimanjaro.

Climate data

Location	Altitude (ft asl)	Average temp. (°F) January	Average temp. (°F) July	Average annual precip. (in.)
Dodoma	3657	75	67	22.4
Dar es Salaam	46	82	74	41.9
Mwanza	3720	74	73	44.1
Tabora	3897	73	71	33.5

Conversion factors: 1 ft = 0.3 m; 1 in. = 25 mm; °C = (°F – 32) × 5/9

Flora and fauna. This same mountain, together with its neighbors (Meru, etc.), is characterized by substantial changes in temperature and rainfall (and thus also in vegetation) with increasing elevation. A steppe environment in the depressions around Kilimanjaro's base is followed by a wet tropical forest (supported by local rainfall levels that can exceed 80–100 in. [2000–2500 mm] per year) that extends between 5000 and 10,000 ft [1500–3000 m]; higher still are alpine meadows extending up to 13,000 ft [4000 m], and then a true tundra, capped by snow and glaciers. The remainder of Tanzania's vegetation landscape is represented by steppes in the eastern part of the plateaus, and savanna to the west, becoming more heavily forested near the lake basins; the coastal strip is covered by a wet tropical forest.

In the past, Tanganyika, as it was then called, was a favorite destination for hunters, with truly devastating consequences for the typical wild fauna that populated the steppes and savannas (elephant, lion, gazelle, rhinoceros, etc.). Many animal species are now given special protection in parks, some of them quite extensive. These include the great Serengeti National Park (with the world-famous Olduvai Gorge that has yielded some of the earliest hominid fossils), the Ruaha National Park, and the Selous, Rungwa, and Moyowosi reserves.

Population. The ancient Khoisan race that formed part of the population of the high plains around the great African equatorial lakes survives on Tanzanian territory as small groups represented by the Sandawe tribe. Most of the population, however, consists of Bantu tribes, such as the Sukuma, and Nilo-Hamitic peoples such as the Masai, a warlike tribe that lives from its flocks but is gradually becoming sedentary. After independence was achieved (1962), the Arab presence, which was very substantial along the coast, decreased considerably, while the Asians (Indians and Pakistanis) who had immigrated during the period of British colonization remained. The European presence has also dropped markedly since independence.

The most densely populated regions of the country are still the coast along with the two islands of Zanzibar and Pemba, where the Muslim and Hindu populations are particularly large. With the exception of the populous coastal cities such as Dar es Salaam, Zanzibar, Tanga, and Mwanza (the last on Lake Victoria), settlements are predominantly rural. The new capital, Dodoma, located on the inland plateau along the rail line from Dar es Salaam to Tabora, is still being settled.

The traditional forms of rural settlement and nomadic herding were reorganized, in the decades following independence, into socialist-inspired cooperative structures, involving the creation of new agricultural villages (*ujamaa*) supplied with technology and health services in order to promote efficient production. However, the social condition of the population has not derived any particular benefit from this program, and remains rather precarious, although the school attendance rate is very high and illiteracy has dropped to extremely low levels.

Most of the population holds animist beliefs, but Christianity is fairly widespread throughout the country. Both Arabic and Bantu languages are spoken along with English; interaction between the latter two has revitalized the ancient Swahili language.

Economic summary. Tanzania's cultural panorama has withstood the transformations experienced by the nation during co-

Administrative structure

Administrative unit (regions)	Area (mi²)	Population (1988 census)	Administrative unit (regions)	Area (mi²)	Population (1988 census)
Arusha	31,690	1,351,675	Tabora	29,394	1,036,293
Dar es Salaam	538	1,360,850	Tanga	10,297	1,283,636
Dodoma	15,946	1,237,819			
Iringa	21,944	1,208,914	Tanganyika	340,970	22,533,758
Kagera	10,984	1,326,183			
Kigoma	14,297	854,817	Pemba (North/South)	380	265,039
Kilimanjaro	5,115	1,108,699	Zanzibar (North/South and Central)	641	375,539
Lindi	25,491	646,550			
Mara	8,399	970,942	Zanzibar	1,021	640,578
Mbeya	23,295	1,476,199			
Morogoro	27,261	1,222,737	TANZANIA	341,991(*)	23,174,336
Mtwara	6,450	889,494			
Mwanza	7,598	1,878,271			
Pwani	12,563	638,015			
Rukwa	26,493	694,974			
Ruvuma	24,576	783,327			
Shinyanga	19,593	1,772,549			
Singida	19,045	791,814			

(*) Excluding 20,644 mi² of inland waters
Capital: Dodoma, pop. 203,833 (1988 estimate)
Major cities (1978 census): Dar es Salaam, pop. 757,346; Zanzibar, pop. 110,669; Mwanza, pop. 110,611; Tanga, pop. 103,409; Mbeya, pop. 76,606.

Conversion factor: 1 mi² = 2.59 km²

lonial domination by Europeans (first Germans and then the British) and Arabs. Production structures, though, have changed profoundly, both because of nationalization and as a consequence of the socialization of agriculture, but the results have not been encouraging; now that former president Nyerere is no longer in power, a return to a market economy is in progress, with greater tolerance for private enterprise and foreign investment.

Due to the lack of mineral wealth, the economy is based primarily on agricultural resources. The most widely grown crops are corn and manioc, which together occupy almost two thirds of the entire cultivated area, and serve mainly to meet domestic needs. Crops grown for export, on the other hand, include coffee, cotton, and sisal, as well as coconut palms (from which copra is obtained); the former are widely grown on the plateau, the latter on the coastal plains. One industry that has been established for centuries on Zanzibar and Pemba is the extraction of clove oil and cultivation of cloves themselves, of which Tanzania is the world's major producer. Other important crops include peanuts, pineapples, bananas, oil palms, sunflowers, sesame, tea, tobacco, ricinus (for castor oil), and sugar cane, as well as a variety of other cereals, tubers, and legumes. Cashew nut cultivation

Socioeconomic data

Income (per capita, US$)	120 (1990)
Population growth rate (% per year)	3.1 (1980–89)
Birth rate (annual, per 1,000 pop.)	47 (1989)
Mortality rate (annual, per 1,000 pop.)	17 (1989)
Life expectancy at birth (years)	49 (1989)
Urban population (% of total)	31 (1989)
Economically active population (% of total)	47.5 (1988)
Illiteracy (% of total pop.)	15 (1991)
Available nutrition (daily calories per capita)	2151 (1989)
Energy consumption (10⁶ tons coal equivalent)	0.94 (1987)

is also typical. Animal holdings include 13 million head of cattle and 11 million sheep and goats (1991). Fishing—in inland waters, especially Lake Victoria—is also widespread.

Mineral resources include small quantities of gold, tin, coal, diamonds, tungsten, and phosphates. Electrical energy production is also modest (885 million kWh in 1989). The predominant industries involve processing of agricultural products: textiles, tobacco, shoes, beer, sugar, and cement are all produced.

The balance of trade is deeply negative. Major exports are cotton, sisal, coffee, tobacco, and cloves; imports include textile products, machinery, food, chemicals, and oil. Considering the size of the country, the communications system is still far from completely efficient. In addition to 37,038 mi [59,738 km] of roads (1984), some 2800 mi [4500 km] of rail lines (1985) are also in operation, linking the Indian Ocean coastline to the inland lakes, and connecting Dar es Salaam to Lusaka in Zambia, and from there to Benguela on the Atlantic. There are major port facilities and airports at Dar es Salaam, Zanzibar, and Tanga.

Historical and cultural profile. *A mosaic of cultures.* During the precolonial period, the territory of present-day Tanzania did not contain any governmental entities with stable, defined borders. Historical development proceeded differently in the two geographical areas that constitute the country (the inland high plains and the region along the Indian Ocean), but both areas were characterized by overlays of different populations and cultures. For example in the continental region, the autochthonous Bushmen, who lived by hunting and gathering, were joined at some time in the past by Bantu populations practicing agriculture, then in turn by groups of Nilo-Hamitic herders.

The coastal strip had a more refined civilization, although it too experienced a melding of disparate elements: these included Arab, Persian, Indonesian, and Indian merchants who followed the monsoon routes and had landed on these shores well before 1000 A.D. in search of gold and ivory. The unity achieved here was not political, but cultural; a variety of Muslim dominions arose along the coast and on the offshore island of Zanzibar, imposing their religion and alphabet on the native black populations; the combination of these two created the cosmopolitan Azanian civilization and the Swahili language.

Zanzibar. It was from this region that foreign penetration into eastern Africa began in the modern era; in the 15th and 16th centuries the Portuguese arrived, although rather than actual colonies, they merely established waystations on the route to India. In the next two centuries the Portuguese were succeeded by Omani Arabs, so numerous that in the first half of the 19th century, both the island of Zanzibar and the mainland coastal strip became subject to Said II, imam of Muscat, who even transferred the capital of his sultanate to Zanzibar and introduced the cultivation of cloves, basis for the island's future prosperity. Said II signed commercial treaties with the European powers, and under pressure from Great Britain, agreed to abolish the trade in slaves involving the mainland, the island, and various foreign companies.

But the slave trade remained active in the interior, and supplied human contraband for export until the end of the century. Through their trading in slaves and ivory, the Arabs maintained close ties with the tribes of the high plains as far inland as Lake Tanganyika; these connections opened the way for the European explorers who very soon penetrated to the heart of black Africa: Speke and Burton in 1857–58, Livingstone in 1886–72, and Stanley in 1871. A struggle to stake out the best territories soon developed between British and German interests, both explorers and private companies.

The German and British occupation. The situation did not become clear until the last decade of the century, when the interior was assigned to Germany, which administered it directly as a colony (German East Africa), while Zanzibar and Pemba were recognized as a British protectorate. Although the sultan maintained his own formal authority over the two islands, actual power was concentrated in the person of the British resident, who functioned simultaneously as his country's consul general and as the sultan's prime minister. After Germany's defeat in World War I, Great Britain was granted a League of Nations mandate over much of German East Africa; the region was given the name Tanganyika, gained relative autonomy after World War II, and became independent in 1961 under the leadership of Julius Nyerere. A few years later Zanzibar also reached the goal of independence, and an anti-Arab revolt by the darker-skinned population (Afro-Shirazi) ejected the sultan and instituted a republic. This allowed Nyerere, in 1964, to achieve a merger between the two nations, forming Tanzania (the name derived from Tanganyika and Zanzibar); the country then embarked on a difficult course of socialist development. With the Arusha Declaration of 1967, Tanzania defined a unique route toward socialism that concentrated on rural development. Its program was as follows:

> *(a) Absence of Exploitation:*
> *A truly Socialist State is one in which all people are workers and in which neither capitalism nor feudalism exists. It does not have two classes of people, a lower class composed of people who work for their living, and an upper class of people who live on the work of others....*

Tanzania is a nation of peasants and workers, but it is not yet a socialist society. It still contains elements of feudalism and capitalism—with their temptations. These feudalistic and capitalistic features of our society could spread and entrench themselves.

(b) Major Means of Production to be under the Control of Peasants and Workers:
The way to build and maintain socialism is to ensure that the major means of production are under the control and ownership of the Peasants and Workers themselves through their Government and their Co-operatives. It is also necessary to ensure that the ruling party is a Party of Peasants and Workers.

These major means of production are: the land; forests; mineral resources; water; oil and electricity; communications; transport; banks; insurance; import and export trade; wholesale business; the steel, machine-tool, arms, motor-car, cement, and fertilizer factories; the textile industry ... [and] large plantations.

Some of these instruments of production are already under the control and ownership of the people's Government.

(c) Democracy:
A state is not socialist simply because all, or all the major, means of production are controlled and owned by the Government. It is necessary for the Government to be elected and led by Peasants and Workers. If the racist Governments of Rhodesia and South Africa were to bring the major means of production in these countries under their control and direction, this would entrench Exploitation. It would not bring about Socialism. There cannot be true Socialism without Democracy.

(d) Socialism is a Belief:
Socialism is a way of life, and a socialist society cannot simply come into existence. A socialist society can only be built by those who believe in, and who themselves practice, the principles of socialism. A committed member of TANU will be a socialist, and his fellow-socialists—that is, his fellow-believers in this political and economic system—are all those in Africa or elsewhere in the world who fight for the rights of peasants and workers. The first duty of a TANU member, and especially of a TANU leader, is to accept these socialist principles, and to live his own life in accordance with them. In particular, a genuine TANU leader will never live off the sweat of another man, nor commit any feudalistic or capitalistic actions.

The successful implementation of socialist objectives depends very much upon the leaders, because socialism is a belief in a particular system of living, and it is difficult for leaders to promote its growth if they do not themselves accept it.

Swahili literature. During the 19th century the Swahili language—a fusion of Arabic, Indian, Portuguese, and English elements within an originally Bantu structure—did not produce original works in Tanzania as it did in Kenya. At the beginning of this century, however, when the Arabic script was abandoned in favor of the Roman alphabet, it emerged as a literary language in which both historically based epic poems (the *utenzi* or *utendi*) and autobiographical stories were written. New genres and topics have been particularly evident in this literature over the past few decades, with forms such as the short novel and the "thriller." The most significant literary personality is Shaaban Robert, a storyteller, poet, and essayist, who has enriched and modernized the Swahili language and is the author of works such as *Utenzi wa vita uya Uhuru*, a long epic poem on the struggle for freedom, published posthumously in 1967.

UGANDA

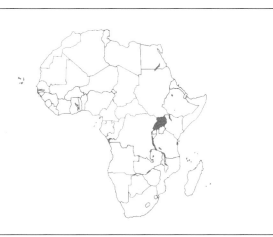

Geopolitical summary

Official name	Republic of Uganda / Jamhuri ya Uganda
Area	93,041 mi^2 [241,038 km^2]
Population	16,582,674 (1991 census)
Form of government	Presidential republic
Administrative structure	10 provinces, including 33 districts divided into 152 counties
Capital	Kampala (pop. 680,800, 1980 estimate)
International relations	Member of UN, OAU, Commonwealth; associate member of EC
Official language	English; Swahili and Bantu languages (Luganda) also spoken
Religion	Catholic 40%, Protestant 20%, Muslim 6%; remainder animist
Currency	Shilling

Natural environment. Uganda is bordered on the north by Sudan, on the west by Zaire, on the south by Rwanda and Tanzania, and on the east by Kenya.

Geological structure and relief. The country occupies a wide stretch of high plateau at an elevation of over 3300 ft [1000 m]; it lies across the equator and is enclosed between the two branches of the great tectonic trench called the Rift Valley, which cut through the structure of east-central Africa. Uganda also includes much of the northern half of Lake Victoria and of other large lakes such as Albert and Edward. As a result, it is contained almost completely within the upper basin of the Nile; the river passes through it from south to north as it emerges from Lake Victoria and continues toward the Sudanese border.

The plateaus that form the framework of Uganda's landscape are cut into the crystalline rocks of the ancient Paleozoic continental basement; in places they are covered by sediments of the Cambrian period. The Rift Valley fractures, which formed in several phases during the Tertiary, not only raised the edges of the plateau—which in some places exceeds 10,000 ft [3000 m] in elevation (10,063 ft [3068 m] on Mt. Kadam near the Kenyan border) and even 17,000 ft [5000 m] near the border with Zaire (17,122 ft [5220 m] on the Ruwenzori massif)—but also allowed acid magma (rhyolitic and nephelinitic) to reach the surface, producing substantial volcanic highlands such as the Elgon massif (15,180 ft [4628 m]) northeast of Lake Victoria, and the Virunga Mountains on the southeastern shore of Lake Edward.

Hydrography. Uganda's hydrographic network, although it is part of the Nile basin, is greatly influenced by tectonics. The part of the Nile that links Lakes Victoria and Kyoga is known as the Victoria Nile, and is interrupted by Owen Falls. The segment connecting Lake Kyoga to Lake Albert is also interrupted by cascades (Kabalega Falls, formerly Murchison Falls). Numerous other watercourses, flowing either directly into the Nile or into Lake Victoria, run through the plateau. The most important of these are the Aswa and the Kagera; a short section of the latter, which is also the spring-fed source of the Nile, forms part of the border between Uganda and Tanzania. The major lake basins, linked to the great African river and its tributaries, occupy the bottom of several deep tectonic trenches, or sit in depressions that in turn are part of various fracture systems. Such is the case with Lake Victoria—whose surface lies at an elevation of 3720 ft [1134 m], and which is nowhere more than 262 ft [80 m] deep—and with Lakes Edward (2991 ft [912 m] in elevation and 850 mi^2 [2200 km^2] in area) and Albert (2030 ft [619 m] in elevation and 2085 mi^2 [5400 km^2] in area). Lake Victoria is the second largest natural body of fresh water in the world (26,287 mi^2 [68,100 km^2]), slightly smaller than Lake Superior. Other lakes include Lake George (95 mi^2 [246 km^2]), which flows into Lake Edward; and Lake Kyoga (3388 ft [1033 m] in elevation and 790 mi^2 [2046 km^2] in area), located between Lakes Victoria and Albert.

Climate. Due to its geographical and topographic situation—traversed by the equator and with average elevations rarely under 3300 ft [1000 m]—Uganda enjoys a typical tropical mountain climate, made considerably more temperate by altitude, that is particularly favorable for human habitation. Average monthly temperatures range from 66–70°F [19–21°C] in January to 66–68°F [19–20°C] in July, with insignificant annual swings but substantial diurnal variations. Nights on the Ugandan high plains can be quite cold. Rainfall is abundant throughout the plateau (annual average of 60–63 in. [1500–1600 mm]), carried by the southeast trade winds from March to June, and by the northeast trades from October to December. Precipitation is even higher in the surrounding mountains, and can even fall as snow; several glaciers exist in the upland valleys of the Ruwenzori, which have an annual precipitation of as much as 157 in. [4000 mm].

Climate data

Location	Altitude (ft asl)	Average temp. (°F) January	Average temp. (°F) July	Average annual precip. (in.)
Kampala	3772	74	70	45.3
Fort Portal	5051	67	66	66.1

Conversion factors: 1 ft = 0.3 m; 1 in. = 25 mm; °C = (°F – 32) × 5/9

Flora and fauna. Uganda's vegetation is quite varied, not least because it has been substantially altered by human activity. Because of the high elevation, the predominant ecosystem is wooded savanna, replaced by typical rainforest along rivers and in lower regions. On higher ground, vegetation develops in altitude bands with alternating forests and temperate meadows, for example on the slopes of the Ruwenzori and Elgon mountains.

The native fauna, typical of the African savanna and including elephant, crocodile, and hippopotamus, was intensively hunted in the past; it is now strictly protected thanks to the creation of several well-known national parks (Kidepo, Kabalega, Ruwenzori).

Population. The demographic picture has been characterized for centuries by an ethnic dichotomy between Bantu farmers and Hamitic herders, with the latter always succeeding in imposing their supremacy. The country suffered an episode of genuine depopulation in the 1970s, when the regime in power brought about the collapse of the economy, both by expelling the Indian and Pakistani immigrants who ran the major economic sectors, and by persecuting—almost to the limits of genocide—any opposition. A semblance of democracy has been restored in recent years and social conditions have slowly started to improve, although per capita income is still among the lowest in Africa.

Most of the population lives on a subsistence basis in small rural villages. The principal urban centers, besides Kampala, the capital, are Jinja and Entebbe on Lake Victoria, Tororo and Mbale in the foothills of Mt. Elgon, Gulu in the north, and Mbarara and Masaka on the plateau west of Lake Victoria.

Due to the labors of early missionaries, Christianity is widely practiced throughout the country, which nevertheless retains animist traditions. Islam is also becoming more widespread.

Economic summary. The principal economic infrastructures are those inherited from the British colonists, who devoted their efforts to plantation agriculture (cotton, coffee, tobacco, sugar

Administrative structure

Administrative unit (provinces)	Area (mi²)	Population (1980 census)
Busoga	7,027	1,221,872
Central	2,494	1,117,648
Eastern	8,399	2,015,530
Karamoja	10,546	350,908
Nile	6,071	811,755
North Buganda	13,028	1,554,371
Northern	16,058	1,261,364
South Buganda	8,222	905,754
Southern	8,290	1,963,428
Western	12,908	1,427,446
UGANDA	93,041(*)	12,630,076

(*) Including 16,960 mi² of inland waters

Capital: Kampala, pop. 650,800 (1990 estimate)

Major cities (1980 census): Jinja, pop. 45,060; Masaka, pop. 29,123; Mbale, pop. 28,039; Entebbe, pop. 20,472.

Conversion factor: 1 mi² = 2.59 km²

Socioeconomic data

Income (per capita, US$)	220 (1990)
Population growth rate (% per year)	3.2 (1980–89)
Birth rate (annual, per 1,000 pop.)	51 (1989)
Mortality rate (annual, per 1,000 pop.)	16 (1989)
Life expectancy at birth (years)	49 (1989)
Urban population (% of total)	10 (1989)
Economically active population (% of total)	44.7 (1988)
Illiteracy (% of total pop.)	51.7 (1991)
Available nutrition (daily calories per capita)	2013 (1989)
Energy consumption (10^6 tons coal equivalent)	0.4 (1987)

cane, tea, etc.) on the high plains, leaving animal husbandry to the local population. Other notable crops include bananas and oil plants such as peanuts and sesame. Cereals (millet, sorghum, corn, rice, and wheat), tubers (potatoes, sweet potatoes, and manioc) and legumes are grown for domestic consumption.

The forests yield rubber as well as mahogany and other valuable hardwoods. Animal holdings comprise over 4 million head of cattle and 4.1 million sheep and goats (1990). Fishing is widely practiced in both lakes and rivers, and supplies the basis for the local diet (265,100 t in 1989).

The economy has recovered from the serious crisis of the 1970s and resumed development, utilizing energy sources provided by hydroelectric plants on the upper Nile (Owen Falls, east of Kampala). There are few important mineral resources, with only modest quantities of tungsten, phosphates, cassiterite, and asbestos. The highland lakes yield small quantities of salt.

The industrial infrastructure is still meager, consisting of a few cement plants (Tororo), a cotton mill, a brewery (Jinja), and a tobacco factory. Steel-making and copper processing plants are in operation at Jinja, and there is a fertilizer factory at Tororo.

Communications are efficient, with approximately 18,600 mi [30,000 km] of roads (1985) and 800 mi [1300 km] of railroad (1986), the latter linked to Kenya's rail system. An active international airport is located at Entebbe. The balance of foreign trade, once positive, is now deeply in the red; principal trading partners are Great Britain, Germany, Japan, and the United States.

Historical and cultural profile. *Warrior kingdoms.* The few remaining traces of prehistoric populations that lived in the region between Lakes Victoria and Albert are rock paintings of animals, stylized human figures, canoes, and concentric circles. The ruins discovered at Bingo and Nturi, in the Masaka desert, have been dated to the late Iron Age, and belong to the Wahima kingdom.

Moving into the historic era, by the 16th century the area that is now Uganda was a mosaic of warrior kingdoms, often fighting with one another, inhabited by a fairly heterogeneous population formed by the slow melding of Bantu groups (originating in Cameroon) and Hamitic and Nilotic peoples. Dominance passed from the Ankole kings to the Bunjoro and Toro kingdoms, and then, in the 19th century, to the kingdom of Buganda, whose king, Mutesa, was the first to be contacted by Europeans. In 1862 he welcomed the British explorers J. H. Speke and Grant, made the acquaintance of C. G. Gordon, governor of the eastern Sudan, and agreed to the establishment of Protestant and Catholic missions, which later became bitter rivals in their zeal to convert the region. However, when Stanley visited the country in 1875 the slave trade was still being pursued, and the Muslims were attempting to spread Islam from their coastal bases.

Stanley described the power of the king of Uganda as follows:

The King of Uganda is greatly dreaded in Karagwé. Before Mwanga was deposed no stranger could pass through the land without obtaining his sanction. The Waganda, after the death of Rumanika, had carried matters with such a high hand that they also taxed Ndagara's Arab guests with the same freedom as they would have exacted toll in Uganda. Two years before our arrival the Waganda were in force at Ndagara's capital, and at Kitangulé to command the ferries across the Alexandra Nile. They found Bakari, a coast trader, occupying the place of Hamed Ibrahim at Kafurro, and demanded from him twenty guns and twenty kegs of powder, which he refused on the ground that he was a guest of the King of Karagwé, and not of the King of Uganda; whereupon he and his principal men were shot forthwith. Considering these things it is not likely we should have had a peaceful passage through Karagwé had we adopted this route for the relief of Emin, with such quantities of ammunition and rifles as would have made Uganda so intractable that nothing but a great military force would have been able to bring its king to reason.

It was clearly demonstrated what hold Uganda maintained in Karagwé, when in obedience to a request from twenty-six of the Pasha's people that I should obtain permission of Ndagara for them to remain in the land until they were cured of their ulcers, I sent word to the king that we had several men and women unable to travel through excessive illness. Ndagara returned a reply stating that on no consideration would he permit the people to stay, as if it once reached the ears of the King of Uganda that he allowed strangers to stay in his country, he would be so exasperated that he would not only send a force to kill the strangers, but that Karagwé would be ruined. His reply was given to the Pasha, and he explained and argued with his wearied and sick followers, but, as he said, they were resolved to stay, as they had only a choice of deaths, and as we were already cruelly loaded, there was no help for it.

Colonial era and independence. In 1890 Germany attempted to impose its own protectorate on the region, but under an Anglo-German agreement of that same year, it was Great Britain that definitively established its rights to Buganda and adjacent lands; this agreement was followed by a long period during which protectorate treaties were signed with local potentates. The colonial period was characterized by considerable economic development centered on the raising of cotton—which made the region one of the richest British possessions in Africa—and by the application of an indirect administration system, which allowed Great Britain to govern Uganda (as the colony was first called in 1902) by way of the local feudal monarchical structure.

Because of this policy, the independence movement that took shape after World War II never conflicted directly with colonial authority, and indigenous elements were gradually incorporated into representative organs. Even so, the indirect British administration, which relied on the survival of figures like the traditional sovereign (*kabaka*), created substantial problems between government and the governed, and even within local communities. British anthropologist Lucy Mair explains the situation:

The history of the kingdoms in Uganda illustrates the difference between the British attitude to kings and that of independent African leaders. Nowhere, except in Basutoland and Swaziland, did a king's dominions extend to the frontier of a colony which was to become independent. In all the rest of the new states, there were populations which owed no allegiance to a king, and very often the new political leaders were drawn from these. The story of Uganda just before and after independence dramatically illustrates the relations between kings and their own subjects, kings and the British authorities, kings and politicians who were not their subjects. Of the four Uganda kingdoms, Buganda was by far the richest and most populous, though of course its population was a minority in the country as a whole. Many Ganda, looking back to their traditions of conquest, assumed that they would dominate an independent Uganda—just what the peoples of the north feared and resented. In the years before independence the kingship and its incumbent had been criticized from within Buganda, and the king was perceived by some of the politically-minded as a fainéant foisted on them by the British. But when a British Governor deposed him because he refused to implement a democratic reform, he became a martyr overnight; and from that time on he was the symbol of his country's prestige. He was restored as a 'constitutional monarch', and as independence approached he and his supporters succeeded in obtaining a large degree of local autonomy for Buganda and a smaller degree of independence from the centre for the other kingdoms. By this time, it is clear, what was at stake was the status of the unit symbolized by the king, not devotion to monarchy as a form of government. The Kabaka of Buganda became President of Uganda, but with narrower powers than most presidents have; there was a moment when he and the elected Prime Minister— Obote, a man from the kingless north—each claimed to have dismissed the other. But the final collapse of kingship was sudden. Obote suspected the Kabaka of plotting a military coup and took the initiative himself; and his army drove all the kings into exile.

This process was completed in 1966, and independence was strengthened a year later with the proclamation of a constitutional republic; the groundwork for this had been laid in 1955 when the country was granted autonomy, and then, in 1962, independence within the British Commonwealth. In 1971 Milton Obote was overthrown by General Idi Amin in a coup supported by the British. Invoking a loosely defined African "authenticity," Amin concentrated power in his own hands. Supported solely by an army composed largely of Nubian and southern Sudanese mercenaries, his rule unfolded into a crescendo of capricious violence. In 1973 he expropriated the assets of British companies, dividing the spoils among himself, his family, and his accomplices. An exile movement gathered under the banner of the National Front for the Liberation of Uganda (NFLU), with military support from Tanzania, deposed Amin in 1973; after a chaotic year that saw three presidents come and go, Obote was once again elected head of government. It must be mentioned that the Commonwealth observers expressed grave doubts about the legitimacy of these elections.

Despite its democratic promises, Obote's regime also embarked on the perilous downhill course of military dictatorship, arousing enormous popular discontent. Once again the army ended up lording it over the country, and once again there was widespread corruption, with patronage positions handed out to members of Obote's own ethnic group. As a result the National Resistance Army waged guerrilla war throughout Uganda in 1985–86, ultimately bringing its leader Yuveri Museveni to power as president.

ZAIRE

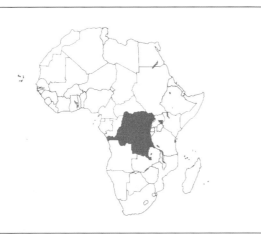

Geopolitical summary

Official name	République du Zaïre
Area	905,126 mi² [2,344,885 km²]
Population	29,671,407 (1984 census); 34,138,000 (1990 estimate)
Form of government	Presidential republic with a one-party legislative council. The head of government is elected every 7 years.
Administrative structure	8 regions plus the capital territory (*ville neutre*)
Capital	Kinshasa (pop. 3,800,000, 1991 estimate)
International relations	Member of UN, OAU; associate member of EC
Official language	French; Bantu and Sudanese languages also spoken
Religion	Catholic 48%, Protestant 28%, Kimbangist (Christian) 17%, Muslim 1.5%; remainder animist
Currency	Zaire

Natural environment. Zaire is bordered on the north by Congo, the Central African Republic, and Sudan; on the east by Uganda, Rwanda, Burundi, and Tanzania; on the west by the Angolan enclave of Cabinda and a short Atlantic coastline; and on the south by Zambia and Angola.

Geological structure and relief. Zaire's territory corresponds almost completely to the left bank of the basin drained by the river of the same name (also commonly referred to as the Congo). The nation's northwestern border runs along the principal right-bank tributary, the Ubangi, and then along the Bomu to the border with the Central African Republic and Sudan. To the east, the Congo river basin is defined by the mountain ranges making up the western margin of the long tectonic furrow occupied by a series of lakes from Albert to Tanganyika, also forming the political boundary of Zaire. The southern border is much more linear, with the upper reaches of many left-bank tributaries

of the Congo remaining in Angolan territory. To the west, Zaire's territory tapers to a narrow corridor following the final stretch of the river down to its mouth and a small portion of the coastal plain.

Geologically and morphologically, Zaire is made up of distinct regional entities: the west-central portion is occupied by the enormous tectonic depression lying at an average elevation of 1000–1300 ft [300–400 m] that contains the vestiges (small lakes and huge marshes) of a large ancient lake basin, and at present is covered by lush tropical forest: it is practically bisected by the Equator. The substrate of this depression, heavily incised and terraced by numerous river valleys, consists of continental Kalahari deposits, including sandstones and both eolian and fluvial sands; above these are the Karroo sedimentary series, also of continental origin, which were deposited after the Hercynian orogeny between the upper Carboniferous and the Triassic, and consist of alternating conglomerates, sandstones, and argillaceous schists. The Karroo series forms the framework of the plateaus which rise to some 3300 ft [1000 m] and define the Congo depression both to the east and to the south of the Kasai-Sankuru valley. The northern high plains, on the other hand, up to the Ubangi-Bomu line, consist of sedimentary formations of Paleozoic age (sandstones, schists, limestones, dolomites, conglomerates), deposited in both marine and continental environments. These facies also crop out to form the high ground closing off the Congo depression to the west—which the river has cut through at Livingstone Falls in its final passage to the Atlantic—and the Shaba plateau in the southeast corner of the country. The Karroo series in turn partially covers the Precambrian endogenous basement rocks (granite, gneiss, schist, etc.) which, along the upper Congo river near the borders with the Central African Republic and Sudan as well as in western Shaba, show traces of repeated peneplanation events; over the entire eastern edge of the country, on the other hand, these rocks are intensely fractured and raised to considerable heights alongside the deep tectonic trench that cradles the great equatorial lakes. The highest peaks here are represented by the volcanoes of the Virunga range (14,783 ft [4507 m]), and farther north, by the Ruwenzori massif (17,122 ft [5220 m]), covered with ice and snow even at these equatorial latitudes. These, like the other peripheral highlands, constitute the boundary of the Congo basin, which often defines the political borders.

To the northwest and north, where the watershed follows approximately the 1650 ft [500 m] contour line, the Congo basin is separated from the basins of the Sanaga and Lake Chad, respectively; while to the northeast, where the adjacent basin is that of the White Nile, the separation is provided by highlands rising to some 6500 ft [2000 m], marking the major continental watershed. To the south, the Congo basin is separated from those of the Zambezi and Kwanza rivers by the Shaba (formerly Katanga) uplands, a succession of tablelands located predominantly at altitudes of 3000–5000 ft [1000–1500 m], which represents the connecting link between the great Zaire depression and the eastern and southern highlands of black Africa.

Hydrography. Zaire possesses an extremely rich hydrographic system, centered on the great Congo basin. The river's perennial flow (approximately 1.4 million ft³ [40,000 m³] per sec) results from the contribution of innumerable tributaries flowing from the south and north toward the equator. Because of its characteristics, it can be said that the Congo's hydrographic system has shaped not only the history of human settlement and European exploration in the area, but also the present-day communications network.

The backbone of this system is the Congo (Zaire), one of the world's longest rivers (2600 mi [4200 km] long), and the second largest, after the Amazon, in terms of the size of its drainage basin (1,425,000 mi^2 [3,690,000 km^2]) and its average flow rate. The initial stretch, from its origin on the slopes of the Shaba highlands near the country's southern border to Kisangani, is called the Lualaba. Along this section it flows generally from south to north, receiving a number of tributaries (Lubudi, Lufira, Luvua, Elila, Ulindi, Lowa), forming several lake basins (Upemba, Kisale), and passing over a series of falls and cataracts such as the Portes d'Enfer and the Boyoma (formerly Stanley) Falls. The river's course downstream from Kisangani (where it receives its definitive name) bends to the northwest and then southwest, crossing the entire equatorial depression and reaching Kinshasa; from there it descends quite rapidly (over the Livingstone Falls) to the Atlantic. It is completely navigable for the 1860-mi [3000-km] stretch between Kisangani and Kinshasa, which Stanley traversed in a steamboat:

The days passed quickly enough. Their earlier hours presented to us every morning panoramas of forest-land, and myriads of forest isles, and broad channels of dead calm water so beshone by the sun that they resembled rivers of quicksilver. In general one might well have said that they were exceedingly monotonous, that is if the traveller was moving upward day by day past the same scenes from such a distance as to lose perception of the details. But we skirted one bank or the other, or steered close to an island to avail ourselves of the deep water, and therefore were saved from the tedium of the monotony.

Seated in an easy-chair scarcely 40 feet from the shore, every revolution of the propeller caused us to see new features of foliage, bank, trees, shrubs, plants, buds and blossoms. We might be indifferent to, or ignorant of the character and virtues of the several plants and varied vegetation we saw, we might have no interest in any portion of the shore, but we certainly forgot the lapse of time while observing the outward forms, and were often kindled into livelier interest whenever an inhabitant of the air or of the water appeared in the field of vision. These delightful views of perfectly calm waters, and vivid green forests with every sprig and leaf still as death, and almost unbroken front line of thick leafy bush sprinkled with butterflies and moths and insects, and wide rivers of shining water, will remain longer in our minds than the stormy aspects which disturbed the exquisite repose of nature almost every afternoon....

Nature and time were at their best for us. The river was neither too high nor too low. Were it the former we should have had the difficulty of finding uninundated ground; had it been the latter we should have been tediously delayed by the shallows. We were permitted to steer generally about 40 yards from the left bank, and to enjoy without interruption over 1000 miles of changing hues and forms of vegetable life, which for their variety, greenness of verdure, and wealth and scent of flowers, the world cannot equal. Tornadoes were rare during the greater portion of the day, whereby we escaped many terrors and perils; they occurred in the evening or the night oftener, when we should be safely moored to the shore. Mosquitoes, gadflies, tsetse and gnats were not so vicious as formerly.... The pugnacious hippopotami and crocodiles were on this occasion well-behaved. The aborigines were modest in their expectations, and

in many instances they gave goats, fowls, and eggs, bananas and plantains ...

From Kinshasa to Matadi, the Livingstone Falls rapids block navigation, which resumes at Matadi for the entire 100 mi [160 km] of the majestic estuary, some 5–10 mi [8–16 km] wide.

The major right-bank tributaries are the Aruwimi and the Ubangi; the left-bank, the Lomami, Kasai, and Kwango. The economic importance of this hydrographic system goes far beyond the fact that it is navigable: it also constitutes an enormous potential source of hydroelectric power, only minimally utilized at present. Also with reference to hydrography, it is worth mentioning that Zaire's territory contains the western sections of Lake Mobutu Sese Seko (the Zairian name for Lake Albert) (2085 mi^2 [5400 km^2]) which belongs to the Nile basin, part of Lake Rutanzige (Edward) farther south (1350 mi^2 [3500 km^2]), as well as portions of Lakes Kivu (1025 mi^2 [2650 km^2]), Tanganyika (12,697 mi^2 [32,893 km^2]), and Mweru (1900 mi^2 [4920 km^2]), which fill part of the long eastern tectonic branch of the Rift Valley.

Climate. Central Zaire, which occupies a huge basin in the central African high plains, has a predominantly equatorial climate, with abundant rainfall and consistently high temperatures. The length of the dry season increases gradually with distance from this equatorial band. Average rainfall for Zaire as a whole is approximately 50 in. [1200 mm] per year, with the maximum about 80 in. [2000 mm] in the equatorial region, and the minimum between 30 and 40 in. [800–1000 mm] on the Shaba highlands. The only area in which annual precipitation reaches a maximum of only 30 in. [800 mm] is Bas-Zaire.

Average annual temperatures are generally between 68° and 77°F [20–25°C], tending to become lower in the upland regions where seasonal variations are also more pronounced; in the southeastern highlands at an elevation of about 5000 ft [1500 m], temperatures are approximately 64–68°F [18–20°C], dropping to 61°F [16°C] higher up. Another significant climatic factor is relative humidity, which reaches very high levels (75–80%) especially in the interior parts of the Congo basin, where days are always hot and humid.

Flora and fauna. These high temperatures, together with abundant and regular rainfall, have encouraged the development of luxuriant vegetation represented by the typical equatorial forest, characterized by stratification of the plant cover into various levels, and generally present within the 60-in. [1500-mm] isohyet. Along watercourses descending from the outer highlands, fringing forests represent true extensions of the evergreen rain forest, which overall occupies an area equal to half the country's territory. It is rich in valuable food and lumber species

Climate data				
Location	Altitude (ft asl)	Average temp. (°F) January	Average temp. (°F) July	Average annual precip. (in.)
Kinshasa	1007	79	73	44.3
Lubumbashi	4034	72	61	48.6
Kananga	2080	76	74	62.4
Kisangani	1404	78	75	67.1
Conversion factors: 1 ft = 0.3 m; 1 in. = 25 mm; °C = (°F − 32) × 5/9				

such as limbal trees, several palm species, *Coffea, Ficus,* mangroves, and—slightly less common—ebony and mahogany, harvesting of which is strictly controlled by the government.

The richness of the sounds and colors of the equatorial forest was also described by Stanley:

> *Burdened with fresh supplies of dried plantains, and guided by the pigmies, we set out from the abandoned grove of Avatiko E.N.E., crossed the clear stream of Ngoki at noon, and at 3 P.M. were encamped by the brook Epeni. We observed numerous traces of the dwarfs in the wilds which we had traversed, in temporary camps, in the crimson skins of the amoma, which they had flung away after eating the acid fruit, in the cracked shells of nuts, in broken twigs that served as guides to the initiated in their mysteries of woodcraft, in bow-traps by the wayside, in the game-pits sunk here and there at the crossings of game-tracks. The land appeared more romantic than anything we had seen. We had wound around wild amphitheatral basins, foliage rising in terraces one above another, painted in different shades of green, and variegated with masses of crimson flowers, and glistening russet, and the snowdrop flowerets of wild mangoes, or the creamy silk floss of the bombax, and as we looked under a layer of foliage that drooped heavily above us, we saw the sunken basin below, an impervious mass of leafage grouped crown to crown like heaped hills of soft satin cushions, promising luxurious rest. Now and then troops of monkeys bounded with prodigious leaps through the branches, others swinging by long tails a hundred feet above our heads, and with marvellous agility hurling their tiny bodies through the air across yawning chasms, and catching an opposite branch, resting for an instant to take a last survey of our line before burying themselves out of sight in the leafy depths. Ibises screamed to their mates to hurry up to view the column of strangers, and touracos argued with one another with all the guttural harshness of a group of Egyptian fellahs, plantain-eaters, sunbirds, grey parrots, green parroquets, and a few white-collared eagles either darted by or sailed across the leafy gulf, or sat drowsily perched in the haze upon aspiring branches. There was an odour of musk, a fragrance of flowers, perfume of lilies mixed with the acrid scent of tusky boars in the air; there were heaps of elephant refuse, the droppings of bush antelopes, the pungent dung of civets, and simians along the tracks, and we were never long away from the sound of rushing rivulets or falling cascades, sunlight streamed in slanting silver lines and shone over the undergrowth and the thick crops of phrynia, arum, and amoma, until their damp leaves glistened, and the dewdrops were brilliant with light.*

The forest thins out on the surrounding high plains, and the vegetation cover takes on the characteristics of a savanna—and, in areas with a longer dry season, a steppe—with typical species such as baobab, sansevieria, aloe, acacia, and other xerophylous grasses and shrubs. The exuberantly lush equatorial forests, the savannas and steppes of the uplands and the mountains are home to the entire range of typical African tropical fauna, including the dreaded tsetse fly, deadly to both humans and animals.

For several years the government has been attempting to protect and develop the nation's wealth of wild animals, creating new parks in addition to those already in existence. These cover a total area of more than 17 million acres [7 million ha], and are located predominantly in the eastern regions, like the Virunga, Maiko, and Kahuzi-Biega national parks. Also noteworthy is the great Salonga National Park in the Congo depression.

Population. Considering the great size of Zaire's territory and the diversity of natural environments with their powerful conditioning factors, the variety of the country's ethnic groups should come as no surprise. The principal group is the Bantu, who have widely settled throughout the northern and southern regions. This group comprises numerous subgroups and tribes, subdivided even further into hundreds of clans, each with its own culture and social structure. Settled in the northern parts of the country are substantial ethnic minorities of Sudanese origin; in the northeast are Nilotic and other groups related to the Sudanese populations of the upper Nile. Small groups of Pygmies, certainly the most ancient inhabitants of this region and now reduced to a few hundred thousand people, still dwell in the depths of the equatorial forest, living by their traditional activities of hunting, fishing, and gathering edible plants.

In everyday life, the population speaks indigenous Bantu and Sudanese languages, subdivided into some 400 dialects spoken primarily in rural areas. The official language, however, is French, used in business and official documents and taught from primary school on. As far as religious beliefs are concerned, only one-fifth of the population follows traditional animist practices, while the other half is Christianized (Catholic, Protestant, and Kimbangist, a syncretistic cult led by the "prophet" Simon Kimbangu). There is also a Muslim minority concentrated in the eastern part of the country.

Demographic structure and dynamics. Zaire's population was more than 34 million in 1990. Relative density is about 39 per mi^2 [15 per km^2], but this drops to 18 per mi^2 [7 per km^2] (1989) in rural areas and rises considerably in urban centers, reaching a maximum of 689 per mi^2 [266 per km^2] in the capital district (1984). Generally, population density and distribution depend on environmental conditions, the incidence of certain diseases (malaria and yellow fever), and on differing levels of economic and urban development. In recent decades the population has more than doubled (from a figure of over 14 million in 1960), thanks to a high birth rate with annual growth consistently on the order of 3.1%, combined with a reduction in mortality rate which, although still high (average life expectancy is 53 years), has dropped substantially. At this rate of increase, Zaire could double its population again by the end of the century; the young (those under 20) already represent more than 50% of the population and their number is likely to increase even further.

Urbanization in the strict sense of the word began with the Belgian colonization (1908) and the almost simultaneous development of mining activities, but has taken on real importance only in the last few decades. Until 1960, any movement of population to urban areas was limited by the Belgian administration, who allowed farmers to buy property in order to promote better land utilization. After decolonization and the end of the war, population shifts became more and more substantial and uncontrollable—the destinations being in most cases the major regional centers (Kinshasa, Lubumbashi, Matadi, Bukavu, and Kisangani)—but were then drastically limited by resolute government action aimed at preventing any further worsening of the social and health-related problems that would have resulted from an ever-increasing mass of people, problems which were already starting to produce marginal conditions in outlying districts. Nonetheless, the urban population has become larger every year due to natural growth and immigration, jumping from 20% in

Administrative structure

Administrative unit (regions)	Area (mi²)	Population (1984 census)
Bandundu	114,124	3,682,845
Bas-Zaïre	20,813	1,971,520
Équateur	155,671	3,405,512
Haut-Zaïre	194,250	4,206,069
Kasai-Occidental	60,589	2,287,416
Kasai-Oriental	64,931	2,402,603
Kivu	99,072	5,187,865
Shaba	191,828	3,874,019
Kinshasa (city)	3,846	2,653,558
ZAIRE	905,126	29,671,407

Capital: Kinshasa, pop. 3,800,000 (1991 estimate)
Major cities (1984 census): Lubumbashi, pop. 543,268; Mbuji-Mayi, pop. 423,363; Kananga, pop. 290,898; Kisangani, pop. 282,650; Bukavu, pop. 171,064; Matadi, pop. 144,742.

Conversion factor: 1 mi² = 2.59 km²

the early 1950s to approximately 39% in 1989, increasing in absolute terms from 2.2 million to about 13 million, one third of which lives in the capital of Kinshasa.

Social conditions. The principal cities of Zaire, with their European-derived structures and architectural forms, are isolated units placed in rural environments at great distances from one another, as often happens in developing countries. The majority of Zaire's population (approximately 61%) lives in small villages in the large expanses between cities; they practice independent subsistence agriculture or work on plantations as wage-earning laborers, and also hunt, fish, and gather wild fruit and other foods to supplement their diets. Rural settlements, although still generally echoing the forms inherited from the "traditional culture" (in terms of village location and organization, type of dwelling, and use of local materials), are experiencing a certain degree of evolution attributable to increased contact between rural populations and the cities, to mass media, and to education.

Economic summary. Since 1972 the economy has been going through a particularly delicate phase, due to a fall in international copper prices and increases in the volume and cost of imports (which led to inflation and thus a deterioration in the balance of trade), and for other reasons (fraud, corruption, and inefficiency among government officials). Little has come of government attempts (1973–1975) to nationalize (or "Zairize") at least 50% of the small and medium-sized mining and agricultural enterprises still held by foreign owners, followed later by the large corporations as well. Many of these enterprises, placed under the control of bureaucrats who often had no managerial talent, experienced a drastic drop in production and were reprivatized in 1976, but without notable results. The economy further deteriorated in 1992, with an almost total standstill of industrial production.

In 1990, gross national product reached US$8.12 billion, equivalent to about US$230 per inhabitant. Farming contributed 38% of GNP, although it employed almost three quarters of the working population; industry and services, employing the remainder of the labor force, contributed 17% and 45%, respec-

tively. To this should be added child labor, which is generally not reflected in the statistics. The International Labor Organization calculates that at least 600,000 Zairian children under the age of 14 work in productive activities.

Agriculture and livestock. Contrasting with a very small percentage (about 3%) of cultivated land and permanent crops is an enormous expanse of territory on which vegetation grows naturally, and which is only partially exploited. The forests, for example, cover more than half the country's total land area, and constitute an extraordinarily rich resource; despite transportation difficulties, exploitation of forest products is constantly increasing (between 1977 and 1989 the figure rose from 459 million to 1.25 billion ft³ [13 million to 35.35 million m³] of felled lumber). The government, by way of a commission established for the purpose, is aiming not only to increase the amount of felling, but also to implement an effective reforestation program which, on a renewal scale of about every 60 years, would support cutting down a volume of industrial timber varying between 318 million and 1.02 billion ft³ [9 and 29 million m³] per year.

The crops most widely cultivated for domestic use are manioc, corn, and millet, which for the most part grow in the less humid regions; the damper areas are used for sweet potatoes, peanuts, and notably rice. Large areas suitable for the cultivation of this cereal exist, and it is becoming a more and more established staple of the local diet, which previously relied mostly on manioc. The most widespread and heavily commercialized plantation crops include cotton, oil palms, and especially coffee. Cotton is widely cultivated in regions with an equatorial climate, both on the savannas north of the equator and in the Shaba and Kasai regions, where several ginning and seed pressing plants are also located. Oil palms are common mostly in the Congo basin and in the area between the Kasai and Kwilu rivers. Coffee is grown especially on the eastern highlands, which are particularly suitable because of their climate and fertile volcanic soil. Despite adverse factors such as the post-colonial crisis, loss of income due to illegal exports (estimated at about 16–22,000 t per year), and fluctuations in world market prices, this crop is always the one that contributes the most to agricultural export earnings. Of lesser importance are rubber-bearing plants, sugar cane, sisal agave, sesame, cocoa, tobacco, and tea (grown especially in the Kivu region). Substantial quantities of fruits and vegetables are grown in the areas surrounding the cities; although they are sold in urban markets, they do not fully satisfy the demand. Notwithstanding recent initiatives, the generally neglected state of Zairian agriculture is obvious. Factors such as the European col-

Socioeconomic data

Income (per capita, US$)	230 (1990)
Population growth rate (% per year)	3.1 (1980–89)
Birth rate (annual, per 1,000 pop.)	45 (1989)
Mortality rate (annual, per 1,000 pop.)	14 (1989)
Life expectancy at birth (years)	53 (1989)
Urban population (% of total)	39 (1989)
Economically active population (% of total)	37.6 (1988)
Illiteracy (% of total pop.)	38.8 (1991)
Available nutrition (daily calories per capita)	2034 (1989)
Energy consumption (10^6 tons coal equivalent)	2.1 (1987)

onization (which always concentrated on plantation production for the European market, expropriating the best land for that purpose), extremely low prices paid to farmers, and an inadequate transportation network, have all slowed the development of traditional agriculture. Only with an enormous quantity of human labor, as the high percentage of the work force employed in agriculture attests (67%, one of the highest in the world), has it been possible to provide a certain meager self-sufficiency of food in villages and rural areas. And it is only by barter, which occurs in every town and wherever the need arises, that the population has succeeded in surviving. Animal husbandry in general, especially involving cattle, could be practiced on a huge scale on Zaire's great expanses of natural meadows and pastures, if it were not limited by trypanosomiasis (sleeping sickness) spread by the tsetse fly. The best animals are those raised on the Ituri, Kivu, Kasai, and Shaba plateaus. Overall animal holdings total 1.55 million head of cattle, 3.97 million sheep and goats, and some 820,000 swine (1990).

Although fishing constitutes one of the sources of subsistence for the population, it is still not being suitably exploited given the enormous fish resources of the country's extensive system of waterways. The use of backward (if ingenious) methods and a lack of processing plants to package and market fish are preventing wider use of this resource. Only a very small portion of the catch (182,600 t in 1989) comes from the sea; the rest is taken from lakes and rivers.

Energy resources and industry. Zaire is particularly rich in mineral resources, to which the country's development is still closely linked. From time immemorial, the indigenous populations have utilized local deposits of iron and copper, but true exploitation did not begin until the beginning of this century under the Belgian colonial administration. The mining organization set up by the Belgians, implemented most intensively in the Shaba region, hinged on a system of stockholding companies which served to gather, on the international money market, the shares needed to start up operations. In October 1906 the Union Minière du Haut-Katanga (UMHK) was established; this company held a monopoly on mining activities in this region, and began industrial exploitation in 1911. Following the events which brought Katanga to the brink of secession, this company was replaced, as of January 1967, by the Société Générale Congolaise des Minérais, 60% of whose stock was held by the government. Exploitation of other mineral resources was also entrusted to companies with a high level of government participation. Zaire's mineral deposits are unevenly distributed within the country: some regions have almost nothing while almost all the resources are concentrated in a few areas, especially Shaba and Kasai-Oriental provinces. This mineral wealth was the stimulus for the separatist movement in these two regions, and the long civil war of the 1960s.

The Shaba mineral region, whose predominant product is copper and which forms part of the "Copper Belt" extending into neighboring Zambia, is some 300 mi [500 km] long and 55 mi [90 km] wide. Open-pit and drift mining, ore-processing installations (separators of the metal from the gangue, foundries, and electric power plants), workers' housing, and road infrastructures are characteristic elements of the landscape on the plateaus of this region. Copper, like the other minerals, is exported both from the Zairian port of Matadi and from other African ports via the continent's rivers and its railroad system, constructed and developed by Belgian engineers during the first decades of colonization. Other products much in demand by modern industry—such as zinc, lead, cadmium, germanium, and cobalt (of which Zaire is the world's leading producer)—are extracted from the copper-bearing ores, which also yield small quantities of platinum, silver, and gold. Shaba also produces almost all the tin and silver extracted in the country, as well as a variety of other minerals (manganese, tungsten, uranium, wolframite) that are increasingly in demand on the world market. The gold deposits, however, are scattered through several regions, although concentrated in the Kivu and Haut-Zaire areas (Ituri and upper Uele basins). Annual production is approximately 70,400 oz [2000 kg]. Zaire is also a world leader in the production of industrial diamonds. The principal diamond pipes, located near Tshikapa and Mbuji-Mayi (Kasai region), had a total production in 1989 of 19 million carats, one-third of the world total.

Industrial installations, which produce goods for domestic consumption, are almost all of small to moderate size and engage either in the processing of agricultural products or in the production of basic necessities (textiles, shoes, etc.). They operate at far less than capacity, and often do not manage to meet domestic demand for various reasons: the most common ones are scarcity of funds and credit, lack of raw materials and spare parts for machinery, and gradual devaluation of investment funds. The country's major processing activities (food processing including breweries, oil processing plants, cigarette factories, grain mills, etc., and the textile sector, shoemaking, chemical and cement plants, etc.) are located only in Kinshasa and a few regional capitals (Lubumbashi, Kisangani, Bukavu). Since 1973, large steel plants, vulcanization and tire production factories, and tractor, truck, and automobile assembly plants have been built in Bas-Zaire between Kinshasa, Matadi, and Maluku, all powered by the hydroelectric plant at Inga.

Despite the enormous potential, electrical energy production in 1989 was some 5.4 billion kWh, most of it supplied by hydroelectric plants (5.3 billion kWh); the remainder came from thermal plants running on coal extracted locally (although some is imported), located for the most part near the country's three major economic development centers: Kinshasa, Kisangani, and Lubumbashi. The highest level of effort in the energy sector is being focused on completion of the Inga project, which involves construction of four hydroelectric plants on the Congo river some 25 mi [40 km] from Matadi; two of these have already been completed by a consortium of Italian companies. Positive results from deep-sea and offshore prospection have led to the identification and exploitation of several petroleum deposits, yielding total production of about 8.2 million bbl [1.3 million metric t] in 1989. Other reserves were discovered in 1981 in the Maidombe subregion. An Italian-built refinery is in operation at Moanda; processing capacity in 1971 was approximately 4.7 billion bbl [750,000 metric t].

Lastly, the world's largest deposits of uranium are located at Shinkolobwe (Shaba). The radioactive material used to manufacture the atomic bombs dropped on Hiroshima and Nagasaki in 1945 was extracted from this mine.

Commerce and communications. Commercial activity is concentrated primarily on exports, which were the basis for the present administrative organization of the country. The Belgian colonizers had designed and implemented Zaire's communication routes and associated infrastructures for the purpose of exporting to world markets the products of the region's mines and plantations, on which the country's economy is still based today. As already mentioned, the majority (80%) of exports consists of mineral resources, three quarters of it copper. Imports, on the other hand, include finished products made of cast iron, iron, and steel, along with machinery and vehicles, consumer goods, some raw materials, and semi-processed products.

The balance of trade maintains a slight excess of exports over imports; most trading partners are Western countries, especially Belgium but also the United States, several European Community nations, South Africa, and Japan. Today as in the past, waterways constitute the foundation of Zaire's communications system. Internal traffic is heaviest on the Congo river between Kinshasa and Kisangani, on the Ubangi as far as Bangui, and on the Kasai up to Ilebo (formerly Port Francqui). In addition to Kinshasa, the center toward which all routes converge, the major ports are those located at the confluence of the major Congo tributaries, where almost all the regional capitals and other smaller centers (Ilebo, Kikwit, etc.) are located; their importance is also due to the fact that they are linked to road and rail lines. These land routes completely replace the river network in those areas where navigation is impossible due to adverse natural conditions. This is particularly the case between Kinshasa and Matadi (Livingstone Falls), and upstream from Kisangani (Boyoma Falls).

Matadi, Zaire's principal seaport, handles approximately half the country's overseas trade. A smaller amount of traffic passes through the port of Boma.

Zaire's single-track, narrow-gauge rail network, designed by the Belgian colonists to service mining activities in the Shaba region, has gradually been extended to peripheral regions that have no waterways suitable for navigation. Its length in 1988 was about 3186 mi [5138 km], with 560 mi [900 km] of it electrified. The railroad lines are connected to those of neighboring countries, particularly Angola, Zambia, and Tanzania, which terminate at the seaports used by Zaire. The road network extends over 28,371 mi [45,760 km] (1980), only a quarter of which is paved; the remainder consists of packed-earth tracks. The main highway is the "Voie Nationale," linking Kinshasa with Lubumbashi in the Shaba region. Zaire has five international airports (Kinshasa, Lubumbashi, Bukavu, Kisangani, and Goma), 40 other large airports, and 150 local airstrips.

One good source of income for Zaire is tourism (39,000 visitors in 1989). Most tourists come to the regions of the major lakes and national parks. Other itineraries follow the main highways, linked to the major airports. Tourism has excellent prospects for expansion, requiring only the development of roads, hotels, and other tourist facilities, and some measure of political stability.

Historical and cultural profile. *The earliest inhabitants.* A number of Bantu kingdoms (Kuba, Luba, Lunda, and Congo) had arisen on the banks of the Congo river before the arrival of the Europeans; their populations practiced agriculture and were characterized by a matrilinear social structure, most typically expressed in artistic terms by sculptures representing a mother with her child in her arms. In addition to this theme, there were also depictions, in the form of both three-dimensional sculptures

and masks, of heroes, ancestors, and rulers, developed by the different tribes in a freely selected variety of styles, ranging from realistic approaches to abstract decompositions with almost "Cubist" results. These local cultures experienced a crisis in the last century with the advance of the white colonizers, who cut off their roots and consequently paralyzed their artistic production.

Belgian colonial rule. The first Europeans who took an interest in these lands were the Portuguese, who by 1483 had already (in the person of Diego Cão) explored the delta of the Congo river.

The description below was written by the Italian Pigafetta after his meeting with Duarte Lopez, who had come to Rome as ambassador of the King of the Congo; it expresses in literary form an image of a country that he had never seen in person, but that he compares with those familiar to him and with the ideas of the classical geographers:

> *In its middle part, the Kingdom of Congo is distant from the Equator in the direction of the Antarctic pole, at the point where the city called Congo is found, by 7 and two thirds degrees, so that it lies in that region which the Ancients believed uninhabitable, which they called the torrid zone, namely the very belt about the Earth, burnt by the Sun. But they were entirely mistaken, for the place is most pleasant, the air temperate beyond all belief; the harsh winter is not felt, but is like the autumn in this region of Rome. They neither use furs nor changes of clothing, nor huddle by the fire; nor is the coolness of the mountain peaks greater than on the plains, indeed generally the winter is a warmer season than the summer by reason of the continuous rains, most heavy two hours before and after midday, such that one may hardly endure them.*
>
> *The men are black as are the women, although some less so, tending toward the olive-skinned, and they have curly black hair, and some also reddish; the men are of middle height and were their color not black they would resemble Portuguese. The pupils of the eyes are of various shades of black, and the color of the sea; and the lips are not large like those of the Nubians and other blacks. Thus their faces are fat and thin and vary as in our own countries, not like the blacks of Nubia and Guinea, who are different. The nights and days there are very similar, so that throughout the year the difference is no greater than a quarter of an hour.*
>
> *Thus in this land the winter, broadly speaking, begins at the time when we have spring, therefore when the Sun enters the northern signs, or the month of March; and when we have winter, as the Sun enters the southern signs in September, their summer begins. In their winter it rains for five months almost continuously, that is April, May, June, July, and August, and there are few calm days as the rain falls so hard and the drops are so large that it is a wonder. These waters soak the earth made arid by the previous hot season, in which it does not rain for six months; thereby it becomes saturated, overflowing the rivers beyond all belief, as they fill with turbid water and spread over the land.*
>
> *The winds that blow in these months in this land are those same which Caesar knew by the Greek word "Etesian," or the ordinary winds of every year, which blow from the North to the West of the compass and to the South-West, and carry the clouds into these very high mountains; which, when they strike them, by their nature they stop and then dissolve into water. From this one sees that when there is rain in the highest mountains, there are clouds upon them.*

But it was Henry Morton Stanley, three centuries later, who explored the river's entire course, revealing the economic potential of its enormous basin, and the way in which it functioned as an access route into the "Dark Continent." The river drew particular attention from Leopold II of Belgium, who in 1876 convened an International Geographical Conference in Brussels, with the stated aim of promoting the exploration of Africa and combating slavery. Financing was provided for a second expedition led by Stanley, who returned three years later to sail up the Congo and establish the city of Vivi. A Research Commission for the Upper Congo was established at the same time; it was later renamed the International Congo Association, and this organization quickly succeeded in changing its nature and gaining recognition as a functioning governmental entity. The Berlin Conference of 1884–85 in fact ratified the existence of an "Independent State of the Congo" with Leopold II as its king; the union between this entity and Belgium resided entirely in the person of the sovereign. He appeared determined, from the outset, to utilize the Congo territory as an inexhaustible source of personal wealth, and indiscriminately exploited indigenous labor in order to squeeze as much as possible out of the trade in rubber and ivory. Thus in 1907, when a groundswell of public opinion forced the king to relinquish "his" possessions, which were then annexed to Belgium as a colony, Parliament passed a series of humanitarian measures against forced labor and other abuses by the concessionary companies, attempting to defend what remained of traditional Congolese society. This was also the context in which many missionaries began studying the oral literature of many ethnic groups, rich in prose and poetry; the transcriptions made at that time would form the basis for analyses by a number of African scholars in the 20th century.

Independence and its problems. Having acquired the status of a colony known as the Belgian Congo, the territory was governed for half a century with no recurrence of the excesses of the preceding thirty years, but the Congolese were allowed no participation whatsoever in their own government; in practice, the constitution of 1908 was not substantially modified until the 1950s. Continued economic development, on the other hand, was not expressed in any improvement in living conditions for the masses, who—especially in the cities—gradually developed a national consciousness and created a broadly based independence front which brought to power the Congolese National Movement headed by Patrice Lumumba. In 1960 Lumumba assumed the duties of prime minister of the new republic that had been created after the proclamation of independence. Independence had been granted by Belgium in great haste, with no time taken to create the essential structures capable of administering the country; there was every hope in Brussels, therefore, that the Belgians could soon intervene in "defense" of this very fragile governing entity. And indeed soon after independence the country was plunged into chaos by a mutiny of the armed forces and the proclamation by Moïse Tshombe of independence for Katanga, a region rich in mineral resources controlled by Western companies. Military intervention by Belgium served only to aggravate anti-foreign sentiments among the population, and only with the arrival of UN troops could the laborious process of normalization begin. In the meantime Lumumba had been assassinated by Katangan forces, but his political legacy was carried on by a revolutionary movement that a few years later, in 1964, embarked on an armed struggle just as the country was experiencing another serious crisis following the appointment of Tshombe, the former secessionist leader, as prime minister.

In his opening remarks to the Panafrican Conference in Leopoldville in 1960, Lumumba spoke in these words to the representatives of the participating nations:

> *You are aware of the causes underlying what is commonly referred to as the "Congo crisis," which in fact represents merely the continuation of a struggle between the forces of oppression and those of liberation.*
>
> *… The classic consequences of colonialism, which we have all experienced and which we are still to some extent experiencing, are particularly deeply rooted here: continued military occupation, tribal divisions which have been promoted and encouraged for years, the deliberate organization of a destructive political opposition, carefully coordinated and supplied with money. You know how difficult it has been in the past for a recently independent nation to rid itself of the military bases installed by the powers which formerly occupied the country. We must affirm, here and now, that as of this moment, Africa will no longer permit imperialist armed forces to remain on its soil. There must be no more Bisertas, or Kitonas, or Kaminas. Now we have our own armies to defend our own countries.*
>
> *… We are here to defend Africa, our heritage, together! We must face the concerted actions of the imperialist powers, of which the Belgian colonialists are merely the instrument, with a common front that unites the free peoples of Africa and those who are still fighting for their own independence. The enemies of freedom must find themselves facing a coalition of free men.*
>
> *The time to make plans has now passed. Today Africa must act. With a single mind, a single will, and a single heart, soon we shall make Africa, our Africa, a truly free and independent continent.*

A bitter civil war, involving both Belgian and U.S. troops, dragged on for another year until General Joseph-Désiré Mobutu mounted a coup d'état in 1965, a turning point in the history of the country, now renamed Zaire after the former name of the river. As spokesman for the most ambitious entrepreneurial middle class in Africa, Mobutu on the one hand embarked on a program of nationalizing certain critical economic sectors in order to regain control of the nation's resources; on the other hand, he strengthened ties with the West rather than breaking them. As a result, the United States in particular made Zaire one of its major strongholds on the African continent. But as the years passed this policy showed signs of strain, stimulating widespread domestic opposition; another factor weakening Mobutu's regime was a flareup of peripheral insurrections in 1978–79. Despite the uncertain domestic situation, Mobutu was elected in 1984 to a third term as president of the republic.

Literature. A large body of literature developed after World War II, written in the local Bantu languages, in French, and in some cases even in Flemish. Its subjects were of course closely linked to the modern realities of Zaire (as in the poetry of P. Kabongo), or to its ancestral memories, as in the case of A. R. Bolama, the most important contemporary Zairian poet as well as the author of sociological and folklore studies. The nationalist tone of Zairian literature became more pronounced over the years, in step with the struggle for independence. It is significant that 1960, the eve of independence, was the publication date of a volume of poetry by Patrice Lumumba.

EQUATORIAL AFRICA

Images

1. *Kenya's animal life is extraordinary and highly diverse. Although conservation legislation dates back to the days of colonial rule, it has not succeeded in preventing serious losses in recent decades. The original African wild fauna now lives almost exclusively in the national parks and game reserves, outside of which these animals are killed for their meat or ivory, or as trophies. Even the national parks are threatened by economic difficulties: farmers see no reason for these vast areas of land to be reserved exclusively for animals, when they could otherwise be utilized for subsistence food production.*

2. *After spreading out into the marshy basin of Malebo Pool, the Congo river narrows considerably and finally becomes impassable to navigation as it reaches the highlands marking the border between Congo and Zaire. It descends quickly to the Atlantic in a series of rapids corresponding to the terraces of the Luanda and Samba plateaus, in a spectacular series of thirty-two cataracts called the Livingstone Falls. The Congo river, which connects to Zaire's entire hydrographic network, is the main transportation route into the interior of equatorial Africa.*

3. *The equatorial forest around the Congo river is home to*

small groups of Pygmies, an African people known since the time of the Pharaohs. Modern anthropologists consider them almost a separate race: their physical structure is not the result of any adaptation to their environment, or crossbreeding, or pathological changes. A typical Pygmy camp consists of a few beehive-shaped huts constructed of branches and leaves, with very meager possessions centered around hunting, their principal means of subsistence.

4. *The Congo river, second longest in Africa (2600 mi [4200 km]) and largest in terms of area drained (1.4 million mi^2 [3.7 million km^2]), begins as the Lualaba river on the slopes of Mt. Musofi on the Katanga plateau in southeastern Zaire. It flows north in almost a straight line, receiving numerous tributaries. Before reaching Kisangani, the river tumbles down the seven cataracts of Boyoma (formerly Stanley) Falls; thereafter, as it flows in a wide and deep bed, it takes the name Congo and becomes navigable. Just upstream from Kinshasa, where it has already reached a width of more than 15 mi [25 km], the river widens into Malebo Pool, a great marsh that fringes the capital city, and then begins its descent to the Atlantic Ocean.*

5. *The dam on the Victoria Nile, built on the northwestern shore of Lake Victoria where the river begins. Lake Victoria, whose surface lies at an elevation of 3719 ft [1134 m] above sea level on the Rift Valley plateau, is the largest lake in Africa. Its catchment basin is some 155,000 mi^2 [400,000 km^2] in size; its average depth is 165 ft [50 m], and the lake's actual area is 26,300 mi^2 [68,100 km^2], 40% of it within Uganda's borders. Its waters are fresh, cool, and shallow, although often whipped up by violent and sudden storms; the lake teems with fish, including enormous tenches weighing up to 650 lb [300 kg].*

6. *Amboseli National Park at the foot of Kilimanjaro, the great mountain in northeastern Tanzania. Kilimanjaro is an extinct volcano standing on a conical base some 55 mi [90 km] in diameter. It is approximately 13,000 ft [4000 m] high, with two main summits: the rocky peak of Mawenzi (17,564 ft [5355 m]), and the dome of Kibo (19,336 ft [5895 m]), the highest peak in Africa. Called "the most beautiful mountain in Africa," it unveils itself only for brief periods: during the day it is often covered by white clouds, which dissipate at sunset, revealing the summit, snow-covered for two or three*

months of the year, and lightly tinged with pink.

7. *Sunset silhouette of a Masai warrior, one of a proud tribe scattered through the vast expanse between Lake Baringo in Kenya and the northern tip of the Tanganyika basin in Tanzania. Long-limbed, with copper-brown skin and curly hair, the Masai dress in a manner adapted to local customs: women generally have shaved heads and wear earrings and necklaces of colored beads, while the men have tightly curled ringlets covered with a paste of red ocher and earth, forming cones that hang down to their shoulders.*

8. *A group of young Masai warriors prepares for a propitiatory dance, wearing their typical costume and accoutrements: the long-bladed spear; short, straight sword; oval leather shield painted with geometric patterns; ocher-colored headdress tipped with feathers; and striped face paint. In recent decades, particularly as a result of contact with the modern world, the Masai are relenting in their age-old avoidance of anything foreign to their own culture.*

9. *A Seychelles native prepares wood and bamboo poles for the framework of a roof. The archipelago, consisting of a total of some 90 atolls (only 36 of them*

inhabited), extends over a wide stretch of the Indian Ocean east of Madagascar. It is composed of four main island groups: to the north are the Seychelles themselves, composed of granite mountains reaching to a height of 3,004 ft [918 m] on Mt. Morne, and the Amirante group, low islets fringed with coral reefs like the two western archipelagos, Farquhar and Aldabra.

10. An inlet on La Digue island, an atoll in the northern Seychelles group near the larger island of Praslin. It still retains its original wild appearance, with enormous blocks of slightly pinkish granite that are scattered like reefs through the Indian Ocean, preventing easy access from seaward.

11. A portion of the tropical forest that extends over 40% of the territory of Zaire, covering the slopes of the Ruwenzori massif; on Margherita Peak (16,794 ft [5120 m]) in Uganda, this mountain marks the highest elevation of the great range that forms the western edge of the great eastern African subsidence trench. Zaire's vegetation varies by region depending on elevation and distance from the Equator, but is primarily tropical with abundant rainfall.

12. Lake Nakuru (Kenya), in the national park of the same name, is famous for its more than 400 species of birds, especially pink flamingos and marabou storks. This is one of the lake basins of the Rift Valley, a huge fissure crossing the western part of the country from north to south, that formed 15 million years ago as a result of intense tectonic activity.

13. Panoramic view of Lake Kivu, on the border between Zaire (on the western shore) and Rwanda. The geomorphology of Rwanda is closely linked to the presence of the great subsidence trench that cuts through much of the African continent, separating the Congo river basin from

the Nile basin and the regions of eastern Africa. Along Rwanda's eastern frontier the enormous tectonic valley is occupied by Lake Kivu, located at an elevation of 4,788 ft [1460 m], and connected to Lake Tanganyika by the Ruzizi river.

14. A view of the modern section of Nairobi, seen from the gardens of Uhuru Park. In the background are the Parliament building with its clock tower, and the circular skyscraper of the Kenyatta Conference Center, obvious signs of the modern European style that characterizes the urban structure of Kenya's capital. Nairobi —the name means "place of water" in the Masai language —was founded in 1889.

15. A street in the modern section of Bangui, capital of the Central African Republic and a major river port at the confluence of the Mpoko and the Ubangi, at the heart of a quadrilateral pattern of rivers that contains the country's highest population density. The city is the jumping-off point for the nation's entire export trade, consisting of diamonds, cotton, lumber, and coffee. Most exports go to France, from which the Republic receives machinery and manufactured cotton goods.

16. View of the modern part of Kampala, capital of Uganda and a commercial center of considerable importance. Its name means literally "place of the impala," a local species of gazelle that is also depicted on the country's coat of arms. Kampala was formerly the official residence of the king of Buganda, and its most recently erected buildings include the graceful city hall (gift of a wealthy Indian merchant), the Kibuli mosque, and the Catholic cathedral of Rubaga, see of Uganda's archbishop.

17. An aerial view of Brazzaville, capital of Congo, stretching in a long arc beside the Congo river where it emerges from Stanley Pool, 650 ft

[200 m] above sea level. A major commercial and industrial center, the city has developed enormously since World War II, thanks to the hydroelectric power station built on the Djoué river and the presence of the river port, which is used only for trade with the interior since the Congo river is not navigable down to its mouth.

18. Harvesting tea on a plantation in Burundi. Burundi is one of the world's poorest nations, with a high population density living on rather barren land with no outlet to the sea; the nation suffers from low levels of per capita income and capital formation, and is highly dependent on exports of a single product, coffee. The foundations of Burundi's economy, even further weakened by its separation from wealthier Rwanda (in 1964), are still agriculture and animal husbandry.

19. A Masai shepherd tends a herd of cattle in a meadow in Amboseli National Park (Kenya). According to the custom of this African people, cattle and other animals are used as trade goods, and may be slaughtered only for ritual purposes. Together with meat and animal blood, their primary food is milk, also regarded as sacred; it is forbidden to drink milk for twelve hours after a meal containing meat.

20. A cargo of fish being unloaded in the harbor at Pointe-Noire, seat of Kouilou prefecture and administrative capital of former French Equatorial Africa, established in 1911. It is Congo's main river outlet to the ocean, and the point through which all overseas maritime trade must pass.

21. The uranium deposits at Mounana, near Lastoursville (Gabon), extend over almost 2000 mi^2 [5000 km^2] and contain reserves of pure ore of at least 4400 t. Precisely because of its substantial mineral reserves, identified but so far only partially exploited, Gabon is potentially one of the richest nations in Africa.

22. A ntad, a stone statuette that guards a tomb according to BaKongo tradition. A Bantu tribe, the BaKongo reached their zenith between the 14th and 17th centuries with the creation of the Kingdom of the Congo; this empire also supported a lively sculptural style inspired by ancestor worship. Typified by statuettes between 20 and 30 in. [50–80 cm] tall, this type of image served not as a portrait of the deceased, but rather as a receptacle for the person's spirit. A figure would also be carved when a chief set out for war, and kept until he returned.

23. The "royal drums" of Gitega (Rwanda) are a phenomenon that is almost unique in African music; these instruments are played during the festivals and the acrobatic ceremonial dances of the BaTutsi, an ethnic minority that has influenced and guided the country's culture and traditions for decades. The drums are symbolic of the former royal period, when they could be played only with the monarch's permission.

24. The ruins of the ancient Arab city of Gedi (Kenya), most probably founded in the 13th century by settlers from Malindi. Most of the remains are of buildings belonging to a city that appears on 17th-century maps with the name of Kilimani; the town was later occupied by Galla nomads and razed to the ground. Buildings include a large mosque, mansions, palaces, and a tomb dated to 1399.

25. The mosque at Bujumbura, present-day capital of Burundi. The city developed around the German military post established in 1890 on the northeast shore of Lake Tanganyika. This photograph shows the considerable influence exerted by the Belgian colonization, which lasted from 1916 until the nation achieved independence in 1962, after separation from neighboring Rwanda.

11

12

SOUTHERN AFRICA

Extending southward between the Atlantic and Indian oceans, this vast region covers an area approximately one fifth the land mass of the entire African continent and embodies all of its most characteristic features, as revealed by both the variety of its landscapes—from the stretches of the Namib and Kalahari deserts to the forests of Madagascar—and its rich human contrasts, with a remarkable mix of peoples, races, and economies: Bushmen, Hottentots, Zulus, Boers, Malagasies, etc., constituting an ethnic and linguistic mosaic that is further complicated by social, political, and economic problems.

Southern Africa is also the Africa of plateaus and mountains, contrasting with the rest of the continent. It has a mean altitude of over 3300 ft [1000 m], mountainous ridges running along the ocean edges, and disappearing rift valleys which on the eastern side have indented its compact coastline. This is also insular Africa, with the large block of Madagascar (which long ago in the Mesozoic separated from the continent and then evolved biologically along entirely original forms), but also with numerous coral archipelagos scattered throughout the warm Indian Ocean, or with the volcanic formations of the Mascarene Islands (Mauritius and Réunion) that emerged from the deep ocean shelves (13–16,000 ft [4–5000 m]). Inscribed in the rocks that form the mountainous skeleton of southern Africa is the entire geological history of the continent: above the vast margins of the pre-Paleozoic crystalline basement, the actual structure of the plateaus consists of a mighty series of stratifications that were deposited during all the geologic eras, from the Cambrian period (Paleozoic) to the Quaternary, spanning a total of about half a billion years. Dating back to the Paleozoic era are the various features of the Karroo tableland (which took its name from the southernmost ranges of Cape Province), including traces of very old glacial formations (such as tillites, morainal and glaciofluvial materials, often intensely metamorphosed) of considerable thicknesses that developed from the Carboniferous until the end of the Triassic. Then, from the middle of the Mesozoic, after a long interval of marine sedimentation, new continental deposits from the Cretaceous period were superimposed over earlier formations, filling in the topographically more depressed areas: that is, the predominantly clayey and sandy Kalahari.

Unlike north central Africa, the southern regions of the continent exhibit a fairly abnormal climate which is determined by both the atmospheric circulation and geographic shape of the coasts and the configuration of the relief, but especially by the influence of external factors, such as marine currents. The abundance of coral formations on the eastern coasts and those of Madagascar is related to the presence of the warm waters of the Mozambique Current, as well as to the latitude itself. On the Atlantic side, it is the cold Benguela Current that predominates, fed by both the Antarctic seas and the deep waters, accounting for a considerable thermal contrast between sea and land, as is further demonstrated by the special isothermal conditions, especially during the summer, along the Namibian coast. The atmospheric circulation exerts its effect primarily on the Indian Ocean side, where the southeasterly trade winds exhaust themselves, leveling off due to a baric depression settled over the Kalahari. This explains the considerable difference in humidity between the arid and desert Atlantic coasts and the eastern coasts, which are richly supplied with rain. In fact, as the southeast trades drift to the Gulf of Guinea, blowing parallel to the coast, they gradually heat up, causing the moisture they contain to move further away from the saturation point, while absorbing more moisture with a thermal ventilation effect that can make the land over which they blow even drier and more arid. This effect is not unknown in other parts of Africa as well, such as the coasts of Somalia and Mauritania, and on other continents. The vegetation of the southern African landscape reflects this contrast. Except for the steppes, the dominant vegetation in the more arid desert areas consists of wooded savanna in all its most characteristic manifestations, assuming the appearance of a moderately dense forest in the more humid areas, with bushes and thorny shrubs in the drier areas. These same elevations contribute to favoring varied deciduous plants on the plateaus of the southern African *veld* which extends unbroken from Cape Province to the Transvaal. Furthermore, the southern end of the continent comes under the decisive influence of the temperate ocean waters and exhibits a Mediterranean-like flora, thus completing the remarkable mirror-image relationship that characterizes the bioclimatic picture of the African continent.

Southern Africa is also the home of two great rivers which have polarized the settlement and economic activities of its people and its countries. There is the Zambezi which, although not very long (1650 mi [2660 km]), does have a very wide basin (500,000 mi^2 [1.3 million km^2])—the fourth largest in Africa—and then the Orange, which also has an extensive basin covering over 400,000 mi^2 [1 million km^2]. In their course from the highlands to the sea, these rivers—like the other African waterways—are often interrupted by rapids and at times spectacular falls (of which the most impressive are surely the Victoria Falls on the border between Zambia and Zimbabwe) corresponding to tectonic and stratigraphic discontinuities. This has made it possible to exploit them for the production of hydroelectric power by the construction of enormous artificial lakes. The large Kariba and Cabora Bassa dams are located on the Zambezi, while the Orange river basin also accommodates many hydroelectric dams.

The European presence has left a profound imprint on southern Africa, as it has indeed on northern Africa. An utterly European state, South Africa, was established here through the enterprise and tenacity of the Dutch settlers who landed in the 17th century (almost contemporaneously with their colonization of the east coast of North America). South Africa today would, in fact, have all the appearance of a European country were it not for the deep ethnic and racial conflicts which have characterized its development and may perhaps reach a logical resolution, as more recent events and the formal abolition of the state of apartheid appear to augur. The structure of two other countries, Zambia and Zimbabwe, the two former Rhodesias which had been spirited British colonies, can also be considered European in make-up. Here, especially in the southern part (called Zimbabwe today in honor of the capital of an ancient native empire), permeated by the pioneering spirit of Cecil Rhodes, the European presence found reason for permanently taking root. The same cannot be said of the two former Portuguese colonies of Angola and Mozambique, the last to achieve independence, which were devastated by slavery, impoverished by centuries of exploitation, and in which the culture of the colonizers bequeathed its very language to what were considered Portuguese provinces. Finally, the ethnic and cultural mosaic of southern Africa is given a finishing touch by the Malagasy presence, where the Asian component predominates and can be considered a gift of the monsoons; it is thanks to the seasonal shifts in these atmospheric and ocean currents that it was possible to establish regular contacts between southeastern Africa and southern Asia (India and Indonesia).

It should also be pointed out that the southern end of Africa was one of the places most frequently visited by European navigators and traders, driven to those distant latitudes in their search for a favorable and easy route to the Indies, those legendary and almost mythical lands of spices on which the heroic exploits of Marco Polo had already opened an inviting window two centuries earlier and direct access to which had been negated by the menacing barrier of Islam. It was especially the Portuguese—guided by the impulse born of the intuitions of Henry the Navigator—whose coastal voyages by Cão, Diaz, and Vasco da Gama contributed to expanding the knowledge of these lands, which were later to become the target of European colonization.

ANGOLA

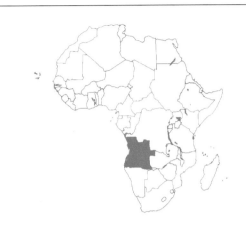

Geopolitical summary

Official name	República Popular de Angola
Area	481,226 mi^2 [1,246,700 km^2]
Population	5,646,166 (1970 census); 9,700,000 (1989 estimate)
Form of government	Presidential republic with a National People's Assembly elected every 3 years and representing a single party. The head of state, who also presides over the executive, is elected every 5 years.
Administrative structure	18 provinces divided into 139 districts
Capital	Luanda (pop. 1,200,000, 1988 estimate)
International relations	Member of UN and OAU; associate member of EC
Official language	Portuguese; Bantu languages are also widespread
Religion	About 65% Roman Catholic, approximately 10% Protestant; remainder of the population is animist
Currency	Kwanza

Natural environment. Angola borders Zaire to the north–northeast, Zambia to the east, Namibia to the south, and the Atlantic Ocean on the west. The enclave of Cabinda, located between the Congo and Zaire, is part of Angola.

Geological structure and relief. The land occupies a vast portion of the southern African plateau facing the Atlantic and has a considerable mean altitude which exceeds 3200 ft [1000 m]. The highest peaks (all over 6500 ft [2000 m]) are found in the west central region of the country (including a series of mountain ranges like the Serra da Chela, Serra da Neve, Serra Upanda, and others, and the Planalto do Bié), which constitutes an orohydrographic node of continental importance because it divides the Congo and Zambezi basins from minor ones which empty directly into the Atlantic. The coast, which runs for about 900 mi

Climate data

Location	Altitude (ft asl)	Average temp. (°F) January	Average temp. (°F) July	Average annual precip. (in.)
Luanda	144	78	69	12.7
Lobito	33	77	68	13.8

Conversion factors: 1 ft = 0.3 m; 1 in. = 25 mm; °C = (°F – 32) × 5/9

[1500 km] between the mouth of the Congo and Cunene rivers (marking the borders with Zaire and Namibia, respectively) is indented, with a narrow flat belt where periodic torrential floods caused by the Benguela current promoted the formation of long sandbars with relative and deep inlets open toward the north, such as the Baía dos Tigres and Baía do Bango.

In geological terms, the entire west central part of the country consists of a body of very old rocks that form the skeleton of part of the plateau and its highest relief (including the Serra Môco, 8594 ft [2620 m], the tallest mountain in the country). The Precambrian crystalline basement, consisting of granite, gneiss, schist, and quartz, crops out directly between the Congo and the Cunene. This ancient layer, often traversed by great granitic intrusions, is partly covered by schist-sand and schist-limestone Paleozoic sediments which form the framework of the Bembe system between the Kwanza and the Congo, as well as the mountains of the Humpata plateau (right bank of the Cunene), largely leveled by erosion. The entire remainder of the country, especially the east central part, consists of sediments of mostly continental origin that form the typical tablelands of the Karroo and the Kalahari of the Carboniferous-Triassic and Cenozoic-Quaternary periods, respectively. Finally, more recent marine sediments, dating back to the Cretaceous-Tertiary period, are buttressed by the ancient basement along the coastal belt.

Hydrography. As mentioned, the hydrographic network of Angola hinges on the coastal elevations and the Bié plateau. Flowing down from these heights toward the Congo Basin are, among others, the Cuango and Cassai, forming a section of the boundary with Zaire; and toward the interior drainage basin of the Kalahari, the Cubango. A tributary of the Zambezi basin (the upper course of which actually flows in Angolan territory) is the Cuando, while the Cuanza and the Cunene flow directly into the Atlantic. At its mouth, the Cuanza forms a wide alluvial plain after a course of 595 mi [960 km], entirely in Angolan territory. All these waterways are often interrupted by rapids and falls which interfere considerably with navigation.

Climate. Although clearly tropical, the climatic conditions of the country reflect a considerable regional diversity, especially with regard to precipitation, as reflected by the fairly high annual mean temperatures, usually above 68°F [20°C] (with a maximum fluctuation in the southern region of about 14°F [8°C], although the monthly means here range from 61°F [16°C] in July to 75°F [24°C] in January). This diversity is determined by pressure systems that settle in the highlands of southern Africa (cyclonic during the summer and anticyclonic during the winter). As a result, precipitation occurs predominantly during the summer half of the year and does not reach a very high median annual rainfall; but in general it is above 39 in. [1000 mm] in the northern regions and 59 in. [1500 mm] in the Bié plateau and the western sierras. Along the coastal strip, however, the rainfall is considerably less (no more than about 12 in. [300 mm] a year) and the climate becomes very dry, with the temperatures tending to be below normal due to the effect of the cold Benguela current which flows down from the southeast along the western coasts of the continent at a temperature that is usually below 59°F [15°C] and is responsible for persistent fog banks which pose a fairly great hazard to navigation. Further north, it flows to the open sea, giving way to a warm current that originates in the Gulf of Guinea. With these temperature contrasts it often happens that the sky is covered with mists (the *cacimbo*) even during the dry season and violent storms are not rare along this coast. Finally, the altitude produces an appreciable increase in the daily temperature ranges, especially in the west central highlands, where nighttime temperatures are especially low compared to temperatures during the day.

Flora and fauna. The vegetation landscape, dominated by shrubby savanna and tall grasses in much of the country, is closely dependent on the thermal and pluviometric conditions. Toward the north, the savanna becomes much more densely wooded, ultimately being replaced by dense tropical forest along the watercourses (fringing forests). Toward the east and southeast, the vegetation becomes poorer, with predominantly steppe-like features. Along the coast, the landscape takes on a typically predesert-like appearance.

Protection of the fauna (still rich in numerous species) and of the natural environment in general is assured by many parks and reservations, some of them very large, such as, among others, the Cameia National Park in the central western highlands, the Lona Park on the border of Namibia, and the Quiçâma National Park on the Atlantic coast south of Luanda.

Population. After independence was gained (1975), almost all Portuguese residents left the country. Consequently, the current population consists of peoples of the Bantu ethnic group, such as the Lunda, Kongo, Chokwe, and Mbundu, who are dedicated to agriculture and cattle raising, and the Herero, who may have originally come from eastern Africa. Population density is rather low and settlements, which are predominantly rural in character (the urban population represents only 28% of the total), are generally located on the coastal strip and in the inland highlands.

The major inhabited centers are distributed throughout the interior of the country and on the coast, as well as along the rail line running from Benguela to Shaba (formerly Katanga) in Zaire and also to Zambia. All the cities were founded by the Portuguese, who came to the area in the 16th century, and these still preserve the typical colonial imprint, with buildings and churches of some magnificence. The colonial heritage is still present, both in terms of the language and the widespread Catholic religion. However, the social conditions of the population are fairly modest and the illiteracy rate is high.

Economic summary. After the achievement of independence and the sudden departure of 300,000 Portuguese residents, the young nation's economy suffered a breakdown due to the shortage of qualified technical personnel to manage productive enterprises and ensure services. The mining of copper, iron ore, and diamonds, and the fishing industry were reduced some 50%, thus forcing the new government to solicit foreign economic aid in order to promote private initiative, while internal warfare raised

Administrative structure

Administrative unit (provinces)	Area (mi²)	Population (1988 census)
Bengo	12,109	154,437
Benguela	12,270	716,506
Bié	27,141	961,691
Cabinda	2,806	114,226
Cuando-Cubango	76,833	173,816
Cuanza Norte	9,337	474,677
Cuanza Sul	21,485	709,679
Cunene	34,486	251,145
Huambo	13,230	1,299,743
Huíla	28,951	835,562
Luanda	933	1,192,242
Lunda Norte	39,674	311,184
Lunda Sul	17,621	149,297
Malanje	37,674	850,727
Moxico	86,087	284,175
Namibe	22,441	79,028
Uíge	22,657	603,914
Zaire	15,490	223,676
ANGOLA	481,226	9,385,725

Capital: Luanda, pop. 1,200,000 (1988 estimate)
Major cities (1970 census): Huambo, pop. 61,885; Lobito, pop. 59,258; Benguela, pop. 40,996; Malange, pop. 31,559; Lubango, pop. 105,000 (1984 estimate); Namibe, pop. 100,000 (1981 estimate).

Conversion factor: 1 mi² = 2.59 km²

serious obstacles to the normal development of economic activity. The negative trade balance with foreign countries was also aggravated by unforeseen meteorological changes (droughts) which caused people to leave the countryside and move to the towns of the capital district. There, the periphery is strewn with *muceques*, as the conglomeration of shacks of immigrants seeking work and a place to live are called.

In terms of its natural resources, Angola is a potentially fairly rich country, although these have been little exploited. Most of the progress in this direction was achieved by the policy of the Portuguese, who had allocated considerable investments for the development of this overseas province. Indeed, adequate utilization of these natural resources offers great possibilities in mining (oil, diamonds, iron, manganese, and phosphates) as well as in

Socioeconomic data

Income (per capita, US$)	750 (1990)
Population growth rate (% per year)	2.6 (1980–89)
Birth rate (annual, per 1,000 pop.)	47 (1989)
Mortality rate (annual, per 1,000 pop.)	19 (1989)
Life expectancy at birth (years)	46 (1989)
Urban population (% of total)	28 (1989)
Economically active population (% of total)	41.4 (1988)
Illiteracy (% of total pop.)	72 (1991)
Available nutrition (daily calories per capita)	1725 (1989)
Energy consumption (10⁶ tons coal equivalent)	0.83 (1987)

agriculture, whether speculative (coffee, sugar cane, and bananas) or traditional (cassava and corn). Agriculture itself still occupies more than half of the actively working population (58%), although it accounts for only about 13% of GNP, 55% of which is derived from manufacturing and mining and 32% from services (with barely one fourth the workers).

Stock raising, predominantly cattle, is widespread in the southern part of the country. The fishing activity in the waters of the Atlantic, with its rich temperature contrasts originating in the Benguela current, supplies numerous canning establishments based in the ports of Benguela, Porto Alexandre, and Porto Amboim.

Industrial activity comprises processing of agricultural products (sugar, cotton, and oil) as well as petroleum refining. Angola exports predominantly to Latin America and the United States, especially petroleum and its derivatives, coffee, diamonds, cereals, etc., while imports from the EC and the former Soviet Union consist of machinery, textiles, and foodstuffs, as well as capital goods. Notwithstanding foreign investments in the petroleum and mining industries, the balance of trade remains constantly negative.

The communications network is still poor. The principal connections follow the coastal strip or penetrate inland, like the trans-African rail line from Benguela to Lusaka (Zambia), and from there to Dar es Salaam on the Indian Ocean.

Historical and cultural profile. On the left bank of the lower reaches of the Congo, in what is today northwestern Angola, the Bantu kingdom of ManiKongo, established in the 14th century, extended well beyond the area of Angola. It was with this kingdom that the Portuguese, who first reached the estuary of the great African river, came into contact. They changed the name of the capital M'Bali to São Salvador, in 1491 began the process of evangelization which induced the local ruler to be baptised, founded small trading bases, and in 1575 established the city of São Paulo de Luanda. In their advance, the Portuguese took advantage of the rivalry among the various tribes and sought to weaken the ManiKongo by allying themselves with the neighboring kingdom of N'Dongo, from whose sovereign N'Gola they then derived the name for their possessions. Together with Guinea, Angola was the African region most affected by the slave trade. Most of those deported were sent to Brazil for the cultivation of sugar cane, which explains why in the 17th century the Portuguese defended this region against Dutch attack, in order to protect the profits derived from a trade which persisted until the end of the 19th century, even after the formal abolition of slavery.

It was only with the end of the slave trade and the necessity for the Portuguese to switch to exploitation of agricultural and mining resources that the conquest of the interior was completed and the Bantu tribe completely subjugated, while the confines of the colony were delimited in the last decades of the century by a series of international agreements. Portuguese exploitation of cotton, coffee, and sugar plantations as well as diamond mines intensified and was based on the widespread practice of forced labor and oppression of the indigenous black population, deprived of its rights.

In 1935, Angola was declared an integral part of Portugal and in 1951 it officially qualified as an "overseas province"; in fact, the Salazar government adopted the guise of a "reformed" colonialism, with the institution of special measures such as formally extending Portuguese citizenship to the Angolans (1961), but these steps were not able to slow an independence movement which, despite strict police surveillance, began to organize in the 1950s and, in 1961, went into armed insurrection. This is how Frantz Fanon recalled those times in his *Les Damnés de la terre* [*The Wretched of the Earth*] :

> We remember that on March 15, 1961, a group of more than two thousand Angolan farmers hurled themselves against the Portuguese positions. Men, women, and children, armed and unarmed, on fire with courage and enthusiasm, then threw themselves in successive waves of compact masses and stormed districts where the Portuguese colonizers, soldiers, and flag dominated. Villages and airports were surrounded and subjected to frequent attacks, but it should be added that thousands of Angolans were mowed down by the settlers' machine guns.

The revolt was initially guided by the Popular Movement for the Liberation of Angola (MPLA) founded by Agostinho Neto and inspired by Marxism and Leninism. It was later joined by the National Front for the Liberation of Angola (FNLA) of Roberto Holden, and the National Union for the Total Independence of Angola (UNITA) of more moderate inspiration. After almost fifteen years of struggle and even before independence was proclaimed, actual war broke out between the MPLA and the other two parties during which an important role was also played by foreign intervention: the Soviets and Cubans backed the MPLA, while Zaire, the United States, and South Africa supported the FNLA and UNITA. The MPLA won because it was politically as well as militarily stronger, and Agostinho Neto became President of the People's Republic that was established (1975). In 1979, upon Neto's death, José Eduardo Dos Santos succeeded him, presenting himself as the defender of ideal continuity and "black" severity against the so-called middle-class yielding by a white or mulatto elite. The MPLA retained power with difficulty, always increasingly threatened by the UNITA guerrillas under Jonas Savimbi, in a country also being severely put to the test by the presence of foreign troops.

It was not until 1988 that an agreement was signed among Angola, Cuba, and South Africa, in which withdrawal of the foreign armed contingents was bound to the goal of recognizing the independence of Namibia. The peace process, although extremely desired by Dos Santos because of the disastrous conditions of the national economy, was actually jeopardized by frequent violations of the armistice.

Literature. Angolan literature includes, above all, an old popular tradition in the various Bantu languages which even in the last century some scholars collected and translated into Portuguese, including anthologies of poems, poetry, lyric ballads, and proverbs. This work of recovery and translation continued also in the 20th century, all the more motivated by the ban against publishing in local languages, which was not lifted until 1975 when independence was achieved. As to publications in Portuguese, after World War II a literature developed with strong civil commitment and denunciation of colonialism, from which emerged the voices of the novelist Castro Soromenio and the poets M. de Andrade, Viriato da Cruz, A. Jacinto, and Agostinho Neto himself.

BOTSWANA

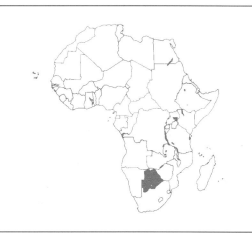

Geopolitical summary

Official name	Republic of Botswana
Area	224,548 mi^2 [581,730 km^2]
Population	941,027 (1981 census); 1,255,749 (1989 estimate)
Form of government	Presidential republic; the National Assembly, only part of which is elected, has a term of office of five years.
Administrative structure	9 district councils and 5 city councils
Capital	Gaborone (pop. 138,471, 1991 estimate)
International relations	Member of UN, OAU, and Commonwealth; associate member of the EC
Official language	English; national language is Tswana (Bantu)
Religion	Animist; 50% Christian (Catholic and Protestant)
Currency	Pula

Natural environment. Botswana borders Namibia to the west and north, Zambia and Zimbabwe to the north and east, and South Africa to the south. The country is a large landlocked area surrounded by relatively high mountains with a mean elevation of 1650–3300 ft [500–1000 m]. The ancient crystalline basement, mostly covered by Tertiary sedimentary deposits, surfaces only in the central western and central eastern regions, where large volcanic (basalt) effusions appear. Southern Botswana is covered by the yellow and reddish sands of the Kalahari Desert (70% of the land), whereas the northern region, the lowest area of the country, is occupied by the Makgadikgadi Salt Pan basin and the endorheic delta of the Okavango (8000 mi^2 [20,000 km^2]).

The hydrographic resources are among the poorest on the continent, consisting of the Okavango river in the north, which flows down from Angola's interior mountains, and the Molopo in the south, which intermittently joins the Orange. In the east, a

Climate data

Location	Altitude (ft asl)	Average temp. (°F) January	Average temp. (°F) July	Average annual precip. (in.)
Gaborone	3329	79	55	21.6

Conversion factors: 1 ft = 0.3 m; 1 in. = 25 mm; °C = (°F – 32) × 5/9

few temporary torrents flow into the Limpopo, the only reliable exorheic basin. Botswana has a clearly arid tropical climate. Temperatures range from 41–72°F [5–22°C] in the winter (April to September) and from 64–91°F [18–33°C] in the summer (October to March), with notably low atmospheric humidity. The limited rainfall (less than 23 in. [600 mm] a year) is concentrated in short, heavy downpours. The Kalahari Desert in the central southern section covers 75% of the country, the remainder of which supports only limited vegetation and a combination of grasses and thorny shrubs. Extensive swamps fed by the waters of the Okavango cover the northwestern region and are home to a great variety of fauna, which is also present in many protected areas of the country, including the Moremi Wildlife Reserve and the Makgadikgadi Pans Game Reserve.

Population. With 1.2 million inhabitants (1989 estimate) and a density of 5 per mi^2 [2 per km^2], Botswana is actually one of the least inhabited countries of the world. Population settlements

Administrative structure

Administrative unit (councils)	Area (mi^2)	Population (1984 estimate)
Districts		
Central	57,024	355,000
Ghanzi	45,513	21,000
Kgalagadi	41,279	26,000
Kgatleng	3,073	49,000
Kweneng	13,854	128,000
North-East	1,976	40,000
North-West:	–	–
Chobe	8,029	9,000
Ngamiland	42,124	75,000
South	10,989	138,000
South-East	687	34,000
Cities[1]		
Francistown	30	36,000
Gaborone	37	79,000
Lobatse	12	22,000
Orapa	4	5,800
Selebi-Pikwe	19	33,000
BOTSWANA	224,548	1,051,000[2]

[1]The area of the cities is included in that of the respective districts.
[2]The general total is rounded off.
Capital: Gaborone, pop. 120,239 (1989 estimate)
Major cities (1989 estimate): Mahalapye, pop. 104,450; Serowe, pop. 95,041; Francistown, pop. 52,725; Selebi-Pikwe, pop. 45,542.

Conversion factor: 1 mi^2 = 2.59 km^2

are primarily rural (76%). The only urban centers are the capital Gaborone and the cities of Serowe, Selebi-Pikwe, Mahalapye, and Francistown. Gaborone, located at 3330 ft [1015 m] on the Notwani river, was established as the capital in 1965; it is an active administrative commercial center with an international airport. The inhabitants of Botswana are predominantly Bantus of the Bechuana group (85% of the population), divided into many tribes, and there is also a minority of Bushmen (4%), the oldest dwellers of the country. The Bechuanas are farmers and cattle-breeders who live in large villages in the country's wetlands. The Bushmen, who are on the way to extinction, still live from hunting and gathering in the dry regions of the Kalahari.

David Livingstone crossed the Kalahari steppes and knew the indigenous populations and their sparse living conditions:

Before narrating the incidents of this journey, I may give some account of the great Kalahari Desert, in order that the reader may understand in some degree the nature of the difficulties we had to encounter.

The space from the Orange River in the south, lat. 29°, to Lake Ngami in the north, and from about 24° east long. to near the west coast, has been called a desert simply because it contains no running water, and very little water in wells. It is by no means destitute of vegetation and inhabitants, for it is covered with grass and a great variety of creeping plants; besides which there are large patches of bushes, and even trees. It is remarkably flat, but intersected in different parts by the beds of ancient rivers; and prodigious herds of certain antelopes, which require little or no water, roam over the trackless plains. The inhabitants, Bushmen and Bakalahari, prey on the game and on the countless rodentia and small species of the feline race which subsist on these. In general, the soil is light-colored soft sand, nearly pure silica. The beds of the ancient rivers contain much alluvial soil; and as that is baked hard by the burning sun, rainwater stands in pools in some of them for several months in the year....

The [Bushmen] are probably the aborigines of the southern portion of the continent.... They are the only real nomads in the country; they never cultivate the soil, nor rear any domestic animal save wretched dogs. They are so intimately acquainted with the habits of the game that they follow them in their migrations, and prey upon them from place to place, and thus prove as complete a check upon their inordinate increase as the other carnivora. The chief subsistence of the Bushmen is the flesh of game, but that is eked out by what the women collect of roots and beans, and fruits of the Desert....

The Bakalahari are traditionally reported to be the oldest of the Bechuana tribes, and they are said to have possessed enormous herds of the large horned cattle ... until they were despoiled of them and driven into the Desert ... [they] retain in undying vigor the Bechuana love for agriculture and domestic animals. They hoe their gardens annually, though often all they can hope for is a supply of melons and pumpkins. And they carefully rear small herds of goats, though I have seen them lift water for them out of small wells with a bit of ostrich egg-shell, or by spoonfuls....

The annual per capita income in 1990 was US$2200 and reflects the economic progress of the country since independence (in 1979 the per capita income was only US$137), although social well-being is enjoyed by a minority, primarily those who work in the mines, while the majority still live from stock raising by primitive methods. In the past few years at least 50,000 people emigrated to South Africa because of the excessive demographic increase (3.7% annually).

Economic summary. At the time of British colonization the country's only resource was livestock, which accounted for 85% of exports. The recent discovery of large diamond deposits (Orapa and Letlakane) has now made Botswana one of the world's major producers (12,146,000 carats in 1990). Mineral resources also include copper and nickel at Selebi-Pikwe, asbestos and manganese at Kwakgwe and Ootse, and coal at Morupule, Mambula, and Palapye. Industry is represented by a number of plants for the processing of minerals and agricultural products, slaughterhouses, and meat-canning facilities.

Agriculture, practiced in rudimentary forms in small traditional holdings on 2.4% of the land, produces primarily subsistence crops that fall short of the country's needs. Production of sorghum was 41,800 t in 1991, that of millet 1100 t and that of corn 4400 t. Moderate quantities of citrus fruit, cotton, and peanuts are raised. Livestock constitutes the principal economic activity, accounting for 8% of exports in addition to providing the food base of the population. In 1990, there were 2.6 million head of cattle, 2.1 million goats, 300,000 sheep as well as donkeys and pigs. Forests cover 18.8% of the land and supply an average 46.6 million ft^3 [1,321,000 m^3] of lumber per year (1989).

The country's trade balance is negative, with diamond, meat, nickel and copper, and hides covering approximately 80% of imports (industrial products, foodstuffs, and fuel), while international aid and remittances from émigrés contribute to maintaining an even balance of trade.

The principal means of transportation is the Bulawayo-Johannesburg railway (444 mi [716 km] in 1984). The network of roads (4976 mi [8026 km] in 1985) is only partly paved. International travel is facilitated by the airports at Gaborone and Francistown.

Historical profile. British colonists gave the inland region of southern Africa located in the Molopo river basin the name of "Bechuanaland" since Bantu tribes of the Bechuana group coming from the south had occupied this territory in the 15th century. Before they could assert their dominion over the area uncontested, the British had to fight the Portuguese and Germans, but in the international agreements of 1885 they were finally able to establish their protectorate over the region north of the Molopo, while the southern part was given the status of a Crown colony (British Bechuanaland) that was annexed ten years later by the Cape Colony. In 1966 the protectorate became an independent

Socioeconomic data

Income (per capita, US$)	2200 (1990)
Population growth rate (% per year)	3.7 (1980–89)
Birth rate (annual, per 1,000 pop.)	48.5 (1989)
Mortality rate (annual, per 1,000 pop.)	11.6 (1989)
Life expectancy at birth (years)	63 (1989)
Urban population (% of total)	23.8 (1989)
Economically active population (% of total)	33.6 (1988)
Illiteracy (% of total pop.)	29.2 (1991)
Available nutrition (daily calories per capita)	2251 (1989)
Energy consumption (10^6 tons coal equivalent)	Included in South Africa

republic and a Commonwealth member, although this constitutional change had no profound impact on the administration of state power which remained in the hands of the Khama dynasty of Bamangwato ethnicity. As the heir to this dynasty and head of the conservatively oriented Botswana Democratic Party (BDP), President Seretse Khama governed the country in a modern, antiracist, and antifeudal spirit until 1980. Here is what Endre Sík wrote about the formation of the Khama government:

After another nine months of procrastination the elections to the National Assembly were held on March 1, 1965. At the elections, in which 189,000 voters (out of the Bechuanaland population of 542,000) took part, Seretse Khama's Bechuanaland Democratic Party won the absolute majority, gaining 28 of 31 seats (the remaining 3 seats went to the People's Party). As a result, the first constitutional government of Bechuanaland was formed with Seretse Khama as Prime Minister. At a press conference the Prime Minister announced that the country's name would be changed to Botswana. As to the relations with the neighbouring countries, he declared that, in spite of the serious differences due to the racial policies of the Republic of South Africa, his country would maintain the commercial relations, because it was poor and needed foreign investments for its own development's sake....

In April 1965 Seretse Khama visited Zambia, and had talks with Kaunda about closer co-operation between the two countries, notably about improvement of communication between them. Upon his return home he declared that, to lessen economic dependence on South Africa, the country had to develop trade relations with Zambia and other African countries.

On July 9, 1965, Philip Matante, the leader of the opposition Bechuanaland People's Party occupying three seats in the legislature, submitted a motion of no confidence. He accused the Seretse Khama government of neglecting to subsidize those suffering from the drought and famine, and said that it had to thank the fraudulent conduct of the elections for his majority won in the legislature. He declared that if his motion should be rejected, he would ask the British government to order a plebiscite to be held. The Prime Minister informed the National Assembly that he was negotiating with the British government for a time-table for independence. He said he was commissioned by the Bechuana people to lead the country to independence without new general elections. Matante's motion was rejected....

On October 13, 1965, the British Secretary of State for the Colonies announced the British government's agreement to Bechuanaland's becoming independent as of September 30, 1966. Two months later, on December 13, Prime Minister Seretse Khama repeated the same announcement at a meeting of the Bechuanaland Parliament, to which most of the chiefs were invited. At the same time the Prime Minister tabled a draft constitution of the independent Bechuana state. Under this act the country should be renamed Botswana and would continue to be a member of the British Commonwealth of Nations, the Prime Minister would be promoted to the office of President of the Republic, and the Deputy Prime Minister would become Vice-President of the Republic.

Economic dependence on South Africa has not prevented Botswana from drawing closer, if cautiously, to the countries of central Africa further advanced in the process of emancipation from colonialism, such as Zambia, Tanzania, and Mozambique. Quett Masire became president after Khama's death in 1980 and continued the previous policy of moderation and nonalignment.

COMOROS

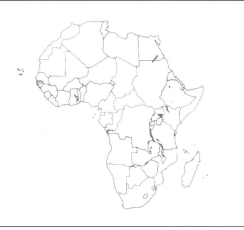

Geopolitical summary

Official name	République Fédérale Islamique des Comores
Area	719 mi^2 [1862 km^2]
Population	346,992 (1980 census); 466,277 (1990 estimate)
Form of government	Federal republic with a president as head of state; the Federal Assembly is elected every 5 years.
Administrative structure	3 islands, each administered by a governor appointed by the president with its own Legislative Council elected directly every 5 years
Capital	Moroni (pop. 22,000, 1988 estimate)
International relations	Member of UN and OAU; associate member of EC
Official language	French and Arabic; widespread use of Bantu dialects and Swahili
Religion	Predominantly Islamic; 1% Catholic
Currency	CFA franc

Natural environment. The Comoros archipelago, which lies between latitudes 11° and 13° at the northern end of the Mozambique Channel between the African mainland and the island of Madagascar, includes four main islands: Grande Comore, Mwali (Mohéli), Anjouan, and Mayotte, which retains political ties to France. They are all of volcanic origin with a fairly mountainous basaltic structure (Karthala, 8397 ft [2560 m], with the largest caldera in the world, and La Grille, 3280 ft [1000 m] on Grande Comore; and N'Tingui, 5234 ft [1596 m] on Anjouan). The coast is predominantly rocky, studded with smaller islands surrounded by coral reefs, and the surface waters are shallow because of the great permeability of the volcanic rock. Where meteoric erosion has produced changes in surface layers consist-

Climate data

Location	Altitude (ft asl)	Average temp. (°F) January	Average temp. (°F) July	Average annual precip. (in.)
Moroni	56	81	75	101.4
Conversion factors: 1 ft = 0.3 m; 1 in. = 25 mm; °C = (°F – 32) × 5/9				

ing of impermeable clay, torrents of water flow through cascades, rapids, and cataracts, especially during the wet season. However, most of the water consumed comes from rainfall collected in reservoirs.

The climate is wet and tropical with two seasons and monsoon conditions. Average annual temperatures are high (77°F [25°C] in Moroni), with a slight temperature swing. The rainy season comes in the summer (southern hemisphere), from November to April, and is due to the northeast monsoon with frequent cyclones, whereas the dry season is caused by the southeast trade winds and runs from May to October; total annual rainfall is slightly over 100 in. [2600 mm]. The vegetation is lush with an exuberance of tropical plants. There are mangroves along the coast; a swath of coconut palms, banana and mango trees in the foothills; and tropical forest in the higher inland areas. The arboreal fauna is similar to that of Madagascar (macaque monkeys, birds, and insects) and there is a great variety of fish.

Population. The most representative ethnic groups are Arabs, Malagasy, Asians, Africans, and people of mixed descent who speak French and Arabic, the official languages, as well as Bantu dialects such as Swahili. In 1975 Islam became the state religion.

Population growth has been especially intense since World War II due to mass Indian immigration and the return of Comorians from Madagascar, as well as the demographic increase. Settlements are concentrated on the coast and 41% of the population lives on Grande Comore (435 per mi^2 [168 per km^2]). The highest population density is found on the second island of Anjouan (1521 per mi^2 [587 per km^2]), whereas only 5% of the people live on Mwali. The major cities are Moroni, the capital (on Grande Comore), Mutsamudu (Anjouan), and Fomboni (Mwali); their architecture is Arabic, with the traditional square and mosque, narrow alleys, and low whitewashed houses, each city with a commercial and fishing port.

Social conditions are aggravated by the very high rate of demographic growth: life expectancy at birth is 52 years and the

Administrative structure

Administrative unit (governorates)	Area (mi^2)	Population (1990 estimate)
Anjouan	164	249,053
Grande Comore	443	192,667
Mwali (Mohéli)	112	24,557
COMOROS	719	466,277

Capital: Moroni, pop. 22,000 (1988 estimate)
Major cities (1980 census): Mutsamudu, pop. 12,518; Fomboni, pop. 5663.

Conversion factor: 1 mi^2 = 2.59 km^2

Socioeconomic data

Income (per capita, US$)	480 (1990)
Population growth rate (% per year)	3.1 (1980-89)
Birth rate (annual, per 1,000 pop.)	47 (1989)
Mortality rate (annual, per 1,000 pop.)	13 (1989)
Life expectancy at birth (years)	54 (1989)
Urban population (% of total)	27.6 (1989)
Economically active population (% of total)	44.9 (1988)
Illiteracy (% of total pop.)	46.3 (1991)
Available nutrition (daily calories per capita)	2059 (1989)
Energy consumption (10^6 tons coal equivalent)	0.02 (1987)

illiteracy rate is 46.3%. Education is provided predominantly by Belgian, Senegalese, and Tunisian teachers.

Economic summary. After gaining independence from France, the country suffered a severe economic crisis and had to reduce consumer goods drastically in an attempt to compensate, at least in part, for the serious deficit in its balance of payments.

Agriculture contributes 37% to GNP, employing 65% of the work force. Although it benefits from a very fertile volcanic soil, subsistence farming is inadequate and almost all foodstuffs must be imported. The chief crops are potatoes, yams, manioc, rice, corn, bananas, and coconuts. Plantation agriculture, however, run by French multinational perfume companies, is well organized. The Comoros are the world's main producer of ylang-ylang, an essential-oil distillate of the *Cananga odorata* flower, and a major producer of *vetiver* (also called "East Indian grass") as well as of citronella, vanilla, copra, sisal, coffee, and cloves. Animal husbandry and fishing play a minor economic role.

Industry accounts for only 13% of GNP and employs no more than 5% of the work force. It is based on small agricultural processing plants, such as distilleries of essential oils, soap-works, mills, and wood-working facilities; tourism, with 50% of GNP, is a major source of income.

The balance of payments is highly negative, with the value of exports covering only half those of imports. The major exports are vanilla, cloves, ylang-ylang, copra, sisal, and coffee. Imports consist of all food products, petroleum, and machinery. Communications include 465 mi [750 km] of roads (1985), shipping connections between the islands and Madagascar and the African mainland, as well as the international airport (Hahaya) of Moroni.

Historical profile. Although Arab traders had been sailing there since the end of the 10th century, Portuguese and then English navigators did not discover the Comoros Islands until 500 years later. French settlement dates back to the 19th century, and in 1912 the archipelago was proclaimed a French colony and administratively joined to Madagascar. Colonial rule brought no modernization and the traditional social structure remained untouched, except for admission of the local elite into the bureaucratic apparatus. During World War II the islands were occupied by the British. In 1946, they separated from Madagascar in a slow decolonization process which culminated in their unilateral proclamation of independence in 1975, from which the island of Mayotte, however, seceded after having voted in a popular referendum to remain linked to France.

LESOTHO

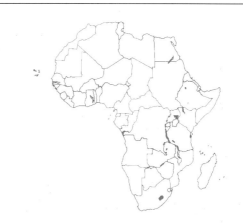

Geopolitical summary

Official name	Kingdom of Lesotho / Muso oa Lesotho
Area	11,717 mi^2 [30,355 km^2]
Population	1,577,536 (1986 census); 1,806,000 (1991 estimate)
Form of government	Constitutional monarchy with a bicameral Parliament (Senate and House of Deputies elected every 5 years by direct vote)
Administrative structure	10 districts
Capital	Maseru (pop. 109,382, 1986 census)
International relations	Member of UN and OAU, and the Commonwealth; associate member of EC
Official language	English and seSotho (Bantu)
Religion	Christian 90% (600,000 Catholics); animist 6%
Currency	Loti

Natural environment. Lesotho is surrounded entirely by the territory of South Africa. It comprises a series of vast tablelands, ranging in altitude from 6500 ft [2000 m] to 9800 ft [3000 m], which consist primarily of large basaltic effusions that cover layers of Paleozoic sandstone and are dissected by watercourses that form the sources of the Orange River. Topographically, it lies between the chain of the Drakensberg (Dragon) Mountains, which is part of the Great Escarpment, to the east, and the Maluti (10,820 ft [3299 m]) and Thaba Putsoa (10,155 ft [3096 m])

Climate data

Location	Altitude (ft asl)	Average temp. (°F) January	Average temp. (°F) July	Average annual precip. (in.)
Maseru	–	20	90	29.3

Conversion factors: 1 ft = 0.3 m; 1 in. = 25 mm; °C = (°F – 32) × 5/9

mountain ranges to the west. All of the hydrographic network is a tributary of the Orange and includes the Caledon, which marks the Western border.

The climate is monsoonlike in the eastern and central sections of the country, which are exposed to the northeastern winter wind (hot and dry from May to September) and to the southeasterly trade which carries the summer rains (annual precipitation of 25 in. [650 mm]), whereas in the western section it is humid and tropical, not unlike that of Sudan.

The vegetation is typical of mountainous grassland with stretches of temperate forest on the inland slopes of the Drakensberg Mountains, and dry savanna to the east. The extensive original fauna, now greatly reduced, is protected in the Sehlabathere National Park.

Population. The predominant ethnic group is that of the BaSotho (80%), followed by the Zulu (20%), and little more than one thousand Europeans.

The population of Lesotho, estimated at 1,806,000 in 1991, has grown more than tenfold since 1875, with the greatest increase occurring after World War II (at an average annual rate of 2.7%). Its highest density (153 per mi^2 [59 per km^2] is the average) is found in the deep valleys of the Orange River and its tributaries and in the northwest region of the country. The predominantly sparse and rural settlements are concentrated around the *kraal* (the native enclosure for cattle), with scant urbanization (20%), the only exception being the capital Maseru, at the border of South Africa.

Lesotho supplies the Republic of South Africa with labor (130,000 wage earners) for its mines.

Economic summary. Lesotho is one of the poorest countries in the world. The economy is based on the starkest subsistence farming and cattle-raising, which account for most of the gross national product and occupy almost the entire active population.

Only 10.5% of the country's land is arable, and agriculture is practiced by generally traditional methods; the chief products are corn, wheat, sorghum, and barley. Livestock resources benefit from the widespread availability of year-round pastures and

Socioeconomic data

Income (per capita, US$)	485 (1990)
Population growth rate (% per year)	2.7 (1980–89)
Birth rate (annual, per 1,000 pop.)	41 (1989)
Mortality rate (annual, per 1,000 pop.)	12 (1989)
Life expectancy at birth (years)	56 (1989)
Urban population (% of total)	20 (1989)
Economically active population (% of total)	46.8 (1988)
Illiteracy (% of total pop.)	26.4 (1991)
Available nutrition (daily calories per capita)	2307 (1989)
Energy consumption (10^6 tons coal equivalent)	Included in South Africa

grazing land (about 66%) in the hill country above the limits suitable for cultivation, and total 1,475,000 head of sheep (1990), 1,065,000 goats, 530,000 cattle, and 253,000 horses and related equines. Wool (1650 t in 1990) is one of the products exported.

There are huge potential hydroelectric power resources (Malibamatso River Project). The only mineral deposits are those of the Letseng-la-Terai diamond mines in the Maluti Mountains. An embryonic industry in consumer foods (breweries) and hide-tanning exists. The disastrous balance of trade (with exports amounting to no more than a fifteenth the volume of imports) is offset by the remittances sent home by workers employed in the South African mines and by international aid. Transportation routes encompass 2923 mi [4715 km] of roads (1988) and 1.25 mi [2 km] of railroad (1983), which connects the capital to the South African rail system. The Maseru airport ensures international communications and holds promise for tourism.

Historical profile. The country derived both its present name and that of Basutoland (by which it was known in the colonial era) from the Bantu tribe of the BaSotho, which had migrated south from eastern Africa at some indeterminate time. To repel the attacks of the Zulu, the BaSotho people coalesced into a unified state and sought the protection of Great Britain, which brought them under the rule of the High Commissioner for South Africa together with Swaziland and Bechuanaland. Chief Moshweshwe was the key ruler through whom this dual process of national consolidation and colonial penetration was achieved. This is how Lucy Mair describes him:

> *According to the old men whose memories were recorded by the missionary Ellenberger, Moshweshwe had had from his youth the ambition to be a great chief; but although he early demonstrated his prowess in successful raids, it was by offering protection to those who sought it that he built up his power.... He consolidated his control over his heterogeneous subjects in part by appointing his brothers and sons as headmen of villages of miscellaneous immigrants, but also by recognizing the chiefs who came with bands of followers as his subordinate authorities. By 1843 he had come to be recognized by the British as a stable element in a still disturbed situation, and he was made responsible for keeping order among his people ...*

Because of its underdevelopment, Basutoland took a long time to achieve its independence, which was not proclaimed until 1966. Three years later, Prime Minister Leabua Jonathan came to power by repressing the internal opposition and with the support of the Republic of South Africa.

Administrative structure

Administrative unit (districts)	Area (mi^2)	Population (1986 census)
Berea	858	194,631
Butha-Buthe	682	100,644
Leribe	8,812	257,988
Mafeteng	818	195,591
Maseru	1,652	311,159
Mohale's Hoek	1,363	164,392
Mokhotlong	1,573	74,676
Qacha's Nek	907	63,984
Quthing	1,126	110,376
Thaba Tseka	1,648	104,095
LESOTHO	11,717	1,577,536

Capital: Maseru, pop. 109,382 (1986 census)

Conversion factor: 1 mi^2 = 2.59 km^2

MADAGASCAR

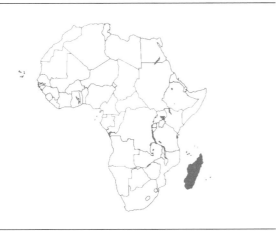

Geopolitical summary

Official name	République Démocratique de Madagascar / Repoblika Demokratika Malagasy
Area	226,598 mi² [587,041 km²]
Population	7,603,790 (1975 census); 11,443,000 (1990 estimate)
Form of government	Parliamentary republic with socialist tendencies; People's National Assembly elected by universal suffrage every 5 years.
Administrative structure	6 provinces
Capital	Antananarivo (pop. 802,390, 1990 estimate)
International relations	Member of UN and OAU; associate member of EC
Official language	Malagasy; French is in widespread use.
Religion	Animist 50%; Catholic 25%; Protestant 20%; Islamic 5%
Currency	Malagasy franc

Natural environment. The island of Madagascar lies in the Indian Ocean off the east coast of Africa opposite Mozambique from which it is separated by the wide Mozambique Channel.

Geological structure, relief and hydrography. According to the theory of plate tectonics, it now seems certain that the island of Madagascar, the world's fourth largest, detached itself from the African continental block during the Mesozoic era. The structure of its eastern strip and central plateaus in particular is constituted by intrusive Precambrian metamorphic rocks and is quite similar to that of the neighboring mainland. The western part of the island is characterized by sedimentary formations comparable to the classic Karroo System (carboniferous schists, sandstone, argillaceous schists, limestone, and conglomerate deposits continuing from the Upper Paleozoic to the Jurassic), which are overlain by other sedimentary elements of ever more recent origin (down through the Tertiary) as we proceed out-

ward, while the inland plateaus dip gently toward the Mozambique Channel, reach an average altitude hovering around 4000–5000 ft [1200–1500 m], and faithfully reflect the typical large tablelands of south-central Africa. Formations of recent volcanic origin (Tertiary and Quaternary) enliven the island's generally mature topography, particularly in the area of the Tsaratanana Massif located in the north, which culminates at 9433 ft [2876 m] and represents the highest elevation, and the Ankaratra mountain group (8670 ft [2643 m]), which lies slightly south of Antananarivo. In the east the plateau rises to a scarp which dominates the coastal plain. The latter consists of a stretch of uniform terrain, generally sandy and dotted with ponds and lagoons, which is so narrow that, except for the area near Farafangana, it is nowhere more than 19 mi [30 km] wide. Virtually in a straight line from Mahavelona to Taolanaro, the east coast affords no natural landing sites because it overlies a large fault that facilitated the reascent of basaltic margins during the Cretaceous. The west coast which faces the African mainland, however, is more indented and often marked by coral reefs which render the possibility of creating convenient landing places difficult. The only natural harbor is Antsiranana (formerly Diégo-Suarez), but its location at the extreme north of the island impedes its development possibilities because it lies so far from commercial centers.

The general westward inclination which characterizes the island of Madagascar explains why its longest water courses are those which flow into the Mozambique Channel (like the Betsiboka, Manambolo, Tsiribihina, and Mangoky). While the rivers of the western section of the country have a very inconstant rate of flow, those flowing into the Indian Ocean, which are much shorter, have a much richer water regimen. In the heart of the depression areas, delimited by the volcanic elevations of the interior, some lake basins have formed, like those of Alaotra and Itasy.

Climate. The contrast between the island's eastern and western slopes is also reflected in the climatic conditions. The eastern strip fronting on the Indian Ocean enjoys the effects of the southeasterly trades, which reach the shores of the island laden with rain (so much so that everywhere on the east coast the annual rainfall exceeds 80 in. [2000 mm], showering as much as 127 in. [3255 mm] of rain in some areas). The central regions, on the other hand, are much drier because by the time the trade winds reach them they have lost much of their humidity. The northern part of the west coast, however, is exposed to the summer monsoon which blows from the northwest and therefore carries much precipitation (over 100 in. [2550 mm] of rain a year at the island of Nosy Be), but as one proceeds southward the rainfall gradually decreases (14–16 in. [350–400 mm] at Toliary). The west coast is also characterized by constantly high

Climate data

Location	Altitude (ft asl)	Average temp. (°F) January	Average temp. (°F) July	Average annual precip. (in.)
Antananarivo	4526	70	59	52.7
Toamasina	16	80	70	126.9
Toliary	33	82	69	13.3

Conversion factors: 1 ft = 0.3 m; 1 in. = 25 mm; °C = (°F – 32) × 5/9

temperatures (ranging at Toliary from 69°F [20.4°C] in July and August to 82°F [27.6°C] in December and January), whereas in the uplands where the altitude moderates the heat the temperature range is slightly more pleasant (59°F [15°C] in July and 70°F [21.1°C] in January at Antananarivo). The east coast, less exposed to the sun because of the frequent rains, enjoys a mean temperature of 80°F [26.7°C] in January and 70°F [21.1°C] in July and August (Toamasina).

Because of the island's persistent isolation, the evolution of living species on Madagascar can be said to have followed a course independent of that of the large continental masses.

Flora and fauna. The presence on the island of distinctive species of flora and fauna (85% of which are endemic to it) is so recurrent as to attribute to the island the characteristics of a subcontinent in this respect. In fact, its spontaneous vegetation embraces some 7000 exclusive species, including a special variety of periwinkle used for the treatment of childhood leukemia, two types of coffee containing no caffeine, and about 700 of the 800 known species of orchids. Unfortunately, the exploitation of timber started during the colonial period; alongside, the intensification of traditional slash-and-burn farming methods led to progressive changes in the original forest, which had previously covered the entire inland plateau area but is now limited to the eastern strip of the island. This thinning out of the rainforest in the central region has resulted in steppe and savannah formations, which are also dominant along the western edge, while a vegetation of underbrush and thickets, highlighted by xerophilous plants, characterizes the drier southern regions. Intensified cultivation also threatens the local fauna, which includes a precious heritage of a great variety of butterflies, over a hundred species of birds that are unique in the world, and about half of all chameleon species. The uniqueness of Madagascar's fauna is further enhanced by the dominant role among mammals of the island's prosimians. Having disappeared almost entirely everywhere else in the world, lemurs have survived on Madagascar (29 species) thanks to the absence of large mammals and poisonous snakes.

However, because of the growing demographic pressure and the limited economic resources of the government, safeguarding this faunal heritage is anything but easy: today only 1.6% of the entire country is protected in the form of national nature preserves (of which there are 11) and national parks (so far only 2 have been established: Amber Mountain National Park at the northernmost tip of the island and Isalo National Park in the massif of the same name).

Population. Fairly complex events have affected Madagascar's human population, which is the result of successive waves of immigrants of various origin, to such an extent, in fact, that today the island is characterized by a considerable ethnic mixture which includes a score of diverse groups. The first wave of people who migrated to the island, presumably in the 2nd century B.C., were of Malayan-Australian ethnic stock. It was followed by immigrants from continental Africa (the Makua who settled in the southern part of the island) and especially Indonesia. Today the largest group (26% of the population) actually consists of the Merina, who came to Madagascar in the 6th century, presumably from Java and Bali, and settled in the central plateaus, from where they contributed to the spread of rice culti-

Administrative structure		
Administrative unit (provinces)	Area (mi²)	Population (1990 estimate)
Antananarivo	22,497	3,811,000
Antsiranana	16,616	715,000
Fianarantsoa	39,516	2,420,000
Mahajanga	57,909	1,253,000
Toamasina	27,758	1,585,000
Toliary	62,302	1,659,000
MADAGASCAR	226,598	11,443,000

Capital: Antananarivo, pop. 802,390 (1990 estimate)
Major cities (1990 estimate): Toamasina, pop. 145,431; Fianarantsoa, pop. 124,489; Mahajanga, pop. 121,967; Toliary, pop. 61,460; Antsiranana, pop. 54,418.

Conversion factor: 1 mi² = 2.59 km²

vation as well as of their language and culture. Subsequent settlers on the island were people of Arab ancestry, natives of the Comoros Islands, Indians, Chinese, and many Europeans (French). Since the 1920s, when 3 million people lived on Madagascar, the country has experienced a considerable population increase thanks to the improvement in sanitary conditions and especially to better control of malaria. The annual growth rate, now stated to be 2.9%, is bound to increase, since a full 40% of today's population is below the age of fifteen years.

Population distribution is highly uneven. The development of rice cultivation has favored population clusters with more than three times the average national density in the central highlands, where Antananarivo is located with about one tenth Madagascar's entire population and its only true urban center. Except for the coastal strip, which is moderately inhabited due to the practice of plantation agriculture and the presence of a few centers devoted to commerce, the western and southern regions appear to be almost entirely unpopulated, notwithstanding the country's vigorous demographic gains of the last several decades.

Despite the modest standard of living, as evidenced by the small per capita income (US$230 per year) and the scant volume of consumer goods (8 television sets and 3.4 telephones per 1000 people), illiteracy is limited to 19.8% of the population thanks to the education reform launched in 1978 in order to make schooling readily available to everyone.

Economic summary. Traditionally based on agriculture, which still absorbs about 75% of the work force, the Malagasy economy has not been able to benefit much from the reforms promised by the military revolution of 1972. The decision to break its formerly very close ties with France and to expand the public sector by nationalizing all French property, in fact triggered a severe economic crisis brought about by the drying up of foreign capital investments and the failure of plans to spur local industrial activity. The resumption of economic relations with the West in the 1980s coincided with a second phase of local economic policy, which recognized the need to develop agriculture and diversify exports in particular.

Although it covers only 5% of the country's land area, agriculture employs the largest number of people and supplies about

80% of the value of exports. The agrarian reform imposed by the revolutionary government did not include collectivization of the land (which today is still managed in such different ways as to permit both traditional forms of itinerant crops and the intensive production of rice and coffee) but simply unification of product marketing activity under state control. Starting in 1983, however, liberalization was introduced in the production and processing of rice, the island's chief crop for which over 40% of the cultivated land is set aside. Rice is grown in every region except the southwest, with intensive cultivation in the inland alluvial basins and especially in the area of Lake Alaotra and near the capital. Manioc, which covers 10% of the cultivated land, is another crop intended for internal consumption. Madagascar also boasts some crops raised for export, the principal of which is coffee (7%), as well as some rather specialized products, such as cloves and vanilla.

There is a fair amount of cattle breeding which, since it is not subject to the limitations imposed by endemic pathogens in various parts of tropical Africa, may perhaps have developed to excess in relation to the size of available pastureland; in 1991 there were over 10 million head of cattle, generally of African stock and limited yield.

The mineral resources of the island have not yet been fully identified but are believed to be considerable. Currently subject to exploitation are deposits of chromite and especially graphite and mica. However, although Madagascar is one of the world's major producers of graphite, the combined foreign sales value of this mineral and mica barely reaches 2% of all exports. Energy production is based on exploitation of the Sakoa coal fields (at Toliary) and that of a few petroleum deposits (small refineries are in operation at Toamasina and Bemolanga). There is a considerable energy potential in the excellent possibilities for the production of hydroelectric power (of the 220,000 kW of electricity produced in 1989, 106,000 was obtained from hydroelectric sources).

Development of the industrial sector is definitely limited and has so far been restricted to only the processing of local agricultural products (production of rice and tapioca, canning of beef, distillation of liquor, and weaving of cotton).

Foreign trade is modest. Madagascar exports primarily agricultural products and imports large quantities of finished goods. However, the country's traditional trade deficit, potentially aggravated by increased internal consumption linked to population gains, is contained by severe import restrictions.

The mountainous and inaccessible nature of much of the

Socioeconomic data

Income (per capita, US$)	230 (1990)
Population growth rate (% per year)	2.9 (1980–89)
Birth rate (annual, per 1,000 pop.)	46 (1989)
Mortality rate (annual, per 1,000 pop.)	16 (1989)
Life expectancy at birth (years)	51 (1989)
Urban population (% of total)	24 (1989)
Economically active population (% of total)	43.7 (1988)
Illiteracy (% of total pop.)	19.8 (1991)
Available nutrition (daily calories per capita)	2101 (1989)
Energy consumption (10^6 tons coal equivalent)	0.42 (1987)

island's terrain partly explains the moderate growth of its ground transportation, which must still rely on infrastructures built during the colonial period, especially with regard to the few existing rail lines (653 mi [1054 km]) in 1988. There is also a shortage of roads (some 5270 mi [8500 km] have been developed), little more than half of which are paved and passable the year round. Decidedly more progress has been made in air communications, with the availability of about a hundred airports of various sizes.

Historical and cultural profile. *Between Africa and Asia.* Known to the Arabs since antiquity and mentioned in the 13th century, the island was settled at some unspecified time by people of Asian origin superimposed on groups that had originated from the African coast. The lack to date of archeological research in depth has made it impossible to formulate more precise theories about Madagascar's first inhabitants. Suffice it to note, however, that Persian and Chinese 11th-century funerary objects have been found in burial grounds in the northwest regions which enable us to affirm that, although the island is geographically close to Africa, it has from the very beginning been open to the most varied influences from the Asian continent with which it has been in contact via the archipelagoes of the Indian Ocean. From a linguistic and cultural standpoint the Paleo-Indonesian contribution has been decisive, and even today there are dwellings and domestic furnishings whose style recalls those of the Fiji Islands. Around the 16th century, Swahili traders brought about a deeper fusion of Asiatic and African elements by establishing trading bases along the northwest coast designed to attract substantial Bantu groups from the neighboring continent.

Local chiefdoms. The Europeans arrived at a later date, some time at the beginning of the 16th century, throughout which the Portuguese, French, English, and Dutch took turns settling on the Malagasy coasts. A century later Richelieu tried to bring the island under French control through the formation of the Société de l'Orient, but this attempt had to be abandoned after several decades of bitter clashes with the local populations. In fact, various ethnically based kingdoms existed or were about to appear which were little disposed to allow themselves to be subjugated by the whites. Attesting to the rule of the Sakalava tribe are the remains of monumental funerary complexes of stone statues surrounded by wooden structures with zoomorphic carvings. Other tribes in the inland of the island, however, used sovereign megalithic-like constructions with carved or fretted steles or posts in their tombs.

European assertion. Early in the 19th century, however, the various ethnic groups were subjected to the Merina warrior tribe, which was the first to establish a unified kingdom to rule over the island. It was through the Merina monarchy and as a result of the continuous intrigues of the courtiers at Antananarivo that the European encroachment made more decisive progress. At first, the British were favored, from whom King Radama I obtained backing for his anti-French position, but their presence became so intrusive as to trigger violent reactions throughout the country, much to the benefit of their European rivals. During the course of the century Great Britain and France played a dangerous game for the control of Madagascar, alternating with open clashes and fragile alliances, even resorting to palace conspiracies. The French won and in 1885 they were granted a protectorate over the island by Queen Ranavalona III;

ten years later they ended the monarchy and annexed the country directly. General Gallieni, who was then appointed military governor in Antananarivo, abolished slavery but also introduced an administrative order reducing the Malagasy to the level of "French subjects." During the entire first half of the 20th century, strong nationalist sentiments developed, expressing themselves in protest riots that exploded after both the first and second world wars. In 1947, when Madagascar was returned to France at the end of the British military occupation (1942–1945), an insurrection led by the Democratic Movement for the Rebirth of Malagasy (MDRM) swept the country. The repression was very harsh, with 80,000 victims, and the state of siege continued until 1956. France was later persuaded to re-examine its administrative methods, granting the country some measure of autonomy inside the French Community (1958) and ultimately independence (1960).

The life of the new republic went through two stages: first, a "moderate" phase under the banner of collaboration with the former mother country and social reconciliation, and then a "revolutionary" phase launched in 1972 by a military coup and characterized by an international alignment within the socialist bloc and a radical reform program to transform the country into a socialist state. This orientation ran into predictable economic obstacles as well as political difficulties because of the highly heterogeneous forces joined in the National Front, which assumed the defense of the Malagasy revolution.

Language and literature. Since antiquity a literature has flourished in the various dialects of the Malagasy language (similar to those of Indonesia, with Bantu and Swahili contributions) in the form of popular songs with musical accompaniment and of rhythmic compositions in irregular verse form (*chain-teni*) recited by two alternating voices. The English and French colonial presence clearly had a negative effect on this traditional literature, and it was particularly France which sought to impose a questionable cultural assimilation. Although already present in the 19th century and the first decades of the 20th century, literary production in both Malagasy and French intensified after World War II, stimulated by the liberation struggle. It was in this context that J. Rabemananyara infused new value into the local cultural heritage by developing the concept of "Malgachitude" on the analogy of the contemporary affirmation of "Negritude." The following poem is taken from the collection entitled *Rites millénaires* [Age-old rites]:

> I bring you, Goddess, a greeting pure as snow.
> A greeting as new as the spring.
> I bring it to you brimming with all my best wishes,
> and all my unconfessed desires.
>
> But what unrevealed message
> will the Oracle of the North bring?
>
> Every word strikes my heart with a glow of certainty
> Most precious manna
> with which the hungry of the desert satisfy
> your mouth of gold
> your mouth of honey,
> promised to the kiss of ecstasy alone ...

MALAWI

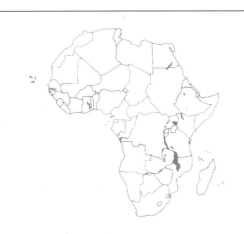

Geopolitical summary

Official name	Republic of Malawi / Mfuko la Malawi
Area	45,735 mi² [118,484 km²]
Population	7,982,607 (1987 census); 8,556,000 (1991 estimate)
Form of government	Presidential republic with a legislative assembly elected every 5 years
Administrative structure	24 districts combined into 3 regions
Capital	Lilongwe (pop. 233,973, 1987 census)
International relations	Member of UN and OAU; associate member of EC
Official language	English; national language is Chichewa (Bantu), other Bantu dialects are widespread
Religion	Majority animist; Catholic 20%; Protestant 8%; Islamic 7%
Currency	Kwacha

Natural environment. Malawi is bounded on the east by Tanzania, from which it is separated by Lake Malawi (formerly Lake Nyasa), on the south by Mozambique, and on the west by Zambia. It is a landlocked country that occupies the entire broad western shoulder sloping toward the tectonic depression washed by the waters of the Malawi lake basin (1551 ft [473 m] above sea level and 2224 ft [678 m] deep) in the southern section of the Rift Valley. Major elevations include the Nyika Massif (8758 ft

Climate data

Location	Altitude (ft asl)	Average temp. (°F) January	Average temp. (°F) July	Average annual precip. (in.)
Lilongwe	3500	73	60	35.1
Zomba	2900	73	63	52.5
Conversion factors: 1 ft = 0.3 m; 1 in. = 25 mm; °C = (°F − 32) × 5/9				

[2670 m]) in the north, which drops suddenly down to the lake, and the Mulanje Peak (10,000 ft [3000 m]), an isolated volcanic structure, in the south.

The country's hydrographic pattern is determined by the presence of Lake Malawi, which drains all the waters flowing down its western side and whose Shire outlet, compensated by Lake Malombe, pours into the Zambezi basin. In the southeast is Lake Chilwa as well as Lake Chiuta, which is a tributary of the Lugenda river.

Malawi is characterized by a wet tropical climate with two seasons, tempered by the altitude and the presence of the wide Lake Malawi basin. Peak temperatures are reached during October and November (86°F [30°C] in the valleys and 75°F [24°C] in the highlands), while 90% of the precipitation collects during the rainy season, which extends from November to April (hot weather), with annual rainfall of 29 in. [750 mm] in the less exposed areas and 58 in. [1500 mm] on the high ground.

As a result of intensive agriculture and overgrazing, the original vegetation cover has been replaced almost everywhere by wooded savanna, characterized by large grassland pastures and dominated by the giant *baobab*. The higher rain-drenched areas are covered by forests of cedar, teak, and mahogany, with gallery forests predominating on the shores of Lake Malawi and along the rivers.

The large carnivores and herbivores typical of Africa are protected in Kasungu, Nvika, and Lengwe national parks.

Population. Malawi is inhabited by Bantu people alongside a minority of Europeans and Asians. The Yao, who dominated the country in the last century, are predominantly Muslim and live in the southern regions; the Ngoni, of Zulu origin and predating the Yao, live in the northern areas; and the Nyanja or Malawi, the lake dwellers, gave the country its name and are considered to be the oldest inhabitants.

Because of its favorable climatic conditions, Malawi has always been heavily populated: in 1939 the population was about 1.6 million, then increased to 2.5 million after the war, and in 1991 totaled 8,556,000, with an average annual gain of 3.4% between 1980 and 1989. The mean population density is 236 per mi^2 [91 per km^2], with the highest recorded in the Shire highlands, where it is 500 per mi^2 [200 per km^2].

Eighty-eight percent of the country is rural and the only large city is Blantyre, which sits at an elevation of 3500 ft [1067 m] on

Administrative structure

Administrative unit (regions)	Area (mi^2)	Population (1987 census)
Central	13,739	3,116,038
Northern	10,395	907,121
Southern	12,257	3,959,448
MALAWI	36,391(*)	7,982,607

(*) Excluding 9344 mi^2 of inland waters
Capital: Lilongwe, pop. 233,973 (1987 census)
Major cities (1987 census): Blantyre, pop. 331,588; Mzuzu, pop. 44,238; Zomba, pop. 42,878.

Conversion factor: 1 mi^2 = 2.59 km^2

Socioeconomic data

Income (per capita, US$)	212 (1990)
Population growth rate (% per year)	3.4 (1980–89)
Birth rate (annual, per 1,000 pop.)	54 (1989)
Mortality rate (annual, per 1,000 pop.)	19 (1989)
Life expectancy at birth (years)	48 (1989)
Urban population (% of total)	12 (1989)
Economically active population (% of total)	43.3 (1988)
Illiteracy (% of total pop.)	58.8 (1991)
Available nutrition (daily calories per capita)	2009 (1989)
Energy consumption (10^6 tons coal equivalent)	0.28 (1987)

the left bank of the Shire. The capital of Lilongwe, which succeeded Zomba in 1975 because of its more central location, is a major agricultural market as well as the seat of some industries.

Considered to be one of the most densely populated regions in Africa, Malawi is a poor country with an annual per capita income of US$212—which has forced part of the population to emigrate to neighboring countries—and a high illiteracy rate of 58.8%.

Economic summary. Malawi's economic situation is very backward, with almost the entire population engaged in poor subsistence farming, while plantation agriculture accounts for over 37% of GNP and 85% of exports. Mineral resources, rich in coal, bauxite, and uranium, have not been sufficiently exploited; electric energy is predominantly based on water power. Industrial activity is very limited, encompassing primarily agricultural food processing and textiles, tobacco manufacturing, and sawmills.

The main food crop is corn (1.75 million t in 1991), grown along the Shire river and on the shores of Lake Malawi, followed by sweet potatoes and manioc. The chief exports are tobacco (100,100 t in 1990) and tea (45,100 t, most of it to Great Britain), sugar cane, peanuts, cotton, and coffee.

Forest covers 31.6% of the country and in 1989 supplied 268 million ft^3 [7.6 million m^3] of lumber, including such precious woods as teak, mahogany and, above all, cedar.

Livestock holdings comprise 1.1 million head of cattle (1990) and 1.2 million sheep and goats. Another resource for the local population is fish, with a catch of 97,000 t (1989).

The trade balance of Malawi is constantly negative, notwithstanding considerable exports of agricultural products. Because of the country's geographic isolation, communications are scarce and transportation difficult, the 514-mi [829-km] railway line (1986) connected to the Mozambique network being the most efficient. In addition, there are some 7440 mi [12,000 km] of roads (1985), only some of them paved, and Lake Malawi, which is navigable all year. International air transportation is assured at the Blantyre and Lilongwe airports.

Historical and cultural profile. As early as 10,000 years ago the western shore of Lake Nyasa was populated by people of Pigmoid origin. These were later joined by other ethnic groups, such as the Mbewe among whom power was inherited through matrilinear descent, and the Maravi, who settled there between the 13th and 15th centuries. This fertile region also attracted numerous Bantu tribes (the Phoka, Nkhamanga, Henga, Ngonde, Tumbuka, and Tonga) dedicated to agriculture and stock raising, who were fated to be decimated by the incursions of Arab and Portuguese slave-traders. During the 19th century, the warlike Nguni and Yao, the first to convert to Islam, also confronted each other along the shores of the lake.

The territory was discovered in 1859 by David Livingstone, converted by the missionaries of the Free Church of Scotland, exploited by the private African Lakes Company—especially for the ivory trade—and contested by Portugal and Great Britain until the latter European power established the British-controlled protectorate of "Nyasaland." From the very outset, the colonial regime aroused strong malcontent among the population which was expressed in the Nationalist Movement that developed between the two world wars and broke out into the open in the 1950s, at the time of the federation with Northern and Southern Rhodesia characterized by the domination of the white elite in Salisbury. In 1953, shortly before the federation was imposed, the chiefs of Nyasaland sent a petition to the British Government and the United Nations, in which they explained the reasons for their opposition to the federation:

> *From our understanding of Article 73 of the United Nations Charter we believe that it would be contrary to the trust accepted through the treaties with our Chiefs ... for the United Kingdom Government to transfer its sovereignty and responsibilities in whole or in part to any other body or persons, or to give up any part of its responsibilities towards the inhabitants of the Protectorate....*
>
> *It is our belief that the progress of the African people ... would be retarded and obstructed by the proposed Federation and the reinforcement of colour barriers which already exist in the political, economic and social spheres, especially in Southern and Northern Rhodesia ...*
>
> *We believe that if the proposed Federal Scheme for Central Africa were imposed on us the powers and responsibilities of Her Majesty's Government in the United Kingdom towards us and our economic, social and political aspirations and interests would be vitally and adversely affected, as indeed would be the mutual good faith that has existed between Britain and the African people of Nyasaland. It would be prejudicial to good relations between ourselves and those in Central Africa who seek to bring about this Federation despite our declared wishes, and, whilst claiming us as their 'partners' in it, relegate us to a position of subordination to themselves.*
>
> *We petition the United Nations, and ask Her Majesty's Government to support us in this ... We ask whether it would be compatible with international law for people who have voluntarily placed themselves under the protection of the Government of the United Kingdom to be handed over, regardless of their views and expressed wishes, to the jurisdiction of another Government; and for territories which are Protectorates and sacred trusts to be handed over by the Protecting and Administering Government to another Government's jurisdiction whether in whole or in part.*

The struggle for independence led by the Nyasaland African Congress (NAC) was also marked by violent clashes with the authorities, but in 1962 Great Britain conceded the territory's right to secede after an electoral consultation gave the separatists 90% of the vote. Self-governing status was granted the following year, and in 1964 full independence. The new state, called "Malawi," was governed along ever more authoritarian lines by Dr. Hastings Kamuzu Banda, the leader of the NAC, which was henceforth the only party sanctioned by the country.

MAURITIUS

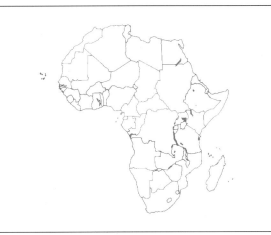

Geopolitical summary

Official name	Republic of Mauritius
Area	787 mi^2 [2040 km^2]
Population	1,000,432 (1983 census); 1,081,669 (1989 estimate)
Form of government	Republic within the British Commonwealth; Legislative Assembly elected by universal suffrage every 5 years
Administrative structure	2 islands and 2 dependencies
Capital	Port Louis (pop. 132,460 1990 estimate)
International relations	Member of UN, OAU, and the Commonwealth; associate member of EC
Official language	English; Creole and Hindi are also widespread
Religion	Hindu 52.5%; Christian 30.7%; Islamic 12.9%; Buddhist 0.4%
Currency	Rupee

Natural environment. The island of Mauritius, 720 mi^2 [1865 km^2] in land area, together with the island of Réunion, is part of the Mascarene Islands located approximately 465 mi [750 km] east of Madagascar in the Indian Ocean at 20° and 22° latitude south. Both islands are of volcanic origin and emerged from a ridge which partly separates the Mascarene basin from that of Madagascar. Mauritius lies 115 mi [185 km] northeast of

Climate data

Location	Altitude (ft asl)	Average temp. (°F) January	Average temp. (°F) July	Average annual precip. (in.)
Port Louis	7	79	68	76

Conversion factors: 1 ft = 0.3 m; 1 in. = 25 mm; °C = (°F – 32) × 5/9

Administrative structure

Administrative unit	Area (mi²)	Population (1989 estimate)
Islands		
Mauritius	720	1,043,631
Rodrigues	40	37,538
Dependencies		
Agalega	27	500
St. Brandon	0.4	–
MAURITIUS	787	1,081,669

Capital: Port Louis, pop. 132,460 (1990 estimate)
Major cities (1990 estimate): Beau Bassin/Rose Hill, pop. 94,236; Curepipe, pop. 66,704; Quatre Bornes, pop. 65,759; Vascoas/Phoenix, pop. 56,335.

Conversion factor: 1 mi² = 2.59 km²

Réunion and, unlike the latter, is no longer volcanically active. The ancient structures that were created by basaltic lava and deposits of tuff and breccia have been completely destroyed by erosion, which is why the highest point (Piton de la Rivière Noire, at the southwestern end of the island) is only 2710 ft [826 m] high, whereas the rest of the island has the appearance of an undulating plateau (from which higher ground emerges from time to time, such as Pieter Both, at 2690 ft [820 m]), that slopes slightly toward the coasts, dissected by waterways running down in a ray-like pattern. Some old craters are still filled by small lake basins, such as the Trou aux Cerfs, the Grand and Petit Bassins, etc. The island is also completely surrounded by coral reefs which rise to a height of 50 ft [15 m] along the northern coasts.

The climate is definitely tropical and especially humid on the east coast and the high ground where there is as much as 200 in. [5000 mm] of rainfall annually; it is drier on the northern and western coasts. The mean monthly temperature fluctuates between 81°F [27.2°C] in January and 73°F [22.8°C] in July. The danger of tropical cyclones is greatest with the intensification of rainfall from January to March.

Little remains of the lush tropical forests that covered the island before its colonization; the rest of the country is given over to extensive plantations of sugar cane and other crops.

The original fauna has also changed considerably, and many native species have been extinct for several centuries, such as the characteristic dodo bird without wings (*Didus ineptus*).

Dependencies of Mauritius include the Agalega Islands, more than 620 mi [1000 km] to the north, and the Cargados-Carajos (Saint Brandon) Islands, 248 mi [400 km] to the northeast; Rodrigues (with an area of 42 mi² [109 km²] and a population of 35,000), lies 372 mi [600 km] to the east and is of volcanic origin, with elevations of no more than 1475 ft [450 m] consisting of basaltic lava. The island of Rodrigues is surrounded by a high (500 ft [150 m]) coral-reef shoreline with many recesses where the remains of a flightless extinct bird known as the "solitaire" (*Pezophaps solitarius*) have been discovered.

Population. The island of Mauritius has had a rather complex settlement history. First settled in 1598 by the Dutch who introduced the cultivation of sugar cane, the island was conquered in 1721 by the French who made it into a base for the Compagnie des Indes, renaming it Île de France and massively importing black slaves from Africa. Following the subsequent British conquest in 1810, the Creole descendants of the earlier European colonizers relied on Indian workers from the Malabar coasts who soon exceeded half a million and were joined by Chinese, Malagasy, and other immigrants. Combined with the high natural increase of 2.5% a year, this rapidly led to overpopulation. Currently the island's population exceeds one million and is increasing at a more measured net rate (1%). The ethnic distribution is 70% Indians and the rest Chinese and Creoles. In addition to Hindi, French Creole languages are widely spoken.

On the whole, social conditions are fairly satisfactory, with a very low rate of illiteracy (17%) and considerable attention paid to education. Curepipe is the seat of a university with an enrollment of a thousand students.

Economic summary. Mauritius' economic mainstay continues to be sugar cane, most of which is exported after being processed in many plants scattered throughout the island. Other crops are palm coconuts, tobacco, coffee, vegetables, bananas, and peanuts, as well as vanilla, sansevieria, and manioc. Livestock includes tens of thousands of cattle, goats, and pigs, and there are also some fishing resources.

Industrial facilities, however, are more modest and include breweries and tobacco processing plants in addition to extensive sugar refineries. An oil refinery and a fertilizer plant are in operation at Port Louis. In order to diversify economic activity and reduce single-crop processing, new industrial initiatives have been undertaken in the textile, electronic, and petrochemical sectors.

Enhanced tourist accommodations are being developed to increase tourism, which has access to the island through the airport of Plaisance in the southeastern part of the island.

Mauritius has a slight deficit in its balance of trade because of the need to import considerable quantities of machinery and industrial products. Major trading partners are Great Britain, France, South Africa, and the United States.

The island has a fairly good road network and excellent harbor facilities: the Mauritian merchant marine possesses thirty or so ships totaling about 100,000 gross register tons (1990). The major port of call is Port Louis.

Historical and cultural profile. An important Arab trading center in the Indian Ocean since the end of the 10th century, the

Socioeconomic data

Income (per capita, US$)	2300 (1990)
Population growth rate (% per year)	1 (1980–89)
Birth rate (annual, per 1,000 pop.)	18 (1989)
Mortality rate (annual, per 1,000 pop.)	6 (1989)
Life expectancy at birth (years)	70 (1989)
Urban population (% of total)	41 (1989)
Economically active population (% of total)	39.1 (1988)
Illiteracy (% of total pop.)	17 (1991)
Available nutrition (daily calories per capita)	2679 (1989)
Energy consumption (10⁶ tons coal equivalent)	0.56 (1987)

island was reached by the Portuguese in 1511. It was initially called Dina Moraze or Dina Meshriq ("Island of the East"), then named Cirne (or "swan"), referring to the large swan-like dodo bird, now extinct, which was widespread in those times and was a symbol of the island. According to some historians, "Cirne" was also said to have been the name of one of the ships of Tristão da Cunha, who came to the island in 1512. Following the Portuguese came the Dutch in 1598, who gave the island its present name in honor of Maurice of Nassau, prince of the House of Orange. After more than a century as a possession of the Dutch East India Company, which introduced the cultivation of cane sugar, Mauritius came under the rule of France, first as a governorate and later as an outright colony. The French settlers first called the island Île-de-France but later renamed it Bonaparte at the time of the first Empire. During the Napoleonic Wars it came under a long siege by the British, which ended when it was ceded to Great Britain in 1814.

But Jules Dumont d'Urville was to write:

> *Although conquered [by the British, ed.] in 1810, Mauritius remains French ... Port-Louis is still a little Paris: pomp, fashions, pleasures of art, need for novelty, political passion—everything comes from our ports, nothing from London or Liverpool. In spite of a 15% surtax, our products are preferred over all others.*
>
> *The fact is that the historic memories of Île-de-France have a place in the annals of the Republic and the Empire. During the war at sea, which over twenty-six years experienced short intervals when our squadrons were decimated in European waters, leaving our colonial possessions without protection, the Île-de-France defended itself for twenty years, went on the offensive with its privateers, and held the wealthy East India company for ransom.... Whereas elsewhere we were subject to the English flag, on Île-de-France it was the English who had to pay heavy duties for passage in the Indian Ocean. Thus, harassed in this extension of the English Channel, the English were determined to drive off our privateers from their sphere of action. They sailed into Port-Louis in such large numbers and so strong that they forced it to capitulate. The peace of 1814 transferred their right of conquest on the map: the Bourbons struck a good bargain with French territory—what they relinquished was restored to them.*
>
> *However, the English got what they stole from the Île-de-France, the roadstead and fortified posts. If the hearts of the islanders continued to belong to France, what did their politics matter!*

In the 19th century, the country's traditional maritime activities, the prime source of its wealth, entered a period of crisis, but the losses were compensated by an extraordinary increase in the production of sugar cane, which doubled in one year from 10,000 t in 1825 and reached 100,000 t by 1854.

Following the abolition of slavery in 1834, the plantations experienced a need for new labor which the British colonial authorities satisfied by allowing large-scale immigration from India. This made the already heterogeneous population of Mauritius even more complex. Independence was achieved in 1968 and the parties which then took over power had to confront not only serious economic problems but a difficult internal situation precisely because of the great ethnic diversity and the contrast between French cultural influence and English politics.

MOZAMBIQUE

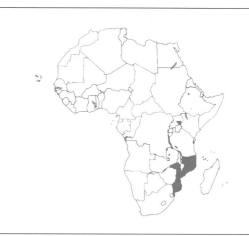

Geopolitical summary

Official name	República de Moçambique
Area	308,561 mi² [799,380 km²]
Population	11,673,725 (1980 census); 14,360,816 (1987 estimate)
Form of government	Presidential republic; the National Assembly is elected by universal suffrage as is the president.
Administrative structure	10 provinces plus the capital
Capital	Maputo (pop. 1,069,727, 1989 estimate)
International relations	Member of UN and OAU; associate member of EC
Official language	Portuguese; Bantu languages widely spoken
Religion	Animist 48%; Christian 19.3% (mainly Catholic); Islamic 16.5%
Currency	Metical

Natural environment. Mozambique is bounded on the north by Tanzania, on the northwest by Zambia and Malawi, on the west by Zimbabwe, on the southwest by South Africa and Swaziland, and on the east by the Indian Ocean.

Geological structure and relief. Mozambique occupies an indented portion of southern Africa's eastern margin, between the mouths of the Ruvuma and Maputo rivers, and consists of three sections: highlands east of Lake Malawi in the north, the low valley of the Zambezi in the center, and mostly alluvial lowlands in the south. The coastal strip, at times fairly wide, is composed almost entirely of conspicuous fluvial alluvia and has a shoreline stretching some 1500 mi [2500 km] and interrupted by many indentations often created by ridges and sand banks. These are responsible for the formation of lagoons and deep coves, usually open to the north, which alternate with the river estuaries and deltas. To the east, Mozambique is separated from the island of

Climate data

Location	Altitude (ft asl)	Average temp. (°F) January	Average temp. (°F) July	Average annual precip. (in.)
Maputo	197	78	65	29.6
Beira	16	82	69	59.3
Tete	574	73	71	30.0
Pemba	13	82	75	29.3

Conversion factors: 1 ft = 0.3 m; 1 in. = 25 mm; °C = (°F − 32) × 5/9

Madagascar by the Mozambique Channel (a sea arm 250–500 mi [400–800 km] wide and up to 1000 ft [3000 m] deep), from which several volcanic and coral islands emerge (Bassas da India, Europa, Juan de Nova). These are dependencies of Mayotte, a French island of the Comoros, which close the mouth of the Channel in the north.

Northern Mozambique, separated from Tanzania by the Ruvuma river, encompasses a large highland with a mean elevation of about 2000 ft [600 m] extending to the shores of Lake Malawi (1551 ft [473 m] above sea level). Geologically, it is composed of a remnant of the continent's ancient crystalline basement complex dating back to the Precambrian era that was leveled by erosion and is frequently traversed by fault lines, especially in the Lake Malawi area. The latter constitutes the southern end of the Rift Valley, that huge tectonic depression which cracked the compactness of the African continent in the Tertiary period. The highest peak near Lake Malawi is Mt. Jeci at 6022 ft [1836 m], while further south, between the depressions of lakes Amaramba and Chilwa and the coast, the Serra Namuli rises to 7934 ft [2419 m]. The central section of Mozambique, including the valley of the Lower Zambezi, is a veritable salient wedged in between Malawi to the north and Zimbabwe to the south, creating a border with Zambia.

The area's geological structure is rather complex: the old continental basement is still present but with many fractures overlaid with effusions of lava and sedimentary materials dating back to various ages. The marginal relief structure rises to over 6600 ft [2000 m], whereas toward the ocean the topography reflects a rapid drop, giving way to a wide alluvial lowland formed partly by the deposits of the Zambezi. South of this extends the southern section of Mozambique, with plateaus and granite mountains such as the Gorongosa range (5990 ft [1862 m]) and Mt. Binga (7990 ft [2436 m]), followed by a larger area of low-lying plains cut out of recent (Pliocene and Quaternary) deposits, partly divided by watercourses such as the Save and Limpopo flowing down from the mountains of Zimbabwe and South Africa.

Hydrography. The country's hydrographic features include the lower basin of the Zambezi and Limpopo rivers, the southeast shores of Lake Malawi, and other waterways which descend from the highlands and empty into the Indian Ocean, such as the Ruvuma, the Lúrio and the Save. The water regimen is governed by the precipitation, namely winter periods of drought alternating with summer floods. The final stretch of the Zambezi is navigable for at least 250 mi [400 km] down to the Cabora Bassa cascades, where a large dam creates an artificial basin in addition to the one upstream from Kariba in Zimbabwe.

Climate. The climatic conditions of Mozambique are typically tropical, with rather high mean annual temperatures (over 68°F [20°C] because of the warm Mozambique Current flowing down in a southwesterly direction along the channel of the same name), appreciable daily temperature swings in the western highlands, and pronounced seasonal precipitation in the form of monsoons. These occur during the summer months (between November and March) and are caused by the southeast trades. Annual rainfall generally exceeds 40 in. [1000 mm] and even reaches 60 in. [1500 mm] in the higher elevations and in the coastal belt opposite Sofala Bay. Precipitation tends to decrease inland and in the southern coastal plains while still remaining above 23–27 in. [600–700 mm] a year.

Flora and fauna. The landscape is not very highly differentiated, with savanna predominating everywhere, more humid (and therefore with more woodland) in some areas and drier in others depending on the intensity of the precipitation, except for the occasionally quite dense tropical gallery forest with mangroves fringing the waterways and along the coast. The fauna consists of the most representative African species, preserved in the large national parks which have been set aside for that purpose, such as Gorongosa north of Beira and Zinave and Banhine parks between the Save and Limpopo rivers.

Population. Mozambique's population is of southern Bantu stock, comprising numerous tribal groups such as the Makonde, Yao, Nyanja, Makua, Sena, Karanga, and others, each with its own dialects. Portuguese and English are also still spoken.

After repatriation of the more than 500,000 Portuguese settlers, who owned about 75 acres [30 ha] of land per family, the number of Europeans dwindled to about 25,000, most of them plantation farmers and technicians. The emigration of native workers to the mines of South Africa gradually decreased after the country gained its independence.

Settlements are predominantly rural and consist of small villages clustered around livestock enclosures (*kraal*), with cylindrical huts at the center of which open space is left to serve as a meeting place. The larger inhabited centers built during the colo-

Administrative structure

Administrative unit (provinces)	Area (mi²)	Population (1987 estimate)
Cabo Delgado	31,893	1,109,921
Gaza	29,224	1,138,724
Inhambane	26,485	1,167,022
Manica	23,801	756,886
Maputo	9,942	544,692
Nampula	31,500	2,837,856
Niassa	49,816	607,670
Sofala	26,255	1,257,710
Tete	38,879	981,319
Zambézia	40,533	2,952,251
Maputo (city)	232	1,006,765
MOZAMBIQUE	308,561	14,360,816

Capital: Maputo, pop. 1,069,727 (1989 estimate)
Major cities (1986 estimate): Beira, pop. 269,700; Nampula, pop. 182,553.

Conversion factor: 1 mi² = 2.59 km²

nial domination have retained their traditional appearance, but are now surrounded by squalid shantytowns. Although the new regimes have maintained the colonial language, they have radically changed the place names (a fairly common practice in all African countries), as in the case of the capital Maputo, formerly named after its founder, the navigator Lourenço Marques. Other major cities are Beira, Quelimane, and Mozambique—all on the coast. Two inland cities are Tete on the Zambezi and Nampula, the terminus of the rail line from Mozambique.

Social conditions, already not brilliant during the colonial era, are still rather precarious because of continuous internal political tensions, so much so that current per capita income is the lowest not only in Africa but in the entire world (about US$85 in 1990)!

Economic summary. Notwithstanding the country's potential wealth, Mozambique's economy is poorly developed because of the lack of infrastructure, the legacy of the colonial period during which no effort was made to train native technicians sufficiently knowledgeable and specialized in various areas of endeavor, and also because of the bloody political and military turns that followed independence. In an attempt to establish a planned economy, the government resorted to nationalizing the major sectors (industry, mining, and agriculture). Agriculture was collectivized and model villages created in which farming families benefited as salaried employees from local educational and health services as they cultivated the plots of land placed at their disposal.

The agricultural sector dominates the productive structure of the country, both in terms of income (accounting for 40% of the total product) and labor (employing 60% of the working population). The input of the other economic sectors is accordingly scant. The land under cultivation amounts to only 3.9% of the total land area, while there are extensive meadows and year-round pastures (about 55%) as well as forest and woodland (about 18%). The chief crops are cereals (rice, sorghum, corn, wheat), industrial crops (cotton, sugar cane, tobacco, peanuts, sunflowers), coconut palms, citrus fruits, and bananas. Animal husbandry is poorly developed and livestock holdings total no more than two million head of large cattle. There is a moderate amount of fishing, primarily in the waters of the Indian Ocean.

Mineral resources are fairly diversified but were never exploited on a large scale, not even during the colonial period. Today, limited quantities of gold, coal, bauxite, copper, asbestos, fluorite, tourmaline, and mica are being extracted, and rich deposits of iron have been discovered near Namapa. The production of electric energy, mostly of thermal origin, received considerable impetus with the construction of the large artificial lake at Cabora Bassa on the Zambezi. Industrial activity includes food processing (sugar), as well as cotton, cement, tobacco and chemical production. Mozambique has always maintained commercial relations with such countries as Portugal, the United States, and South Africa. A special collaboration has been established with South Africa in connection with the construction of the Cabora Bassa Dam for the use by South Africa of the ports of Maputo and Beira, while in the past many Mozambique miners had emigrated to that country.

Exports have improved in recent years, but continue to lag behind imports, resulting in a constant and growing indebtedness on international markets. The still very modest network of roads (24,287 mi [39,173 km] in 1985) and rail lines (2383 mi [3843 km]) is inadequate for a developing country. Shipping is of greater importance and includes access to good ports of call (Maputo, Beira, Pemba, Quelimane, Inhambane) which are also available to the landlocked countries of Zambia and Malawi. Major airports are located at Maputo, Beira, Quelimane, and Nampula.

Historical and cultural profile. Inhabited during earliest times by people who have left us cave paintings of scant artistic interest, Madagascar was populated during the first century of our era by Bantu groups which came from the region of the "great lakes," settling initially on the coast and later in the interior, where the Empire of Monomotapa, among others, was established between the Zambezi and Limpopo rivers. Starting in the 7th century, navigators from various Asian areas landed on the Mozambique coast and clusters of Swahili traders founded trading posts which were soon destined to become important centers: the fame of Sofala, Mozambique, Chinde, and Quelimane was to spread as far as India and China. The route to the Indies also led the Lusitanian navigators, such as Covilhão (1489) and Vasco da Gama (1498), to disembark at these ports, and so it was actually the Portuguese who came to impose their dominion over the region—first on the coastal belt and then in the interior—by sailing up the Zambezi and founding the river ports of Sena and Tete. Here is a description of the Sofala and Mozambique kingdoms, rich as the travel account which the Italian Filippo Pigafetta wrote for Charles V:

> Thus the Kingdom of Sofala lies below the Magnice and Cuama rivers and the coast of the Sea. It is small in size and has villages and lands, the chief of which is an island that lies in the river itself, called Sofala, from which that entire country derives its name; it is inhabited by Mohammedans and the King himself is of the same sect. He pays allegiance to the Crown of Portugal in order not to be subject to the Monomotapa Empire. There, at the mouth of the Cuama river, the Portuguese have a fortress, trading in these lands in gold aplenty and ivory and amber, which are found on that coast, as well as in slaves, in exchange for Bombay cloth and silk from Cambay which is what those people wear. The Mohammedans who now live in these lands are not native to the country, but before the Portuguese reached these regions, they plied their trade there in small boats from the coast of Arabia Felix. When the Portuguese came to rule these lands, those Mohammedans who were there remained, and today they are neither heathen nor of the Mohammedan sect.
>
> A little beyond we suddenly come upon the Kingdom of Mozambique, located at 14.5 degrees [latitude] south, which takes its name from three islands lying at the mouth of the River

Socioeconomic data

Income (per capita, US$)	85 (1990)
Population growth rate (% per year)	2.7 (1980–89)
Birth rate (annual, per 1,000 pop.)	46 (1989)
Mortality rate (annual, per 1,000 pop.)	17 (1989)
Life expectancy at birth (years)	49 (1989)
Urban population (% of total)	26 (1989)
Economically active population (% of total)	52.9 (1988)
Illiteracy (% of total pop.)	83.4 (1991)
Available nutrition (daily calories per capita)	1632 (1989)
Energy consumption (10^6 tons coal equivalent)	0.48 (1987)

Megincate, where there is a large and safe harbor capable of accommodating any manner of craft. The kingdom is small but abounds in every kind of food and is a port of call for all ships that sail to that country from Portugal and India. On one of these islands, called Mozambique, which is the principal and ruling one that gives its name to all the others as well as to the kingdom and the above-mentioned port, stands a fortress garrisoned by the Portuguese on which all the other fortresses on that coast depend and from which they draw their supplies. The fleets which sail from Portugal to India, if they cannot complete their voyage, winter in Mozambique, and those from India which sail for Europe must necessarily anchor at Mozambique to replenish their provisions.

Having exhausted the country's large mineral resources, the Portuguese forgot about gold, silver, copper, and iron to dedicate themselves to the slave trade, which became especially lucrative in the 18th century when France began to require labor for the plantations introduced on the Indian Ocean islands conquered at that time. In order to better control Mozambique, which at first was a dependency of Goa, Portugal introduced an autonomous colonial government in 1752 and in the course of a century reinforced its domination, defending the country against the attacks not only of the local tribes but also of the Dutch and Arabs.

In 1891 the Berlin Conference recognized their dominion and defined the colony's borders, while at the same time negating Portugal's aspirations to unite its possessions on the Indian Ocean with those on the Atlantic in a single great Portuguese Central Africa. During the decades that followed the colony remained a theater of conflict between the colonial troops and the populations of the interior which had been swollen by mass migrations of Nguni from the south, and it was not until 1915 that the area could be said to have been "pacified." Portuguese rule proved to be ruthless, especially during Salazar's regime, characterized by racial segregation, police rule, general forced labor practices, and the "export" of Mozambique workers to the mines of South Africa. For these reasons the nationalist movement for some time found it difficult to organize, but when it did come out into the open in the 1960s as the Front for the Liberation of Mozambique (FRELIMO), it immediately embraced revolutionary ideas and soon dedicated itself to armed struggle.

Mozambique achieved independence in 1975, along with the other Portuguese colonies, after the "revolt of the captains" in Lisbon the previous year put an end to the Salazar regime. FRELIMO then came to power as head of the People's Republic that was established. The new state was actively opposed by both the Rhodesian settlers (defeated after the independence of Zimbabwe) and South Africa, which supported the dissident internal guerrilla movement RENAMO (Mozambique National Resistance). After more than 10 years of bloody civil war, with over 600,000 dead, an agreement was reached in 1992 between the insurgents and the government.

The Mozambique region is devoid of long-standing artistic traditions, but various Bantu tribes have developed fairly significant forms of art in recent centuries. In the field of wood-carving there are the anthropomorphic and zoomorphic statues of the Thonga and Makonde peoples. This latter group also produced objects for daily use modeled in clay by a skillful technique and decorated the facades of their houses with stylistically original ornamental geometric elements, figures from the animal world, and family scenes.

NAMIBIA

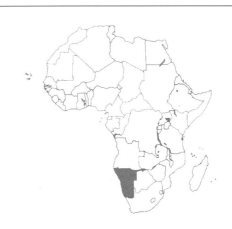

Geopolitical summary

Official name	Republic of Namibia
Area	317,734 mi^2 [823,144 km^2]
Population	1,009,900 (1982 census); 1,184,000 (1987 estimate)
Form of government	Constitutional republic
Administrative structure	26 districts
Capital	Windhoek (pop. 114,500, 1988 estimate)
International relations	Member of UN and OAU
Official language	English and Afrikaans; Bantu, Hottentot, and Bushman languages widespread
Religion	Predominantly Christian (around 200,000 Catholics) and animist
Currency	South African rand

Natural environment. Namibia is bounded on the north by Angola, on the northeast by Zambia, on the east by Botswana, on the southeast and south by South Africa, and on the west by the Atlantic. The country's orographic structure is characterized by an extensive highland complex composed of ancient crystalline basement outcrops, in large part covered by Paleozoic and Tertiary sediments, as well as many basaltic effusions of an average altitude of 3300–5000 ft [1000–1500 m] and mountains often over 6500 ft [2000 m] in elevation, such as the Auasberge at 8148 ft [2484 m], Brandberg at 8561 ft [2610 m], and Kransberg at 7708 ft [2350 m]. The coastal strip along the straight Atlantic coastline, interspersed by lagoons and sandbanks formed by accumulations of the Benguela Current, is a desolate desert.

The climate is tropical in character: arid and desert-like on the coast and semiarid with a short rainy season inland. Annual rainfall is less than 4 in. [100 mm] in the coastal region and over 23 in. [600 mm] in the northeast. The temperatures are moder-

Climate data

Location	Altitude (ft asl)	Average temp. (°F) January	Average temp. (°F) July	Average annual precip. (in.)
Windhoek	5674	73	55	8.8
Swakopmund	16	64	60	0.4

Conversion factors: 1 ft = 0.3 m; 1 in. = 25 mm; °C = (°F − 32) × 5/9

ated by the Benguela Current and inland by the high elevations.

Vegetation is limited to xerophytous species in the Namib Desert along the coast, and steppelike species on the inland plateaus, with grasslands, acacia, and euphorbia; savanna and remnants of the tropical forest survive along the river banks. The typical desert fauna is scarce, consisting of small rodents, scorpions, reptiles, and gazelle in addition to tapir and baboons, while large herbivores like the antelope and giraffe still roam the steppe. Wildlife is protected in the Etosha and Namib reserves.

Population. The population of Namibia is 88% Bantu, with a minority of Bushmen and Hottentots. Representing the largest Bantu ethnic group are the Ovambo, who eke out a meager living as farmers in the northern part of the country or work in the mines. The Herero in the north-central regions are engaged in stock raising. The white minority, a third of whom are descendants of the German colonists, occupies a privileged political and economic position compared to the native peoples who live under very backward social conditions, with a literacy rate of only 45%.

Namibia is one of the least populated countries in Africa, with a population of some 1.2 million and a density of about 2.6 per mi² [1 per km²]. In recent years the demographic growth rate has increased to 3.2%. The highest population density is in the central plateau regions, where the major cities are located, inhabited predominantly by whites. Windhoek, the capital, lies at an altitude of 5674 ft [1730 m] and is an important trade, rail, and road junction. Walvisbaai (Walvis Bay) and Lüderitz are the coun-

try's principal ports. Other towns are Keetmanshoop, Rundu, Rehoboth, and Swakopmund.

Economic summary. The mainstays of Namibia's economy are its mineral resources (which represent 90% of its exports), stock raising, and fishing. Crops are grown on only 0.8% of the territory and include corn, millet, and sorghum, but in insufficient amounts to meet the country's needs. Stock raising, which occupies much of the indigenous population, totaled 2 million head of cattle in 1990 and 9.3 million sheep and goats (including about 1.3 million Karakul sheep that supply the pelts for astrakhan coats). Fishing is a rich economic resource (catch of 22,327 t in 1989).

Diamonds are among Namibia's chief mining resources with an average annual yield of one million carats, exploited by multinational companies, most of which are South African. Other minerals are zinc, lead, copper, and small quantities of silver and cadmium; also extracted are manganese, vanadium, tungsten, pyrite, gold, salt (along the coast), phosphates, and uranium. Industrial activity includes the processing of metals, fish (sardine and crayfish canning), and cattle products (butter and cheese production, meat canning, and preparation of pelts).

Although no official data are available, it is estimated that the export of diamonds, cattle, and fish—mostly to South Africa—covers the cost of imports, mainly petroleum and food products.

Transportation facilities encompass a rail system of 1477 mi [2382 km] (1985) which runs across the length of the country, with branch lines and about 26,000 mi [42,000 km] of roads (1987). An international airport is located at Ondelkaremba, near the capital.

Historical and cultural profile. In ancient times the present-day territory of Namibia was the site of migrations of peoples which involved all of southern Africa, and were characterized by Bantu-speaking peoples supplanting Bushmen and Hottentots. Relationships among the various ethnic groups were not always

Administrative structure

Administrative unit (districts)	Area (mi²)	Population (1987 estimate)	Administrative unit (districts)	Area (mi²)	Population (1987 estimate)
Bethanien	6,950	3,000	Namaland	8,152	15,000
Boesmanland	7,129	3,000	Okahandja	6,809	15,000
Caprivi-Oos	4,452	44,000	Omaruru	3,252	6,000
Damaraland	17,972	28,000	Otjiwarongo	7,932	19,000
Gobabis	15,999	25,000	Outjo	14,947	10,000
Grootfontein	10,237	25,000	Owambo	19,995	520,000
Hereroland-Oos	20,052	22,000	Rehoboth	5,474	33,000
Hereroland-Wes	6,369	18,000	Swakopmund	17,253	18,000
Kaokoland	22,461	20,000	Tsumeb	6,338	22,000
Karasburg	14,713	11,000	Windhoek	12,927	129,000
Karibib	5,107	10,000			
Kavango	19,669	122,000	**NAMIBIA**	317,734	1,184,000
Keetmanshoop	14,785	20,000			
Lüderitz	20,482	16,000			
Maltahöhe	9,871	6,000			
Mariental	18,408	24,000			

Capital: Windhoek, pop. 114,500 (1988 estimate)

Major cities (1988 estimate): Swakopmund, pop. 15,500; Rehoboth, pop. 15,000; Rundu, pop. 15,000.

Conversion factor: 1 mi² = 2.59 km²

Socioeconomic data

Income (per capita, US$)	1342 (1990)
Population growth rate (% per year)	3.2 (1980–89)
Birth rate (annual, per 1,000 pop.)	44 (1989)
Mortality rate (annual, per 1,000 pop.)	12.1 (1989)
Life expectancy at birth (years)	55 (1989)
Urban population (% of total)	27.8 (1989)
Economically active population (% of total)	29.7 (1988)
Illiteracy (% of total pop.)	27.5 (1991)
Available nutrition (daily calories per capita)	2183 (1989)
Energy consumption (10^6 tons coal equivalent)	Included in South Africa

SOUTH AFRICA

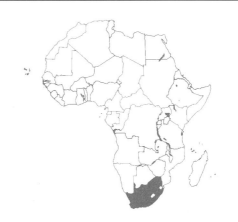

peaceful and even as late as the 19th century there were fierce encounters between the Herero and Hottentots. The European presence, which was varied and without deep roots, did not extend inland: Portuguese navigators, Dutch traders, Nordic whalers, and English and German missionaries stayed in the coastal plain known as the Namib, and colonization began particularly late. In 1883 Germany obtained the bay at Angra Pequeña (which they later renamed Lüderitz) from a Hottentot chief and the explorer Gustav Nachtigal extended the protectorate to the interior of the country, which was renamed South-West Africa. After World War I the Union of South Africa, which had militarily occupied the territory, received a League of Nations mandate to administer the country, and the end of World War II gave the Pretoria government the pretext for claiming annexation of the former German colony. South-West Africa was thus incorporated into the Union of South Africa as its fifth province. Notwithstanding the UN refusal to recognize the *de facto* situation and despite the struggle of the South-West Africa People's Organization (SWAPO), it took several decades before independence was achieved. The SWAPO leadership appealed to the UN on several occasions, requesting it to intervene in South-West Africa and expel the South Africans on their behalf. In October 1969, a SWAPO delegation led by Gottfried Hage Geingob, the party's representative in the United States, participated in the meeting of the UN's Fourth Committee and presented the following requests by his organization: recognition of the struggle in Namibia and granting of material aid; imposition on the government of South Africa to withdraw from the Namibian state; respect for the laws governing the Territories, with the resulting need to tax companies doing business in Namibia, and granting of "a greater number of scholarships to Namibians." The country obtained its independence in 1990 under the name of Namibia. What made this aspiration possible was, above all, the new, more relaxed international climate in which the 1988 negotiations were held in New York among Angola, Cuba, and South Africa to define the country's new order.

Geopolitical summary

Official name	Republic of South Africa / Republiek van Suid-Afrika
Area	433,565 mi^2 [1,123,226 km^2]
Population	23,385,645 (1985 census); 30,796,000 (1990 estimate)
Form of government	Presidential republic with a Parliament comprising 3 Houses elected by the respective ethnic communities (Whites, Coloureds, Asians)
Administrative structure	4 provinces and 10 autonomous Bantu territories (Black Homelands), of which 4 are recognized as independent
Capitals	Pretoria (pop. 822,925, 1985 census) with administrative functions; Cape Town (pop. 1,911,521, 1985 census) with legislative functions; Bloemfontein (pop. 232,984, 1985 census) with judicial functions.
International relations	Member of UN
Official language	Afrikaans and English; numerous Bantu languages are widespread
Religion	1980 census showed 2.36 million Catholics, 13.34 million Protestants, 2.3 million other Christians, 119,000 Jews, 512,000 Hindus, 318,000 Muslims, 5.9 million of various beliefs (animist, etc.)
Currency	Rand

Natural environment. The territory of South Africa occupies the southern extremity of the continent with a series of highlands washed by the Atlantic Ocean on the west and the Indian Ocean on the south and east, and bounded on the north by several rivers, including the Orange, Limpopo, and the gullies of the Molopo and Nossob. The borders of South Africa are delimited to the north by Botswana and Zimbabwe, to the northeast by Mozambique and Swaziland, and to the northwest by Namibia.

Geological structure and relief. The average elevation of South Africa is above 3300 ft [1000 m], dropping below that only in the inland steppe regions (Grootvloere Hakskeepan) and along the coastal strips, where rocky promontories and hazardous peninsulas (such as those of Cape Columbine and the Cape of Good Hope) alternate with small alluvial lowlands (Groot Berg, Sundays, Umfolozi, and Pongolo), not infrequently surrounded by lagoon-like bodies of water. The coastal profile is fairly rectilinear on the eastern side between Cape Padrone and the Mozambique border.

Geographically, the continuity of the South African territory is interrupted by the completely enclosed enclaves of Lesotho (on the Basuto plateau in the upper basin of the Orange) and Swaziland. Moreover, there is a long and narrow salient thrust between the eastern border of Namibia (coinciding with the meridian at 20° east of Greenwich) and the southwest border of Botswana (along the course of the Nossob). South Africa also owns the enclave at Walvis Bay facing the inlet of the same name on the Atlantic Ocean, inside the territory of Namibia, as well as the group of Penguin Islands which rise in front of its southern coast. The two islands of Prince Edward and Marion, which lie in the Indian Ocean some 1200 mi [1900 km] southeast of the Cape of Good Hope, likewise belong to South Africa.

In geological terms, the overall structure of the country represents a fairly simple model, even if particularly complex in detail. It is a large remnant of the Precambrian continental slope, metamorphosed and repeatedly leveled by erosion, over which a thick nappe of sediments of marine and continental origin was deposited during the entire Paleozoic era. These sediments, in turn, were subjected to the effects of the Caledonian and Variscan orogenies which produced huge fold structures, as still evidenced by the mountain chains and raised margins (Drakensberg Mountains) of the scarps (Great Escarpment) which delimit the entire South African territory, both at its periphery and inland. Outcrops of the ancient basement, predominantly formed by metamorphic rock rich in granite intrusions often accompanied by auriferous mineralization (as in the Witwatersrand), are especially prevalent in the Orange and Vaal basins, as well as in the northeastern part of the country. Paleozoic sedimentary series (sometimes predominantly dolomitic limestone, sometimes argillaceous schist, and elsewhere composed of materials of continental origin such as typical glacial deposits like tillites), formed between the Cambrian and Carboniferous, constitute the framework of the principal mountains that brighten the South African highlands, such as those of

Nomakwaland (5600 ft [1707 m]), Cape Province (over 6500 ft [2000 m]), the northern part of the Orange Free State (6084 ft [1855 m]), and the Transvaal (over 6500 ft [2000 m]), as well as those which overlook the southern coasts (Swartberge, 7629 ft [2326 m], Langeberg, etc.). These sediments are in turn covered—over a vast area of 231,500 mi^2 [600,000 km^2] throughout the east central part of the country—by the typical Karroo formations of southern Africa (ranging in age from the Upper Carboniferous and Upper Triassic) with its rich beds of coal in addition to fossil deposits (plants and vertebrates) of great interest. Tectonic movements toward the end of the Triassic period caused the effusion of considerable quantities of basaltic lava as much as 6500 ft [2000 m] thick in the area of the Basuto highland (which has peaks over 10,000 ft [3000 m] high, the highest point being Thabana Ntlenyama at 11,421 ft [3482 m]). Other Mesozoic sediments of marine origin produced minor elevations along the ocean slopes, whereas the Cretaceous was probably responsible for the formation of those volcanic chimneys (also found in various localities of Cape Province, Orange, and Transvaal) that formed the famous Kimberley diamond deposits. Finally, the depressions of the Molopo basin, as well as some coastal plains, are covered by sediments of continental origin of the Kalahari series.

Hydrography. The country's water resources are determined by the relief and structure of its hydrographic network, characterized by the two major river basins of the Orange (1153 mi [1860 km] long), which flows into the Atlantic, and the Limpopo (1000 mi [1600 km]), which empties into the Indian Ocean, in addition to many shorter watercourses that flow down from the highlands to the ocean. The main right-bank tributary of the Orange river is the Vaal (750 mi [1200 km]), which crosses the Witwatersrand region, while the Olifants (350 mi [560 km]) descends from the northern slopes of the Drakensberg Mountains into the Limpopo. There are no large natural lakes because of the intense evaporation and lack of rainfall, but this is compensated for by many artificial basins.

Climate. South Africa's climatic conditions are greatly influenced by its geographic position (the Tropic of Capricorn crosses the northern regions), altitude, and the presence of the sea. Monthly mean temperatures in January exceed 77°F [25°C] in the interior regions which look out onto the Kalahari and stay around 68°F [20°C] on the southern coasts, whereas in July—the coldest month—they are almost always 50–59°F [10–15°C]. Temperature swings are, however, greater in the interior plateaus. Rain falls predominantly during the summer months and affects the entire eastern margin of the plateau, being carried there by the southwest trade winds, and can at times exceed 39 in. [1000 mm] a year. In the Cape area, rainfall is concentrated during the winter months, induced at this latitude by the cold fronts coming from the Antarctic. Finally, the entire western central area is characterized by a considerable shortage of precipitation, which falls to very low levels (sometimes even less than 4 in. [100 mm] a year) on the Atlantic coast, influenced by the cold Benguela Current in contrast to the warm Agulhas or Mozambique Current off the east coast. Both of these ocean currents and their thermal contrasts are responsible for the frequent formation of fogbanks which endanger navigation. On the whole, the climate of the southern tip of South Africa is of the hot temperate or Mediterranean type, whereas inland and along the northwestern

Climate data				
Location	Altitude (ft asl)	Average temp. (°F) January	Average temp. (°F) July	Average annual precip. (in.)
Pretoria	4,592	70	52	30.6
Cape Town	39	69	54	19.9
Bloemfontein	4,569	73	47	22.0
Durban	16	75	62	39.4
Johannesburg	5,750	68	51	27.7
Conversion factors: 1 ft = 0.3 m; 1 in. = 25 mm; °C = (°F − 32) × 5/9				

coasts it is more typically tropical, with a tendency to be continental and arid, especially in the northern central regions.

Flora and fauna. The vegetation of the country, also in connection with the barometric effects, is quite varied. In the Cape region Mediterranean species are the predominant flora (maqui and shrub), followed along the eastern coastal belt by wet tropical forest interspersed with dense grassland which also covers the slopes of the Drakensberg Mountains and, with temperate characteristics, extends to the Basuto plateau as well. Much of the area drained by the Limpopo and its right-bank tributaries is covered by scrub and tree-lined savanna, with acacia and baobab, which often give way to stretches of gallery forest along the waterways. On the interior plateaus the forests are gradually replaced by temperate meadowland, then shrubby and arid steppes, and, finally, in the lower basin of the Orange river, an actual desert environment which, in this case, is the southern extremity of the Namib Desert.

After having been intensively hunted for centuries, the big-game South African wildlife is now confined in numerous protected areas and especially in extensive national parks, the largest and most famous of which is the Kruger National Park, at the northeastern end of the country. Other noteworthy wildlife preserves are the Addo Elephant National Park, which encompasses the evergreen forest of the same name north of Port Elizabeth, and the Kalahari Gemsbok National Park on the border with Namibia and Botswana, which is home to many species of antelopes. Among the protected species, worth mentioning is the geometric tortoise (*Psammobates geometricus*), which lives in the Elandsberg Reserve in Cape Province.

Population. The most recent events that have marked the public life of South Africa following the release of Nelson Mandela and other representatives of the black opposition and the formal abrogation of the policy of *apartheid*, or racial segregation, are bound to have a profound effect on the development of the richest nation in Africa and, above all, its social and demographic conditions. At present these are characterized by the fact that the majority of the population that is two thirds Negro of Bantu stock, has so far been confined to passive participation in the economic and social life of the country, which is firmly in the hands of the white population. The latter represents a little less than 20% of the total, while the balance comprises Indians, Bushmen, and Hottentots in very small number (3.5%) and so-called "Coloureds" (approximately 11%), the descendants of unions between whites and natives. The whites, in turn, include two major groups, the "Afrikaners," descendants of the Dutch settlers who were the first to develop the region after landing there in the 17th century (Cape Town was founded by them in 1652), and the English, who subjugated them at the end of the Anglo-Boer wars, a bloody conflict that raged from 1899–1902.

South Africa was originally inhabited by Bushmen and Hottentots, nomads dedicated to hunting and gathering who were chased toward the arid regions of the Kalahari and Namibia by successive Bantu immigrations from the northeast. The largest Bantu groups were the Zulus, who had strenuously opposed white expansion and now live in Natal province, and the Xhosa, who settled in the Transkei. Other Bantu groups, who now populate Orange, Transvaal, and Cape provinces, were the Sotho, the Bechuanas, the Tsonga, the Swazi, and the Venda. They were

Administrative structure		
Administrative unit	Area (mi^2)	Population (1985 census)
Provinces		
Cape	247,572	5,041,137
Natal	23,297	2,145,018
Orange Free State	49,152	1,776,903
Transvaal	88,532	7,532,179
Black Homelands		
Gazankulu	2,534	497,213
KaNgwane	1,476	392,782
KwaNdebele	355	235,855
Kwazulu	11,966	3,747,015
Lebowa	8,428	1,835,984
Qwaqwa	253	181,559
SOUTH AFRICA	433,565 (*)	23,385,645

(*) Excluding 4 independent Bantustan regions: Bophuthatswana (16,984 mi^2, pop. 1,959,000 in 1990); Ciskei (3281 mi^2, pop. 860,000 in 1990); Transkei (16,058 mi^2, pop. 3,000,000 in 1990); Venda (2860 mi^2, pop. 560,000 in 1990).

Capitals (1985 census): Pretoria, with administrative functions, pop. 822,925; Cape Town (urban agglomeration), with legislative functions, pop. 1,911,521; Bloemfontein, with judicial functions, pop. 232,984. Major cities (1985 census): Johannesburg, pop. 1,609,408; Durban, pop. 982,075; Port Elizabeth, pop. 651,993.

Conversion factor: 1 mi^2 = 2.59 km^2

artificially settled in the so-called "Bantustans," units inhabited exclusively by blacks which the South African government considers independent and self-governing. In practice, these are actually reservations, pure and simple, where black workers are housed and considered to have immigrated, having no longer retained South African citizenship. There are currently ten such South African Bantustans: Bophuthatswana, Ciskei, Transkei, and Venda were the first, to which were later added Kwazulu, Lebowa, Qwaqwa, Gazankulu, KaNgwane, and KwaNdebele, covering an overall area of 64,194 mi^2 [166,306 km^2] and home to about 12.5 million people. It should be noted that the independence of the first four has been challenged by the United Nations as well as the Organization of African Unity (OAU). It is predictable, however, that once racial segregation is abolished the existence of these fictitious political entities will also be questioned. Recently, in fact, they have been the scene of inflamed popular claims for reintegration into South Africa.

Today, both the white and nonwhite populations of South Africa are experiencing a substantial growth rate, averaging 2.4% a year (very restrained among whites and rather higher among blacks). The Bantu people in particular are increasing at a much greater rate than the others. In the past, especially during the last decade of the 19th century and the first decade of the 20th, the white population experienced a marked increase because of a mass influx of immigrants from various parts of Europe attracted by the illusion of easy gains following the discovery of rich gold deposits in the Witwatersrand. Earlier, when

England assumed possession of the colony (1814), the old Dutch settlers (Boers) were driven inland where, after a long migration known as the Great Trek (1834–44), they established the republics of the Orange Free State and the Transvaal.

In geographic terms, the density of the population of South Africa is still rather low (70 per mi^2 [27 per km^2]) and its distribution somewhat irregular. The most densely populated areas are found around Cape Town and in the southeast coastal belt, between Port Elizabeth up to the Mozambique border (Natal), as well as in the districts of Bloemfontein and Johannesburg in the highlands. Approximately 60% of the population live in urban centers, and are predominantly white, but also Indian and Coloured. The blacks, however, are restricted to outlying shanty-towns or live in the agricultural regions of the veld dedicated to farming and cattle-raising; their villages are very poor and made up of houses crowded around a central area (kraal) set aside as a shelter for the livestock and as a meeting place.

Situated in the heart of the Witwatersrand (near Pretoria, the administrative capital of the state) is the highly modern metropolis of Johannesburg, South Africa's most active and liveliest city and its main industrial center. Other cities on the plateau are Bloemfontein, the capital of the Orange Free State; Kimberley, near the world's most famous diamond deposits; Vereeinging, also in the Witwatersrand, and many others. On the ocean is Cape Town, the legislative capital of South Africa and the most populous city in the country as well as one of the most beautiful of the continent because of its magnificent and picturesque view of False Bay, dominated by Table Mountain and the legendary Cape of Good Hope.

Other major cities on the southeast coast are Durban, the maritime terminal of the rail line linking it to Johannesburg; Port Elizabeth, nestled in wide Algoa Bay; and East London.

South Africa, with its variegated ethnic composition, is also a mosaic of languages and religions. The most widespread languages are Afrikaans, derived from Dutch and spoken by the Afrikaners, and English, imposed by the subsequent conquerors. Both of these languages are spoken by the majority of whites and Coloureds. The blacks are also familiar with English, but generally use Bantu dialects. The Asian population speaks various Indic and Dravidian languages (Tamil, Hindi, Gujarati, etc.). Christianity is fairly widespread among the whites as well as among the Coloureds and Bantus, with a predominance of Protestant denominations, among which the Dutch Reformed Church holds a prominent position. There are slightly more than two million Catholics. The Hindu and Muslim religions are generally more widespread among the people of Asian origin. Animist beliefs are still deep-rooted among the Bantus and completely preserved by the Bushmen and Hottentots.

Economic summary. In spite of the contradictions inherent in the social framework of South Africa, the country's economy is no doubt the most thriving on the African continent, favored by the wealth of its natural resources—mineral as well as agricultural, including livestock—but above all by its especially efficient and flourishing commercial relations with the rest of the world, although these have been seriously impaired by the economic sanctions imposed by the UN to counter its apartheid policy. The fact is, however, that a differentiated reading of the country's economic picture in terms of the ethnic realities involved reflects

contrasting aspects: individual income, which in 1990 was about US$2680, rises to US$8500 if only the white population is considered, whereas for the Asians and Coloureds it drops to US$1800 and is as low as US$700 for the black population. Structurally speaking, the economy of South Africa can be regarded as industrial since industry accounts for about one third of GNP, while mining activity is still substantial (14%) and agriculture fairly limited (5%). On the other hand, the agricultural sector is heavily burdened because 14% of the work force, mostly black, is dedicated to the sole purpose of self-subsistence. Also weak is the share of the service sector, in terms of both work force and income, because of a basic failure to develop the domestic market.

In recent years, the effects of the antiapartheid economic sanctions, combined with the crisis in raw material prices, especially of gold and petroleum, has led to a notable drop in foreign investments, with disastrous consequences on unemployment, which now affects fully a third of the working population, and on depreciation of the national currency (the rand), which over ten years lost almost four times its value, posing serious problems for foreign debt repayment.

Agriculture and livestock. The agricultural sector—which involves altogether about one tenth of the country's land area, two thirds of which consist of meadows and pastures, and only 4% of forests—is characterized by major contrasts. On the one hand, the most fertile land is set aside for plantations and farms directly managed by whites, using extensive equipment and modern methods. On the other hand, the land farmed by the Bantus is very unproductive and is often combined with cattle-raising, in their villages usually surrounding the *kraal*, the typical enclosure where they keep their livestock. South African agricultural development benefits from a system of irrigation consisting of many artificial basins and an extensive network of canals. Cultivation encompasses a rather broad panorama of crops, comprising almost every kind of cereal (wheat, corn, rye, barley, oats, millet, sorghum), tuberous plants, garden vegetables, and legumes. There is also considerable orchard produce, including apples, pears, and other pome fruits, in addition to citrus and tropical fruits, and particularly grapes around Cape Town. Industrial crops occupy a special place in the agricultural economy and include sugar cane (Natal), tobacco (Transvaal and Cape Province), and cotton, as well as many oil and fiber products (peanut, sunflower, castor, sisal, etc.). Forestry products include primarily tannery products and construction lumber from the forests that cover the Drakensberg Mountains.

Chiefly raised by white farmers are the animals that supply

Socioeconomic data	
Income (per capita, US$)	2680 (1990)
Population growth rate (% per year)	2.4 (1980–89)
Birth rate (annual, per 1,000 pop.)	34 (1989)
Mortality rate (annual, per 1,000 pop.)	10 (1989)
Life expectancy at birth (years)	62 (1989)
Urban population (% of total)	59 (1989)
Economically active population (% of total)	36.3 (1988)
Illiteracy (% of total pop.)	20.7 (1991)
Available nutrition (daily calories per capita)	3035 (1989)
Energy consumption (10^6 tons coal equivalent)	107.6 (1987)

meat, wool (angora) and prized pelts (karakul). In 1991, the country's livestock included 13.5 million head of cattle, raised in the more humid regions, 6 million goats, and 32.6 million sheep, which graze in the drier interior highlands. Fishing, carried out largely in the cold, oxygen-rich waters of the Atlantic where fish are more plentiful, is particularly well-developed, with a catch (sardines, cod, anchovies, etc.) totaling 966,438 t in 1989. Whaling is still practiced off the southern coast.

Energy resources and industry. South Africa is pre-eminently a mining country: it is among the world's top ten producers of silver, coal, copper, iron, phosphates, and uranium, as well as by far its largest producer of gold (1.32 million lb [601,524 kg] in 1989), which is 40% of the world output and accounts for half of the value of South African exports. The major deposits are found in the Witwatersrand. The production of diamonds, including industrial diamonds, is also considerable (9,116,000 carats in 1989). Iron and coal are extracted primarily in the Transvaal, together with other minerals such as platinum, manganese, chromium, antimony, vanadium, lead, tin, and nickel. There are also substantial deposits of nonmetallic ore, including asbestos, magnesite, mica, sulfur, and pyrites.

South Africa has virtually no hydrocarbon deposits, although it does produce some energy synthetically by fractional distillation of coal; it therefore has to import considerable quantities of fuel to meet its domestic needs. Most of the electric energy (162 billion kWh in 1990) is produced by thermal plants, with approximately 4 billion kWh by nuclear power.

Industrial activity is highly diversified and primarily concentrated in the Transvaal, the Cape Town area, and along the southeast coastal belt (Durban-Pinetown and Elizabeth-Uitenhage areas). It includes mechanical engineering (automobile, aircraft, radio equipment, weapons, etc.), chemicals (various acids, ammonia, plastics, fertilizer, synthetic fiber, etc.), metallurgy (steel and cast iron, copper, aluminum, tin, zinc, etc.), rubber (tires), textiles (wool and cotton yarn and fabrics), paper (newsprint, mechanical and chemical pulp), and cement. The food-processing industry is also highly developed.

Commerce and communications. South Africa's balance of trade is slightly negative. Its major trading partners are the United States, United Kingdom, Japan, and Germany. Its dense and especially well-integrated communications network includes 14,644 mi [23,619 km] of railroads (1989), largely electrified, and 113,440 mi [182,968 km] of roads (1987). The country's merchant fleet of 236 ships has a displacement of some 350,000 gross register tons, and its major ports are Cape Town, Durban, Port Elizabeth, East London, Mossel Bay, and Richard's Bay, the latter handling an annual volume of over 55 million t of freight. The main international airports are located at Cape Town, Johannesburg, Pretoria, Durban, and East London. A large volume of tourists (about one million a year) is attracted by the first-rate accommodations and sights of considerable interest.

Historical and cultural profile. *Indigenous population.* The southern extremity of the African continent did not represent a particular goal of European conquest for a long time, and until the 17th century it was inhabited by Hottentot and Bushman tribes—the former pastoral and skilled in copper-working, the latter hunters and gatherers since the stone age. It is particularly to these two peoples that South Africa owes the unique forms of its indigenous art which contrasts with the expressive art forms of the Bantu groups who settled there fairly recently. Their cave paintings and carvings, which were still being done in the last century, bear remarkable witness to a tradition that dates back to the 1st century A.D. Peaceful and devoid of means of defense, these first inhabitants were driven toward the interior of the region and dispersed by the Dutch, who in the middle of the 17th century founded the Cape Colony which was only formally under control of the Dutch East India Company.

The Boers. Taking advantage of the fertile soil and propitious climate, the Dutch settlers dedicated themselves to agriculture and animal husbandry and were therefore called *boeren* (farmers). It was not the Hottentots or Bushmen who resisted their advance but rather the various Bantu tribes (Nguni, Zulu, and Xhosa) who had for several centuries migrated from the north and to whom the Boers referred by the pejorative name *kaffir* (from the Arabic *kafir*, infidel). The "Kaffir wars" kept the Dutch colonists busy for a very long time (from 1779 to 1850), preventing them from realizing the portent and possible consequences of the contemporaneous settlement of the English in the Cape region. After having driven the so-called "natives" into the interior, the Boers were thus forced to withdraw during the Great Trek (1834–1839) to the east-central African territories, where they established the Republic of Natal (later wrested from them by the British), the Orange Free State and the Transvaal. The latter two states, whose independence Great Britain recognized in the mid-1850s, constituted the bulwarks of Boer national consciousness.

For fully half of the 19th century the Dutch settlers fought a relentless struggle to defend their cultural and political autonomy against the English. This was also the background for the defense of the Afrikaans language, a Dutch dialect profoundly modified by several centuries of contact with the various African, European, and even Asian languages spoken along the coasts of southern Africa. An important literary tradition, still vital today, was to be based on this claim that Afrikaans was a written as well as a spoken language.

English and Boers. Between the English colonies of the Cape and Natal, which were established with a modern capitalist outlook, and the Boer states of Orange and the Transvaal, where agricultural and animal husbandry interests predominated, conflict was inevitable. Extensive mining activity soon developed in the English possessions and when, at the turn of the century, diamond deposits and gold mines were also discovered in the Transvaal, British plans of conquest were kindled. The bloody Anglo-Boer War (1899–1902) ended with the military defeat of the Boers, but they achieved a substantial political success: at the national constitutional convention held in Durban and Cape Town in 1908–1909, the British themselves proposed that the Transvaal and Orange Free State not be admitted as possessions of the Crown in a subordinate role but be included with equal dignity in a federal state, which was in fact constituted the following year as the Union of South Africa. Although the Boer element ended up being integrated with the British and ultimately becoming predominant in the new state, a member of the Commonwealth, the real problem from the start was the opposition between the white elite and the great mass of the black, mulatto, and Asian population. Afrikaans became widespread in the schools and in civilian and religious life as a second official language, but far from being an expression of cultural autonomy, as in the previous century, it gradually imposed itself as the language of the dominant classes.

The racial problem. The party which very soon asserted itself on the South African political scene and came to power, first in 1924 and again in 1948, was the Nationalist Party (NP), representing the European community of Dutch origin and speaking Afrikaans, which was responsible for the policy of racial segregation that was to characterize South African society from then on. This policy represented a response to the problems created by the living together of various ethnic populations which, however, sounded a jarring note when viewed against the background of the parallel advance of the African people on the road to emancipation. In practice, the blacks were deprived of political and civil rights, subjected to salary discrimination, forbidden to intermarry with whites, and not allowed to own land outside of limited reservations, resulting in territorial segregation that restricted them to ghetto cities which they could leave only with a pass. Popular protests, such as those of Sharpeville and Langa (1960), which were crushed by bloody repressions, and even the condemnations of the UN and the other Commonwealth countries failed to correct the Pretoria government's orientation. In 1961, the only response of the Union of South Africa was to sever its ties with the Commonwealth and transform itself into the Republic of South Africa under the presidency of Pieter W. Botha, thus leaving it free to make apartheid even harsher. In recent decades, however, this policy increasingly met with a vast movement of internal protests (leading in 1976–1977 to the mass uprisings of Soweto and Guguleto, which ended in a bloodbath) as well as with international condemnation that paved the way to the application of economic sanctions by the UN.

In 1986, the NP proclaimed a state of emergency and only in 1989, when F.W. de Klerk became president of the republic, were there beginning signs of a possible reversal of direction, such as restoration of freedom of the press, suspension of the death penalty for opponents of the regime, and a return to legality of the African National Congress (ANC), the largest anti-racism party in South Africa. A further clearing on this horizon was the release in 1990 of the ANC leader Nelson Mandela, who had been imprisoned for over 27 years.

Twentieth-century literature. South African literature developed under constant confrontation with the most dramatic problems born of racial segregation. Even the Afrikaans literature, more involved with stylistic experimentation than with content and concerned primarily with achieving a literary dignity of its own, proved in more than one case that it was able to express the country's reality, with such significant voices as those of D. J. Oppermans, S. J. Pretorius, Olga Kirsch, Ina Rousseau, and Breyten Breytenbach, who in 1957 was sentenced to nine years in jail for his stand against apartheid. Both white and Coloured authors writing in English have addressed the racial question and social problems, in particular, William Plomer, Laurens van der Prost, Herman Charles Bosman, and Nadine Gordimer. Writers of African ethnic and cultural background are the Dhlomo brothers, R. Mazisi Kumene, Oswald Mbuyiseni Mtshali, and Noni Jabavu, of Zulu origin. Other South African writers dealing with secular subjects (such as Peter Abrahams and Ezekiel Mphahlele) chose to go into exile or were forced to do so by repression.

SWAZILAND

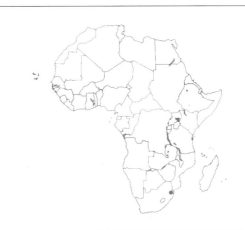

Geopolitical summary

Official name	Kingdom of Swaziland / Umbuso weSwatini
Area	6703 mi² [17,364 km²]
Population	681,059 (1986 census); 768,000 (1990 estimate)
Form of government	Constitutional monarchy with bicameral Parliament (Senate and House of Deputies)
Administrative structure	4 regions
Capital	Mbabane (pop. 38,290, 1986 census)
International relations	Member of UN, OAU, Commonwealth, associate member of EC
Official language	English and siSwati
Religion	Animist 40%, Protestant 30%, Catholic 7%
Currency	Lilangeni

Natural environment. Swaziland is bounded on the north, west, south, and southwest by the Republic of South Africa, and on the east by Mozambique. Its territory occupies a section of the eastern edge of the large South African plateaus (constituted here by pre-Paleozoic crystalline rocks), intersected by many waterways which drop directly down to the Indian Ocean.

From the higher areas of the Drakensberg Mountains the topographic features slope down stepwise toward the east in distinct altitude zones with their individual morphological, climatic, and vegetation characteristics. The first such zone consists of the

Climate data

Location	Altitude (ft asl)	Average temp. (°F) January	Average temp. (°F) July	Average annual precip. (in.)
Mbabane	3739	68	54	54.6

Conversion factors: 1 ft = 0.3 m; 1 in. = 25 mm; °C = (°F – 32) × 5/9

Administrative structure

Administrative unit (regions)	Area (mi²)	Population (1986 census)
Hhohho	1,378	178,936
Lubombo	2,296	153,958
Manzini	1,570	192,596
Shiselweni	1,459	155,569
SWAZILAND	6,703	681,059

Capital: Mbabane, pop. 38,290 (1986 census)
Major cities (1986 census): Manzini, pop. 18,084; Big Bend, pop. 9676; Mhlume, pop. 6509.

Conversion factor: 1 mi² = 2.59 km²

Highveld (with elevations of 5900–2900 ft [1800–900 m]), followed by the Middleveld (3450–1000 ft [1050–300 m]), and finally the Lowveld (1000–500 ft [300–150 m]), which is characterized by a valley depression running from north to south and delimited on the east by the short Lebombo (2300 ft [700 m]) mountain range. The Umbeluzi, Usutu, and Ingwavuma rivers and their numerous tributaries cut deeply into the region after flowing down from the Great Escarpment into the Indian Ocean.

The climate is subtropical—humid and monsoon-like—in the Highveld and tropical like that of the Sudan in the Lowveld. Annual rainfall in the higher altitudes ranges from 40 to 90 in. [1000 to 2300 mm] between October and March, but in the lower interior region it is reduced to 20–27 in. [500–700 mm].

The natural temperate forest vegetation has been transformed into meadows and plantations, except in the Usutu valley and the savanna of the Lowveld, and the fauna has been reduced to a few specimens that can still be seen in the Milwane Reserve south of Mbabane.

Population. The Swazi (AmaSwazi), who belong to the southern Bantu group related to the Zulu, give the area its ethnic homogeneity. The population of 768,000 (1990 estimate), having tripled since the end of World War II because of the high growth rate (4.5%), with a mean density of 114 per mi² [44 per km²] lives mostly (69.6%) in rural settlements. The chief cities are the capital Mbabane and Manzini, which are inhabited mainly by whites. The low mortality rate (1.25%) and high birth rate (4.68%) account for the country's marked demographic

Socioeconomic data

Income (per capita, US$)	789 (1990)
Population growth rate (% per year)	4.5 (1980-89)
Birth rate (annual, per 1,000 pop.)	46.8 (1989)
Mortality rate (annual, per 1,000 pop.)	12.5 (1989)
Life expectancy at birth (years)	53.7 (1989)
Urban population (% of total)	30.4 (1989)
Economically active population (% of total)	40.3 (1988)
Illiteracy (% of total pop.)	32.2 (1991)
Available nutrition (daily calories per capita)	2554 (1989)
Energy consumption (10⁶ tons coal equivalent)	Included in South Africa

growth. Life expectancy at birth is 54 years and the illiteracy rate is 32.2%.

Economic summary. Swaziland's economy is characterized by substantial mineral reserves and a dualistic production process, with a prevalence of modern methods. The principal mineral resources are diamonds, asbestos, iron, coal, tin, and gold. Industrial activity consists primarily of wood-working, cotton-ginning, brewing, and cattle-slaughtering. Industry and modern plantation agriculture (sugar cane, tropical and citrus fruits, tobacco, and cotton) account for 56% of GNP. Traditional agriculture (potatoes, corn, sorghum, millet, manioc, rice, and vegetables) is practiced exclusively by the Swazi population, with cattle-raising (662,000 head of cattle in 1990) a very important element of indigenous economic activity. In the 1990s Swaziland has maintained a positive balance of payments, to which a moderate tourist flow from South Africa has contributed. Transportation facilities are meager: 1723 mi [2779 km] of packed dirt roads (1988), 229 mi [370 km] of rail (1987) linking the country to Mozambique, and one international airport at Matsapa.

Historical and cultural profile. In the second half of the 18th century the Nguni tribe of the of the Nkosa-Dlamini peoples was driven north by the displacement of the Zulu groups until it occupied the area between the Usutu and Komati rivers which was to become the nucleus of the nation of Swaziland.

Sobhuza I, who ascended the throne in 1815, is regarded as the nation's founder; he was succeeded by Mswati I, who ruled from 1840 to 1868 and gave the country its name. The need to defend itself from constant Zulu attacks induced the country to accept colonial influence, and starting in the 20th century Great Britain administered it together with Bechuanaland and Basutoland through its "High Commission." Though smaller in size and population than the other two protectorates, Swaziland was of far greater economic importance because of its rich mineral resources (asbestos, iron, tin) and fertile soil. Britain congratulated itself in referring to this possession as a happy island sheltered from social and racial tensions, to such a degree that, even in 1959, the Labourite John Hatch was still able to claim that

Swaziland has more hopeful economic prospects than the other two territories and this may in part explain the happier atmosphere in race relations. Certainly considerable confidence has been established between the administration and the leaders of Swazi opinion.

Actually, the Swazi were subject to forms of economic and political oppression clearly stamped in the colonial mold that were aggravated by the traditional ethnic conflicts. Yet the monarchical form of government was basically unaffected by the Anglo-Saxon presence and was not abolished even at the end of the road when, after a series of post-World War II stages, independence was achieved in 1968. Swaziland then became part of the Commonwealth as a constitutional monarchy, but in 1973 Sobhuza II repealed the constitution and assumed full power, dissolving the political parties. Both before and after independence, the life of Swaziland came under the influence of the neighboring Republic of South Africa, which determined the country's economic choices and, after the death of Sobhuza II (1982), intervened directly in its political life. After a few years of regency, Mswati III became king in 1986.

ZAMBIA

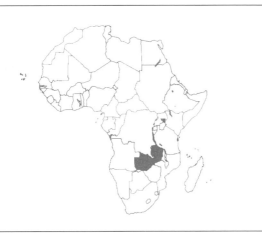

Geopolitical summary

Official name	Republic of Zambia
Area	290,509 mi² [752,614 km²]
Population	7,818,447 (1990 census)
Form of government	Presidential republic; Head of State and the Legislative Assembly are elected every 5 years
Administrative structure	9 provinces
Capital	Lusaka (pop. 982,362, 1990 estimate)
International relations	Member of UN, OAU, and Commonwealth; associate member of EC
Official language	English; Bantu languages spoken
Religion	Animist majority, Catholics 13%, Protestants
Currency	Kwacha

Natural environment. Zambia is bounded on the north by Zaire and Tanzania, on the east by Malawi, on the south by Mozambique, Zimbabwe, Botswana, and Namibia, and on the west by Angola.

Geological structure and relief and hydrography. The landlocked territory of Zambia, which comprises most of the area to the west of the Zambezi, is rather irregular in contour with a deep constriction in the middle that divides it into two distinct and separate sections. The northeastern part is composed of the basin of the Luangwa, the major left-bank tributary of the Zambezi, and by a series of plateaus of the Congo River basin, delimited by one of its spring-fed branches (the Luapula) and including part of Lake Mweru and the wide marshy depression of Lake Bangweulu which, in turn, is separated from the Muchinga mountain chain (6061 ft [1848 m]), the Luangwa tectonic trench, and the southwestern ridge (Mt. Sunsu, 6783 ft [2068 m]) of the Chingobo Mountains by the southern shores of

Lake Tanganyika. Because it lies adjacent to the great East Africa Rift Valley, this section of Zambia is very much subject to tectonic forces, with the result that fairly diverse rocky formations come into contact, such as conspicuous remnants of Precambrian crystalline basement and the limits of the sedimentary series of the Karroo System.

The west-central part also consists of vast plateaus extensively intersected by the course of the Zambezi and its tributaries, often with the formation of rapids and cascades such as the famous Victoria Falls, on the boundary with Zimbabwe. These plateaus rise to an average elevation of more than 3300 ft [1000 m] above sea level (with peaks no higher than 6600 ft [2000 m]) and are largely patterned on the sedimentary series of the continental environment of the Kalahari formed during the Upper Tertiary which is present in the regions further to the west. These sediments are covered by a substrate of Precambrian rocks, the result of extensive peneplanation by erosion, which are often dissected by swamp and lake depressions connected by the course of the Kafue, a left-shore tributary of the Zambezi. From its headstreams on the boundary with Zaire, the Zambezi leaves the territory of Zambia to flow across Angola for 125 mi [200 km], then comes back into Zambia, crosses the plateaus of Barotseland, and at the border with Namibia and Zimbabwe enters into a complex system of basalt gorges inside which it forms the Victoria Falls (354 ft [108 m] drop and a flow rate that varies from 6000 ft³ [10 million l] per min in November to 265,000 ft³ [450 million l] per min in March and April). After following the border with Zimbabwe for a long stretch, the Zambezi enters Mozambique, where it empties after a course of 1649 mi [2660 km] into a wide delta on the Indian Ocean. The river has a huge catchment basin of 514,500 mi² [1,333,000 km²] and is fed by many tributaries, chiefly the Kafue and Luangwa.

David Livingstone discovered the Zambezi in 1851:

> *At Sebituane's death the chieftainship devolved, as her father intended, on a daughter named Ma-mochisane. He had promised to show us his country and to select a suitable locality for our residence. We had now to look to the daughter, who was living twelve days to the north, at Naliele. We were obliged, therefore, to remain until a message came from her; and when it did, she gave us perfect liberty to visit any part of the country we chose. Mr. Oswell and I then proceeded one hundred and thirty miles to the northeast, to Sesheke; and in the end of June, 1851, we were rewarded by the discovery of the Zambesi, in the centre of the continent. This was a most important point, for that river was not previously known to exist there at all. The Portuguese maps all represent it as rising far to the east of where we now were; and if ever any thing like a chain of trading stations had existed across the country between the latitudes 12° and 18° south, this magnificent portion of the river must have been known before. We saw it at the end of the dry season, at the time when the river is about at its lowest, and yet there was a breadth of from three hundred to six hundred yards of deep flowing water. Mr. Oswell said he had never seen such a fine river, even in India. At the period of its annual inundation it rises fully twenty feet in perpendicular height, and floods fifteen or twenty miles of lands adjacent to its banks.*

Climate. Zambia is generally affected by the tropical and continental latitudes, although their excesses are moderated by the fairly high average altitude. The mean monthly temperatures

Climate data

Location	Altitude (ft asl)	Average temp. (°F) January	Average temp. (°F) July	Average annual precip. (in.)
Lusaka	4,251	70	61	32.6
Livingstone	2,998	75	61	26.3
Ndola	4,159	70	59	50.5

Conversion factors: 1 ft = 0.3 m; 1 in. = 25 mm; °C = (°F – 32) × 5/9

range from 61°F [16°C] in June and July to 82°F [28°C] in December; the continentality of the climate, however, makes fairly wide daily temperature swings possible. Although rainfall can vary considerably from year to year, it is seasonal and generally coincides with the hottest period of the year; it represents the principal element of climatic differentiation, diminishing from 47 in. [1200 mm] annually in the north to 23 in. [600 mm] in the south.

Flora and fauna. The type of vegetation varies similarly, from the equatorial forest in the northern regions and along the course of the Zambezi to the savannah in the southern and eastern regions. Zambia is endowed with a remarkable system of national parks (totaling 18 protected areas which cover approximately 8% of the national territory). Noteworthy among these, because of its characteristic large herds of elephants, is the one at South Luangwa covering 3500 mi² [9000 km²] in the central portion of the Luangwa river valley. Another is the great Kafue national park (8650 mi² [22,400 km²]) in the river basin of the same name.

Population. Because of the essentially mild climatic conditions and favorable environment of these regions, they were settled in prehistoric times, the earliest human presence having been traced to the Paleolithic. Identified among the oldest tribal stock are some groups of Bushmen and Hottentots, who settled the area during the Mesolithic, and some Pygmy tribes, which still dwell around Lake Bangweulu. But, as throughout south central

Administrative structure

Administrative unit (provinces)	Area (mi²)	Population (1980 census)
Central	36,436	513,835
Copperbelt	12,093	1,248,888
Eastern	26,675	656,381
Luapula	19,519	412,798
Lusaka	8,452	693,878
Northern	57,061	677,894
North-Western	48,569	301,677
Southern	32,919	686,469
Western	48,785	487,988
ZAMBIA	290,509	5,679,808

Capital: Lusaka, pop. 982,362 (1990 estimate)
Major cities (1990 estimate): Kitwe, pop. 348,571; Ndola (1987 estimate), pop. 418,142; Mufulira, pop. 175,025; Chingola, pop. 186,769; Luanshya, pop. 147,747.

Conversion factor: 1 mi² = 2.59 km²

Africa, the dominant people today are the Bantu, who in much more recent times (early 19th century) overran the local resident populations and drove them into more inhospitable areas.

Although the territory of Zambia was settled very long ago, it is today one of the least populated regions of Africa, with a density of 26 people per mi² [10 per km²]. This population distribution is highly unbalanced. More than one fifth of the total population (with well over half the white inhabitants) is crowded into the province of Copperbelt alone, which is dotted by many cities, notably Ndola, that have developed around the major mineral deposits. There is also some concentration along the rail line linking this highly populated area with the capital of Lusaka and the tourist center of Livingstone (Maramba), at the Zimbabwe border. Outside this development center there are vast areas which are almost totally unpopulated.

Starting in the 1950s, when Zambia had barely 2 million people, there has been a phase of considerable demographic growth (currently stated to be 3.7% a year), made possible by a sharp decline in mortality as a result of increased food supplies and a measure of endemic disease control, although malaria and sleeping sickness continue to plague large regional areas. The rather low annual per capita income of US$420 (1990) accounts for the extremely lean standard of living of the majority of the population, as well as the increasingly meager level of consumption even among the people living in the cities (49%), due to the import restrictions decreed by the government.

Economic summary. After having been extensively characterized in colonial times by a single dominant activity (copper mining), the economy of Zambia today is extremely vulnerable and dangerously dependent on world market fluctuations, notwithstanding attempts by the government to diversify industrial activity and expand agriculture.

Agriculture absorbs over 65% of the working population while occupying only 7% of the land area and supplying barely 14% of the gross national product. The agricultural sector is characterized by a sharp dualism: on the one hand, the large agricultural holdings managed by the state or former European colonists, who have never had any possibility of withdrawing their investments, produce 50% of the meat and milk and over 30% of the corn, although they occupy a limited portion of the cultivated area; and on the other hand the farms under direct management are still operating at subsistence levels in spite of the recent investments made by the state.

Finally, the more remote areas are still engaged in itinerant farming, with cultivated land being abandoned and villages moved every 4 or 5 years.

The mining sector continues to represent the key sector of the national economy; in addition to copper (with an annual total in 1990 of about 550,000 t), there is also considerable extraction of cobalt (of which Zambia is the world's second biggest producer, after Zaire), lead, zinc, tin, and manganese. The mining industry, after having experienced a boom during the 1970s, when it contributed 30% to the GNP, entered an extremely critical phase in the decade that followed because of the collapse in world prices of copper and other major local products, resulting in its share of the GNP dropping to only 15%. This had a negative effect on Zambia's entire economy. Except for the coal that is mined in modest quantities in the Maamba basin and used in metal pro-

Socioeconomic data

Income (per capita, US$)	420 (1990)
Population growth rate (% per year)	3.7 (1980–89)
Birth rate (annual, per 1,000 pop.)	49 (1989)
Mortality rate (annual, per 1,000 pop.)	13 (1989)
Life expectancy at birth (years)	54 (1989)
Urban population (% of total)	49 (1989)
Economically active population (% of total)	32.3 (1988)
Illiteracy (% of total pop.)	31.4 (1991)
Available nutrition (daily calories per capita)	2026 (1989)
Energy consumption (10^6 tons coal equivalent)	1.9 (1987)

cessing, power is derived from a few large hydroelectric plants erected near the Zambezi to meet the energy requirements of the mining sector.

Industry, which is largely managed by the state, absorbs about 8% of the work force and contributes 24% to GNP. Processing activities, concentrated for the most part in the Copperbelt around the city of Lusaka, are primarily earmarked to satisfy local consumption. In addition to production intended for the manufacture of copper, the principal products are foodstuffs, beverages and tobacco derivatives.

Following the drop in the price of copper, which accounts for 90% of the country's export earnings, Zambia's foreign trade has recently experienced a downturn. The foreign-exchange reduction brought about by the copper crisis has prompted the government to institute fairly restrictive measures with regard to imports of consumer goods. This has essentially enabled Zambia to maintain a stable balance of trade.

The country derives a limited income from international tourism, the development of which is hampered by a lack of local infrastructure. Not only are tourist facilities inadequate, but so are transportation and communications. Although the 1342 mi [2164 km] rail system (1988) is fairly well developed because of the need to transport copper to the Indian Ocean ports, the road network other than the Lusaka-Ndola development hinge is rather limited (23,163 mi [37,359 km]), especially with regard to paved stretches (3817 mi [6157 km]). Air transportation is better developed, based as it is on a few major airports (Lusaka, Livingstone, and Ndola) and a few small service stopovers.

Historical and cultural profile. *Zambia in ancient times.* The inland region of southern Africa contains a wealth of important paleontological fossils. *Homo erectus*—halfway on the road of human lineage between the ape-like hominids and those with typically human features—was actually discovered in Zambia. In some populations of this region the stone age extended fairly close to our own era, in others the passage to the iron age was relatively recent. But next to communities that remained marginalized alongside the historical process were others who experienced a rich and intense history. The Bantu kingdoms of Luba and Lunda, which had their center in the Shaba area of Zaire, extended their offshoots to today's territory of Zambia. Here too came the Barotse, still from the north, whose king was deemed to have descended from the Nyambe, or Supreme Being, as well as Nguni peoples, this time from the south. Here, between the 17th and 18th centuries, came the Portuguese, Arabs, and

Swahilis, who were chiefly interested in the slave trade and quite ready to ignite the conflicts between the various tribes to that end. The white presence increased as the ancient kingdoms declined.

Colonization. Britain's interest in the region of Zambia arose in the second half of the 19th century after David Livingstone had ascended the course of the Zambezi and Cecil Rhodes, president of the British South Africa Company (BSAC), had included it as a possession. Having overcome the resistance of the Nguni and Lunda, the British embarked on colonization with the establishment of the two protectorates of Northern Rhodesia, corresponding to the present territory of Zambia, and Southern Rhodesia, today's Zimbabwe. In the northern protectorate colonization was less systematic than in the southern part; it lacked the large white population which was to create so many racial problems in the south. The two possessions followed different political and constitutional courses when the BSAC was dissolved in 1924 and Northern Rhodesia came under the direct administration of the Colonial Office, while Southern Rhodesia obtained the status of an autonomous colony.

The new state. Northern Rhodesia seemed to be destined to remain one of the poorest and most barren British possessions until rich deposits of copper were discovered in the Copperbelt. This was followed by a constant influx of whites from the south, while pressure was simultaneously exerted by Salisbury to obtain control of the region. It was to that end that the Federation of Central Africa was established in 1953 which forcibly linked the two Rhodesias to neighboring Nyasaland for a decade to the advantage of the government in Salisbury. From the very outset Northern Rhodesia was hostile to the Federation and played an important role in the process that led to its dissolution in 1963, prior to independence which was achieved in October of the following year. The new nation, which adopted the name Zambia, was headed by President Kenneth Kaunda, a charismatic leader who embodied the most advanced objectives of African nationalism in the program of "Zambian humanism," to which he sought to adapt the country's policies in the years to follow— with considerable difficulties. Starting in 1977–1978, the left-wing minority of the country's single party began to press for introduction of scientific socialism as the dominant ideology, replacing Kaunda's humanism. Elected president again in 1978, Kaunda was confronted two years later by a failed attempt by the right wing to overthrow his government. Re-elected once more by an overwhelming majority, Kaunda further strengthened his position. Notwithstanding his renewed authority, Kaunda's strong personality gradually began to decline with time due to the emergence of a bureaucratic middle class and the persistence of relations with Zimbabwe, but prevented Zambia from losing contact with "revolutionary" Africa.

ZIMBABWE

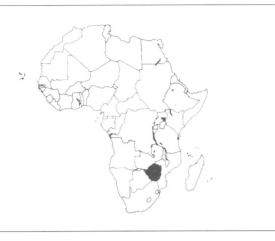

Geopolitical summary

Official name	Republic of Zimbabwe
Area	150,833 mi^2 [390,759 km^2]
Population	7,546,071 (1982 census); 9,369,373 (1990 estimate)
Form of government	Presidential republic with socialist tendencies; Parliament is composed of two branches: House of Representatives and Senate; Head of State remains in office for 6 years
Administrative structure	8 provinces
Capital	Harare (pop. 863,000, 1987 estimate)
International relations	Member of UN, OAU; associate member of EC
Official language	English; Bantu languages also widespread
Religion	Animist majority, Christian 20%
Currency	Dollar

Natural environment. Zimbabwe is bounded on the north by Zambia, on the east by Mozambique, on the south by South Africa, and on the west by Botswana.

Geological structure and relief. The landlocked territory of Zimbabwe occupies a complex system of highlands between the Zambezi valley to the north and the Limpopo valley to the south. Its morphology is typically that of a high plateau linked with the presence of wide remnants of the continental crystalline basement characterized by extensive rises and faults and serving as a watershed between the two river basins. The topography is streaked with granite intrusions which have left many residual mountains, among them the Great Dyke which crosses the plateau from north to south for 300 mi [500 km]. This range contains fairly important mineral resources, chiefly chromium, and is composed of basic effusive (predominantly doleritic) rocks.

On its northwest and southeast margins, the crystalline base-

ment is covered by sedimentary Karroo formations deposited in a continental environment and containing extensive remains of vertebrates dating back to a period between the Upper Carboniferous and Upper Triassic. The repeated tectonic upheavals that dislocated the basement and its covering layers resulted in intensive explosive volcanic activity producing numerous diamond-bearing conduits. Late Triassic eruptions were responsible for the expansions of basaltic lava on the banks of the Zambezi, particularly in the area of Victoria Falls, where erosion reveals thicknesses of as much as 2300 ft [700 m]. Of special interest are the ridges of the Mavuradonha range in the north (peak elevation of 5783 ft [1763 m]) and those of the Inyanga range in the northeast, which rises to 8502 ft [2592 m], the highest elevation in the country, before sloping abruptly toward Mozambique. The country can be divided into three topographic zones (or velds). The high veld (over 4000–4600 ft [1200–1400 m] in altitude), which serves as a watershed between the Zambezi basin and those of the Limpopo and the Save, encompasses the middle of the country from the northeast to the southwest between Harare and Bulawayo and is the most important area because of its natural resources and favorable climate for human habitation. The second zone, or middle veld, roughly 2000–2300 to 4000–4600 ft [600–700 and 1200–1400 m] in elevation, extends to the northwestern and southeastern margins of the high veld before gently sloping down to the low veld. This third zone comprises the low river valleys and continues into the desert expanses of the Kalahari to the northwest and the coastal plains of Mozambique to the southeast.

Hydrography. The Zambezi, one of the world's biggest rivers in terms of both its length (1650 mi [2660 km]) and size of its drainage basin (513,400 mi^2 [1,330,000 km^2]), collects the waters from about two thirds of the country. It flows into Zimbabwe from Zambia, defines the border between the two countries, and boasts the spectacular Victoria Falls (a mile [1600 m] wide with a 400-ft [122-m] drop). Among the many right-bank tributaries are the Shangani and Umniati, which supply the huge hydroelectric Kariba Dam basin (175 mi [280 km] long). The western part of the country is drained by the Nata-Monzamnyama river, which ends in the endorheic salt-pan basin of Lake Makarikari in Botswana. The watercourses of the southeastern part of the country flow together into the Limpopo, Save, and other lesser rivers which cross Mozambique and, like the Zambezi, empty into the Indian Ocean.

Climate. Zimbabwe's climate is tropical, with two seasons: one dry, from April to October (winter), and the other rainy, from November to March (summer). The temperature, especially in the middle and high velds, ranges from a minimum of 57°F [14°C] in July to a maximum of 73°F [23°C] in January. Temperature swings are greater in the depressed and interior areas where the anticyclonic air masses of the Kalahari dominate the atmosphere. Mean annual rainfall is 27 in. [700 mm], but this level is about 4 in. [100 mm] higher in the high veld and on the slopes facing the Indian Ocean.

Except for the Limpopo and Zambezi valleys, which are warmer and more humid because they are subjected to the weather perturbations coming from the east and southeast, the territory enjoys moderate air currents, which favor human settlement.

Flora and fauna. The vegetation cover of Zimbabwe is typical of tropical climates. There is a preponderance of thin forests ("tree veld") of deciduous trees and savanna, grasslands inter-

Climate data				
Location	**Altitude (ft asl)**	**Average temp. (°F) January**	**Average temp. (°F) July**	**Average annual precip. (in.)**
Harare	4,825	69	57	32.4
Bulawayo	4,451	71	57	23.2
Victoria Falls	3,001	78	61	27.7
Conversion factors: 1 ft = 0.3 m; 1 in. = 25 mm; °C = (°F – 32) × 5/9				

spersed with small trees and shrubs (baobab, acacia, etc.), and a prevalence of xerophylous plants, especially in the region facing the Kalahari. In various sections of the tree veld, the presence of the non-native eucalyptus tree attests to ongoing human influence, and along the river banks gallery forests thrive.

The plantations of the European colonists, which are mainly located on the ridges facing the Indian Ocean, and the itinerant farming of the indigenous population have had the effect of restricting the natural vegetation, though it is still well preserved in the less populated northwestern section of the country.

The fauna, particularly rich and varied prior to colonization, has been sharply reduced because of the large game hunts carried out to resupply the world's circuses and zoos. Recent legislation has regulated such hunting and established national parks for the protection of wildlife, such as the Hwange National Park near the border with Botswana.

Population. The population of Zimbabwe has a density of approximately 62 per mi^2 [24 per km^2]. It consists of 96% blacks, 3% whites, and for the balance Asians and people of mixed race (Coloureds). The majority is of Bantu stock and belongs to the Matabele and Mashona peoples.

The annual demographic increase is approximately 3.5%, due almost completely to the natural growth rate of the indigenous population, with hardly any contribution by the other racial groups. Altogether, these never even reached 300,000, or about 5% of the population, and today their number is declining because of the large emigration of the Europeans which began after the country's independence. Approximately 60% of the black population live in the "tribal trust lands" (45% of the country), which are actually reservations, in villages dispersed in the "tree veld," where subsistence agriculture, extensive cattle-raising, and especially hunting are practiced. Some of these people periodically leave their villages (*kraal* or *umzi*) to work in the efficient agricultural enterprises operated by the white settlers. A good many of the native people, having assimilated European living habits more extensively, live in the so-called "white" rural areas (38% of the country) and in the urban agglomerations, where they have integrated into the modern economy.

Part of the white population is made up of descendants of the English colonists who, after Cecil Rhodes, settled in the vast territory between the Zambezi and Limpopo rivers obtained in concession from the indigenous tribal chiefs in 1890. The original 800 settlers grew to 280,000 by 1975, due above all to waves of immigrants from the Union of South Africa and the United Kingdom. Then, because of political instability and guerrilla attacks, an exodus of the white community started which reduced the number of its members to 63,000 between 1976 and 1979 (in 1983 there were 180,000).

The urban centers developed along communication lines, especially the railroads. Major cities are Harare (formerly Salisbury) and Bulawayo, which together contain about 70% of the urban population. Other urban centers are Mutare (formerly Umtali), Hwange (formerly Wankie), Chinhayi (formerly Sinoia), Gweru (formerly Gwelo), and Masvingo (formerly Fort Victoria).

The official language is English, but the indigenous population speaks the traditional native languages Shona and Ndebele. More than half of the blacks are animists, while 30% are Protestants (Anglicans and Presbyterians), as are the majority of whites. Catholics number about 750,000 (90% of them black), and there are also Jews (6000) as well as, among the Asians, Muslims and Hindus. Harare is the seat of a university founded in 1970.

Economic summary. The economy of Zimbabwe, still managed according to free-market criteria, is strongly based on a dualistic principle. Operating in contrast to the traditional subsistence economy, which encompasses the great majority of the black population, is an economic system of commercial agriculture, mining and light industry, and an efficient service sector managed by whites with cheap labor.

Still of primary importance is the agricultural sector which in 1990 occupied 45% of the total work force, although cultivated land (6.9 million acres [2.8 million ha]) represents no more than 7.4% of the country's total area. Five thousand white farmers, who employ a considerable labor force (some 250,000 salaried workers), own 35 million acres [14 million ha] of land located in areas where there is abundant rainfall and little erosion of the soil. The black population, on the other hand, owns individual properties evaluated at 3.7 million acres [1.5 million ha] of collective land known as "tribal trust lands," which are overpopulated and impoverished.

The "land question" has always been a source of conflict between the European and African ethnic groups and was the cause of the recent civil war. The Constitution of 1980 prohibited the confiscation and general expropriation of land without adequate compensation. The present government has instituted a cautious land reform, however, settling black farmers on lands

Administrative structure		
Administrative unit (provinces)	**Area (mi^2)**	**Population (1982 census)**
Manicaland	13,460	1,099,202
Mashonaland Central	10,532	563,407
Mashonaland East	9,625	1,495,984
Mashonaland West	23,340	858,962
Masvingo	17,104	1,031,697
Matabeleland North	28,385	885,339
Matabeleland South	25,627	519,636
Midlands	22,761	1,091,844
ZIMBABWE	150,833	7,546,071

Capital: Harare, pop. 863,000 (1987 estimate)
Major cities (1982 census): Bulawayo, pop. 414,800; Chitungwiza, pop. 175,000; Gweru, pop. 70,000; Mutare, pop. 70,000.

Conversion factor: 1 mi^2 = 2.59 km^2

Socioeconomic data

Income (per capita, US$)	640 (1990)
Population growth rate (% per year)	3.5 (1980–89)
Birth rate (annual, per 1,000 pop.)	37 (1989)
Mortality rate (annual, per 1,000 pop.)	7 (1989)
Life expectancy at birth (years)	64 (1989)
Urban population (% of total)	27 (1989)
Economically active population (% of total)	39.6 (1988)
Illiteracy (% of total pop.)	24 (1991)
Available nutrition (daily calories per capita)	2232 (1989)
Energy consumption (10^6 tons coal equivalent)	6.5 (1987)

underutilized or abandoned by whites. About 4000 families received land during the first year of this program. Although the "plan" does not fully satisfy the expectations of the ethnic African majority, which would like "everything at once," it seeks to preserve the efficiency and productivity characteristics of the preceding white management. However, there have been cases, especially along the eastern border, of abusive occupation of land by blacks which neutralize the constitutional terms and weaken the reassuring stand of the government, concerned about not losing the highly skilled labor of the whites and the capital investments from the Western world, which the country absolutely needs.

Agriculture and livestock. Agriculture, faced with satisfying domestic needs, has made substantial progress with regard to previously little cultivated crops (1,744,600 t of corn in 1991 and 278,300 t of wheat, as well as sorghum, sugar cane, cotton, silk, coffee, etc.), while reducing land planted with tobacco to 153,000 acres [62,000 ha] on which 152,900 t were grown (in 1964–65, tobacco alone accounted for 55% of the agricultural output).

Specialized modern farms, some of which are operated by ethnic Africans and use irrigation equipment, produce a variety of fruit (pears, apples, peaches, and citrus). State and producer organizations operating in various production sectors provide helpful domestic and foreign marketing facilities. The 666,000 families (totaling 4 million people) that live on the tribal trust lands and practice typical subsistence farming, where millet and corn constitute the basic foods, have remained largely untouched by this progress.

In spite of the damages caused by the war, animal husbandry is fairly widespread on the tribal trust lands, especially of cattle, which accounts for 60% of the total (6,711,000 head in 1990). In general, cattle-raising as practiced by the blacks is much less profitable than that practiced by the whites, who have adopted modern methods and use choice animals. Thanks to its mild climate and lush pastures, the high veld is the most suitable area for the livestock industry.

Forestry has benefited from the substantial expansion (about 49% of the total area) of natural forests rich in prized hardwoods (teak and mahogany) and from plantations of cellulose-producing species (pine and eucalyptus), which are processed primarily in Mutare. Annual timber production is estimated at some 275 million ft^3 [7.8 million m^3].

In 1989 freshwater fishing produced 22,000 t.

Energy resources and industry. Zimbabwe's subsoil contains a great variety of minerals and metals much sought after by modern industry. The country's mineral fortune is primarily derived from four products: gold, asbestos, nickel, and chromium, of which Zimbabwe is one of the world's leading producers. The mining industry is managed by the Harare government.

Very important to the economy is the production of electric energy (approximately 7750 million kWh in 1988), primarily of hydroelectric origin. The principal hydroelectric power plant (633,000 kW of installed capacity), that of Kariba on the Zambezi, was built by Italian companies and inaugurated in 1960.

Also of some importance are the thermoelectric power plants in Harare, Bulawayo, and Umnyati, each of which has an installed capacity of about 200,000 kW. The remaining energy needs are met by imports of electricity from Zambia.

The most industrialized areas of the country are those of Harare and Bulawayo, which account for 76% of total production and also employ three quarters of the industrial work force (150,000 people, including 130,000 ethnic Africans and 20,000 ethnic Europeans, corresponding to 14% of the colored labor force and 18% of the white). Recent industrial decentralization has been undertaken at Que Que, Redcliff (seat of a pilot iron-and-steel plant), and Gwelo, a major center of textile and shoe manufacturing, and chromite refining. The manufacturing industries are predominantly run by multinational companies and large corporate subsidiaries of foreign firms.

Zimbabwe's industry is quite diversified and can satisfy almost all domestic needs, especially with regard to food (sugar, canned goods, flour, beverages, fats and vegetable oils, coffee, and tea), which is remarkable for a Third World country.

After a decline that bottomed out in 1979, tourism experienced a rapid recovery thanks to increased political stability. Hotel facilities (over 500,000 in 1989) are considered to be satisfactory, especially in the cities of Harare and Bulawayo. Other tourist attractions are Victoria Falls, Kariba Dam, the archeological treasures of ancient Zimbabwe, and a number of national parks.

Commerce and communications. The country's balance of trade has remained positive in spite of considerable fluctuations. Exports consist almost entirely of agricultural and mineral commodities. In addition to the export of tobacco which, although quantitatively cut back, is still considerable, there has been a constant increase in exports of corn, cotton, coffee, nickel, and ferrochromium. Imports include a variety of consumer and capital goods as well as raw materials, primarily in the areas of transportation and petroleum, chemical, and pharmaceutical products. Zimbabwe's major trade partners are South Africa, the EC countries, the United States, Japan, and Malawi.

Deprived of an outlet to the sea, Zimbabwe has developed a transportation network since the early decades of European colonization which links it to the ports on the Indian Ocean (in Mozambique) as well as those on the Atlantic (in Angola). Rail service, covering 2104 mi [3394 km] (1983), is efficiently maintained by the National Railways of Zimbabwe, and the road system comprises 28,431 mi [45,856 km] of highways in 1985. There are about 6200 mi [10,000 km] of paved roads. Inland water navigation is of no economic importance, except for the ships operated on the artificial Kariba lake for tourism and fishing (sardines) purposes. The country has good domestic and foreign aviation facilities and many of the ethnic Europeans operate private planes. The major airports are at Harare and Bulawayo.

Historical and cultural profile. *Ancient kingdom.* When it achieved independence, Southern Rhodesia chose the ancient name Zimbabwe, significant because it dates back to a long period of prosperity. Zimbabwe (which means "large stone house" and indicated the residence of the ruler) was called the capital of the great Monomotapa Empire which flourished in the region between the Zambezi and Limpopo rivers, reaching the height of its glory in the 15th century. This kingdom, in turn, was the culmination of an historical process rooted in earliest antiquity, for as early as the 5th century B.C. this area was the site of a farming and iron-age culture in which were assimilated Bantu groups like the Karanga (Shona), Mbere, and Rozwi peoples. This indigenous civilization, which before adopting a unified order was structured in a network of developed city-states, founded its wealth on mining, working, and trading metals, notably gold.

Colonization. This wealth which was at the root of the ascent of the Monomotapa Empire also brought about its decline when it came into contact with the Europeans. It was the Portuguese who, attracted by the Kingdom's prosperity, intruded into its dynastic feuds and ultimately made it their vassal in the 17th century.

Filippo Pigafetta, without ever having seen it, described the fabulous Monomotapa Empire in the late 16th century as follows:

> *The Monomotapa Empire, extending inland from the shores below these two rivers, the Magnice and Cuama, has a vast number of caves of gold that is transported to all the neighboring regions and to Sofala as well as other lands of Africa. Some say that it is from these countries that gold was brought by sea to Solomon for the temple in Jerusalem, which is not improbable, for in the countries of Monomotapa there are many old edifices of great workmanship and fine architecture fashioned of stone and lime and wood—something that is not seen in the surrounding provinces.*
>
> *The Monomotapa Empire is large and its many people are pagan heathens, who are black, of medium stature, very brave in war, and swift of foot. There are numerous kings, vassals of the Monomotapa ruler, who often rebel and plot against him. Their arms consist of bows and arrows and light darts. This Emperor maintains many armies, divided into legions in the provinces in the manner of the Romans, for being a man of great rank he must wage war constantly to maintain his standing. Among the warriors we have mentioned, those reputed to be the most valorous are the female legions, which are greatly valued by the King and are the backbone of his military forces: they burn their left breasts with fire so that they will not get in the way when shooting arrows, in the custom of the Ancient Amazons so celebrated by the historiographers of early secular accounts. The arms they use are bows and arrows; they are very slender and swift, courageous and daring, masters of the bow and arrow; above all, skilled and steady in war. In battle they resort to great warlike cunning, habitually feigning retreat, acting as if put to rout, and taking flight, but often turning around and assailing their opponents with a hail of arrows; and when seeing their enemies elated by victory, already dispersing, they suddenly fall on them with great daring and kill them. Thus, because of their speed in ambushing the enemy and their other fighting skills they are held in great dread in these parts. From their King the women receive grants of land where they live alone and after some time they unite with men they have chosen themselves at their pleasure for procreation, and any*

> *male children born to them are sent to these men's houses but the females they keep for themselves to raise in the arts of war. The Empire of Monomotapa thus lies on an island formed by the seacoast, by the River Magnice, by a portion of the lake from which the latter flows, and by the River Cuama. Toward the south it borders of the territory of the Rulers of the Cape of Good Hope, already described, and toward the north it is bounded by the Empire of the Monenugi, as we shall come to see. Now returning to our subject, which is to follow the seacoast, and having crossed the River Cuama, we find a small kingdom on the sea called Angoche, which takes its name from some islands of that name, located directly opposite to it and inhabited by the very same Mohammedans and heathens as are in the country of Sofala, traders who sail along this coast in small boats and traffic in the same goods as do those of Sofala.*

The final collapse came in the beginning of the 19th century, when the Anglo-Boer settlement in South Africa caused by way of repercussion a series of migrations from the south, including that of the Nguni and Matabele, who swept away the Monomotapa Empire and established a new kingdom. In the meantime, the inland area of South Africa had assumed a definite strategic importance: for Great Britain it was an important link in the chain that could join Egypt and the Cape, and for the Portuguese it could constitute the connection between Angola and Mozambique.

Located at the crossroads of such ambitious imperial plans, the region saw Great Britain prevail: the artifact of colonization was Cecil Rhodes, who founded Salisbury in 1890 and from whom the territories took their name. His British South Africa Company (BSAC) obtained from London the authorization to administer the region. According to the terms of this mandate (1923), the northern part remained tied to Britain as a protectorate and the southern part became a self-governing Crown colony with the name of Southern Rhodesia, whose new status granted much greater freedom of action to the white colonists who came to settle in large numbers and took advantage of the situation to exclude the black population in the most rigid manner from the exercise of power. The country adopted legislation which in racial matters directly echoed that which prevailed in South Africa. In fact, the Salisbury government always maintained close contacts with the authorities in Pretoria as well as with the Portuguese colonial regimes in Angola and Mozambique.

Between 1953 and 1963 Southern Rhodesia sought to extend its influence to Northern Rhodesia and Nyasaland by joining them in the Central African Federation, but the attempt failed precisely because of the resistance of the African elements, who were hostile to the racist policy of Salisbury.

Toward independence. Even Great Britain opposed this policy, and Jan Smith, who became head of the government in 1964, reacted to British and American pressure by tightening the screws against the non-white population and unilaterally proclaiming independence. This ushered in a period of international tension and armed struggle by the liberation movement which was eased in 1980 by elections that gave a majority to the Zimbabwe African National Union (ZANU) of Robert Mugabe, who became Prime Minister with a majority of seats in the Parliament. That same year the country became independent and adopted its present name.

SOUTHERN AFRICA

Images

1. *Night view of a modern street in Johannesburg, the Transvaal. In the background is the J. G. Strijdom communications tower inaugurated on April 17, 1971, the tallest building (882 ft [269 m] high) ever built on the entire African continent. The city owes its very foundation and also its economic development and demographic growth to the discovery of large gold deposits in 1886.*

2. *The Namib Desert in the coastal region of Namibia extends from the Cunene to the Orange rivers. About 800 mi [1300 km] long and 30–90 mi [50–150 km] wide, it is the most arid and inhospitable region of the country because of the lack of precipitation (0.6–0.8 in. [15–20 mm] a year). The landscape is desolate and the only undulations are the diffuse sand dunes sculptured by the southwestern winds.*

3. *As their name indicates, the Bushmen used to live in the thickets of the northeast from which they were driven toward the arid regions of the Kalahari. Many aspects of their culture (such as their rock drawings and paintings) link this ethnic group—probably the oldest and most primitive in all of Africa—to the inhabitants of the Upper Paleolithic period.*

4. *Semidesert stretch of savanna in the southeast of Zimbabwe, which geologically consists largely of an Archeozoic crystalline mass over which Carboniferous and Cenozoic sedimentary rocks were later deposited and superimposed. Completely different from these formations is the absolutely unique phenomenon of the Great Dyke: an aligned and continuous outcropping of intrusive, eruptive, basic and ultrabasic rock which "cuts" the territory from north to south over a length of 298 mi [480 km]. This is of relevant economic interest because of the presence of considerable concentrations of extractive ore.*

5. *The Orange river as it flows through a canyon in South Africa's Augrabies National Park. About 1300 mi [2080 km] long, the Orange is among the largest tributaries of the Indian Ocean. It rises in northeastern Basutoland from the slopes of the Mont aux Sources and then runs toward the southwest, is joined by the Vaal, and marks the southern boundary between Namibia and South Africa.*

6. *Dusk on the Zambezi, the fourth largest river in Africa (1650 mi [2660 km] long and with a basin about 513,400 mi^2 [1.33 million km^2] wide). It rises in Zambia near the border where Angola and Zaire meet, then describes a wide curve representing the boundary between Zambia and Rhodesia, and forms the Victoria Falls. At the end of its course it empties into the Indian Ocean in a wide delta with several branches often quite distant from one another and all obstructed by sand, except for the navigable channel.*

7. *The Victoria Falls, discovered in 1855 by the Englishman David Livingstone, who named them in honor of Queen Victoria, are formed by the Zambezi at the boundary between Rhodesia and Zambia. At this point in its course, the river plunges over a huge basalt shelf 400 ft [122 m] high and 5580 ft [1700 m] wide, narrowing into a chasm barely 250 ft [75 m] across. Its catchment basin is important to the area's economy because in 1938 it became the site of a hydroelectric power station.*

8. *Madagascar monkey. Southern Africa is endowed with remarkable morphological diversity and is the home of hundreds of animal and floral species unmatched by any other region of the world.*

However, colonization of this area by the European powers has progressively dimmed the picture of this part of Africa as a "Paradise on Earth" by reducing the number of large carnivores and herbivores for food, sport, and commercial reasons.

9. *Flowery plain and tree-lined hills in Namaqualand, a vast region of Southwest Africa inhabited by the Hottentot Namaqua people from whom it derives its name. It is divided into two areas by the Orange river—Great Namaqualand and Little Namaqualand to the south, which is part of South Africa.*

10. *The traditional economy of Madagascar is essentially based on the dual elements of rice and the zebu. Rice is the staple food of most Malagasy and therefore the most typical crop of the island, where the rice paddies surrounding the villages create a landscape similar to that of Southeast Asia. A constant presence in the wide-ranging rice fields of Madagascar is the zebu, or Malagasy ox. This bovine animal has a strong body, is accustomed to being mounted and working in the fields and, moreover, constitutes a good source of meat. Zebu ownership is a sign of wealth and well-being.*

11. *Native woman of Mauritius. Indians represent the most substantial ethnic element on this island state east of Madagascar. They are the last descendants of the large groups which emigrated there during the latter years of the 19th century. Unlike the other ethnic populations of the island, the Indians have preserved their traditions and customs, even their religious practices, like the famous "marching on fire," a sort of penitential act in which Indian believers are driven to walk miraculously on burning coals.*

12. *Member of the highly heterogeneous population of the Comoros, a group of islands located at the northern mouth of the Mozambique Channel. Their ethnic composition is the result of a series of intermarriages among Arabs, blacks, Malagasy, and Asians. One of their noteworthy traditions, a legacy of the archipelago's most pervasive economic activity, is the procession of the canoes leaving to go (or returning from) fishing; it honors the fearlessness of those who go out to sea and boldly challenge the ocean currents in their slender canoes hollowed from tree trunks with no other floats than empty gourds.*

13. *Family group from the highlands of Madagascar. Notwithstanding the migrations and complexity of the population, the result of constant ethnic intermarriage, a score of distinct groups can be singled out on the island, each embodying a certain unity of culture and tradition. The largest of these are the Merina, who may have come from Java in the 16th century and settled in the interior plateau, bringing with them rice, cotton, perhaps the zebu, and their dominant culture.*

14. *South African woman of the Ndebele tribe, a Bantu people and one of the Rhodesian principal ethnic groups that immigrated into the area in 1830. Thanks to their strong sense of social identity and the high value they place on family*

ties, the Ndebele maintain their traditions intact; their dress, especially when at work, follows European styles, but reverts to that of earlier times during festivities.

15. *Nighttime view of the skyscrapers of Cape Town against the suggestive natural scenery of Table Bay. The city was founded in 1652 by Dutch settlers led by Jan van Riebeeck who chose it as a fortified base of the Dutch East India Co. In addition to being the political capital of the Republic of South Africa, Cape Town is the country's chief port, serving to resupply ships on the India route and as the point of departure for its exports.*

16. *The monument to Paul Kruger in downtown Pretoria, South Africa. Kruger was president of the Transvaal at the time of the Boer rebellion against the English, who were guilty, in the agreement of 1881, of having turned their promise of self-government into political submission to the queen. The Afrikaners, today's descendants of the Boers, venerate Kruger as a national hero.*

17. *View of Windhoek, capital of Namibia, located at an elevation of 5500 ft [1680 m] in the heart of a natural amphitheater framed by the country's highest peaks (the Erongoberg, 7700 ft [2350 m], and the Brandberg at 11,828 ft [2606 m]). Seat of the major administrative and political bodies, Windhoek is a flourishing market for the domestic trade in cattle and karakul pelts.*

18. *Palace of the People in Moçâmedes, capital of the province of the same name and active fishing port on the southern shore of the estuary of the Bero river in southwest Angola. Former Portuguese West Africa has a very low average population density, with most of the inhabitants concentrated in the coastal zone along the rail link with Zaire, of which Moçâmedes is the terminus.*

19. *Aerial view of Soweto, the urban suburb of Johannesburg set aside as a black township. Since 1948, when the repressive and segregationist laws promulgated by the white minority for the overwhelming black majority became even more restrictive of individual and collective liberty, Soweto has been a symbol of the South African problem. Recent abolition of all major racial laws opens up new and difficult prospects for the country, the development of which is still hampered by the climate of violence and political instability created by apartheid.*

20. *Panoramic view of a diamond mine located near the Atlantic coast of Namibia. As in neighboring South Africa, often called "the land of gold and diamonds," one of the most consistent sources of Namibia's income are its mineral resources; exploitation of the diamond deposits, among others, found mostly in the subsoil along the coastal zone north of the Orange river, accounts for approximately 45% of the entire national income.*

21. *The big Kariba Dam arched over the gorge of the Zambezi represents one of the great technical achievements of modern Africa: 420 ft [128 m] high and 1900 ft [580 m] long, it is one of the world's largest artificial basins. Capable of storing about 5.65 trillion ft^3 [160 billion m^3] of water, its construction required the evacuation of over 50,000 Batonga. It feeds two large hydroelectric stations that supply power to both Zambia and Zimbabwe.*

22. *The South African city of Durban in Natal Province was named after the British administrator Benjamin D'Urban, who was governor of Cape Province from 1834 to 1838. The city's most important resource is without doubt its port, the busiest and best-equipped of South Africa.*

23. *Preserved in the locality of Noittgedacht, city of Kimberley*

(South Africa), are many rock paintings and carvings which some scholars, while unable to date them, attribute to the Bushmen. These petroglyph paintings represent animals and plants according to exact stylistic canons: absence of background, preference for profile design, rareness of perspective.

24. *The inland stone structures of "Great Zimbabwe," the seat of the glorious Bantu Empire of Monomotapa between the 12th and 14th centuries, is the reference point for the ancient history of former Rhodesia. These ruins include an elliptic temple, an acropolis about 2300 ft [700 m] high, and a valley rich in archeological finds. In the 12th century the Bantu peoples in the area of today's Zimbabwe formed a very loose confederation whose head bore the title Monomotapa and was regarded as a kind of divine ruler or priest-king.*

25. *Funeral statue attributed to the Sakolava tribal confederation that originated in southern Madagascar. Malagasy art reflects the special ethnic situation of the island on the basis of which its inhabitants can be considered to have originally come from Southeast Asia at the beginning of our historic era. The most common figures are nude women or bodies in erotic poses, 20–25 in. [50–60 cm] high, placed at the four corners of the tomb.*

26. *Mask representing the face of a master circumciser. The ancient history of Angola records the presence, alongside some hunting and herding tribes which had no artistic tradition whatever, of peoples in the northeast of the country who were influenced by the more highly developed Congo and Luanda kingdoms and expressed themselves in attractive artistic forms, of which the evidence of greatest interest are the wooden masks used for the initiation rites of young people, with facial features modeled in clay.*

3

4

11

12

13

23

24

POSSESSIONS

FRENCH AFRICA

France participated in the colonization of Africa along with the other European powers, with voyages, exploration, and the establishment of private trading companies (subject to government control); by the 17th century, it had thereby gained a foothold on the western coast of the continent. Colonial expansion resumed vigorously two centuries later, and Algeria, conquered in 1830, became the point of departure for French military expeditions that later set out for Morocco and Tunisia, so that the entire Maghreb came under French influence. From there, France's expansion continued to the south, crossing the Sahara, encompassing part of the Sudan, and extending to the Gulf of Guinea. In 1895, French West Africa was created (comprising Guinea, Senegal, Sudan, Upper Volta, Ivory Coast, Dahomey, Niger, and Mauritania), followed in 1910 by French Equatorial Africa (Ubangi-Shari, Chad, Gabon, and the Middle Congo). It was in the Maghreb countries, after World War I, that the nationalist movement began which eventually spread out to the entire continent. This movement led to proclamations of independence in Morocco and Tunisia in the 1950s, while the nations of black Africa achieved that status in the following decade through a more gradual process, marked by the intermediate stages of the Franco-African Union and Community of Nations, federalist experiments that were based on relative autonomy for each member state. All that remained to France, besides a few Indian Ocean islands, was Algeria, which gained its freedom after a long and bloody war that cost the Algerian people more than a million dead.

Island of Réunion. After the loss of Algeria (1962), all that France retained in Africa were a few island dependencies in the Indian Ocean around Madagascar, specifically Mayotte in the Comoro group, and La Réunion in the Mascarene archipelago.

Situated some 450 mi [700 km] east of Madagascar (21° S latitude), the island of Réunion, formerly known as Île Bourbon, has a land area of 969 mi^2 [2510 km^2]; it is of volcanic origin and is entirely mountainous. It consists of two juxtaposed volcanic structures, the Piton de la Fournaise (8630 ft [2631 m]), and the Piton des Neiges (10,066 ft [3069 m]), the latter completely extinct and covered with basaltic lava. Erosion has effaced the main crater, but a number of temporary cones still survive on the slopes. Piton de la Fournaise, located southeast of the first peak, is still active: the circular traces of old craters are clearly evident, as is the central crater known as the *cratère brûlant* [burning crater]. The eruptions of this volcano are generally of mixed type, part Hawaiian and part Strombolian, characterized by effusions of highly fluid lava and copious ejections of ash and debris. The last major eruption occurred in 1977. Between the two volcanic structures stretches a connecting ridge that comprises the Plaine des Palmistes and the Plaine des Cafres.

There are many watercourses descending from these mountains directly into the sea; they are very active during the rainy season (from November to April), and especially when the island is struck by violent tropical cyclones. Precipitation is extremely high, especially on the east coast exposed to the moist southeast trade wind, with annual totals well in excess of 160 in. [4000 mm]. The west coast is more sheltered: about 60 in. [1500 mm] of rain falls annually at St. Denis. Average monthly temperatures, again at St. Denis, range between 79°F [26°C] in January and 61°C [16°C] in July.

When the island was discovered in 1513 by the Portuguese navigator Pedro Mascarenhas, its mountainous slopes were covered with a dense tropical forest of tall trees. After the French conquest (1642) and centuries of intense exploitation, the biological face of the island has changed completely. Sugar cane plantations have expanded enormously, and groves of tamarind (*Acacia heterophila*) flourish where the ancient forests once stood. A number of species of exotic birds still survive.

The population of about 600,000 (1990) consists mostly of mulattoes, but includes blacks, Malagasies, Europeans, Indians, and Chinese. Creole dialects and Gujarati are widely spoken alongside French. The capital, St. Denis, has a population of more than 100,000. After World War II, the island became an overseas *département* of France, with considerable autonomy and regular financial assistance from the mother country. As a result, environmental and social conditions, which had seriously deteriorated during the last world conflict, have now perceptibly improved.

The economy is based principally on the cultivation of sugar cane and related activities: sugar factories, rum distilleries, etc. Other well-known crops include vanilla and geraniums, as well as tobacco and essential oils. Livestock (cattle, goats, and pigs) are raised, and fishing is practiced. The main harbor is Pointe-des-Galets on the northwest coast; there is an international airport (Gillot) at St. Denis. A tourist industry is also being developed, for which a suitable hospitality infrastructure already exists. The island's dependencies include a series of coral islets scattered around Madagascar: Tromelin (370 mi [600 km] north of Réunion), the Îles Glorieuses in the Comoro group, and the islands of Juan de Nova, Bassas da India, and Bassas da Europa in the Mozambique Channel.

Mayotte. Also located in the Comoro group is the island of Mayotte (Mahoré), which opted out of the archipelago's declaration of independence in 1975 and thereby acquired the status of a "territorial community." The total area of the island and its nearby dependencies is 144 mi^2 [374 km^2], with a population of about 67,000 in 1985. The island, of volcanic origin, is entirely surrounded by coral reefs. The population consists of Arabs intermingled with black, Malagasy, and Indian elements. The predominant religion is Islam. The capital is the town of Dzaoudzi (pop. about 6,000), which also has an airfield. The island's economy is essentially based on exotic produce (vanilla, ylang-ylang, sugar cane, coconuts, etc.), and on animal husbandry and fishing. There is negligible tourism. French sovereignty is disputed by the nearby republic of the Comoros, to which the island physically belongs. It also houses a French military base.

BRITISH AFRICA

Extending itself not only toward the lands beyond the Atlantic but also toward India and the Orient, Great Britain participated along with the other European powers in the conquest of Africa. That conquest proceeded in well-defined stages, first in the form of a commercial presence on the coasts of a still unknown continent, then by means of the great exploratory expeditions of the 18th and 19th centuries, and finally as a campaign of genuine military subjugation, in which the continent itself was divided up. Gambia, Sierra Leone, the Gold Coast, and Nigeria were the first British colonies in Africa, together with the Cape Colony wrested from the Dutch which took on such historic importance in the last years of the 19th century. During this same period, after the opening of the Suez Canal (1869), Britain's interest was piqued by Egypt; then it turned to the southeast, from Sudan to Somalia, Uganda, Kenya, and Nyasaland, creating a link between its northern possessions and the southern territories at the Cape. This vast empire was ruled under a variety of constitutional systems that were felt at various times to be suitable for different situations: Crown colonies governed directly from London by the Colonial Office, protectorates over essentially independent states, and "dominions" that enjoyed relative autonomy. It was the dominions that gradually drained the vitality from the imperial formula during the years just before and after

World War II, ultimately emerging as independent nations within the British Commonwealth. Today, the island of St. Helena, with its dependencies Ascension, Tristan da Cunha, and Gough, represents the last African survivor of what was once the world's largest colonial empire.

The volcanic islands that rise from the Atlantic south of the Gulf of Guinea and west of the western coast of Africa (Ascension, St. Helena, Tristan da Cunha, and Gough) are part of the British colony of St. Helena, with its capital at Jamestown.

The total area of these islands is 162 mi^2 [419 km^2], and their population is approximately 7000.

Ascension. The northernmost of the group is the island of Ascension (34 mi^2 [88 km^2]), with about a thousand inhabitants, located 465 mi [750 km] south of the coastline of the Gulf of Guinea (8° S latitude) and 250 mi [400 km] west of the mouth of the Congo. The island was discovered on Ascension Day in the year 1501 by the Portuguese navigator João de Nova Castella. Consisting entirely of volcanic rock, its highest point is Green Mountain (2828 ft [859 m]), and the capital is Georgetown on the west coast. The island became a dependency of St. Helena in 1922. Abounding with sea turtles, wild donkeys and rabbits, and "sea swallows" (terns) which nest there, the island supports a few crops. At present its most important function is as a connecting point for undersea cables between the coasts of Africa and South America.

Saint Helena. The island of St. Helena, 560 mi [900 km] from the coast of Angola (16° S latitude), has an area of 47 mi^2 [122 km^2] and a population of about 5600 in 1987, mostly black and mulatto. Its climate is fairly mild, with temperatures that vary each month between 75° and 84°F [24–29°C] in summer and 64–75°C [18–24°C] in winter. Precipitation is not especially plentiful (13–36 in. [325–925 mm] per year depending on exposure). St. Helena was also discovered by the Portuguese navigator João de Nova Castella, on August 18, 1502. Occupied temporarily by the Dutch (1651), in 1673 it became a British dominion, used as a guard post on the route to the Indies and therefore equipped with a fortified citadel at Jamestown. Here, at a place called Longwood, Napoleon Bonaparte was confined from 1815 until his death in 1821. St. Helena later housed prisoners from the Boer War (1899–1902). The island's shores are steep and rocky; its highest point is Diana's Peak at 2827 ft [862 m]. It is connected by undersea cable to Cape Town and to Ascension island. The economy is based on a few crops, livestock, and fishing.

Tristan da Cunha. The southernmost of these islands is Tristan da Cunha (40 mi^2 [104 km^2]), located almost 900 mi [1400 km] west of the Cape of Good Hope (37° S latitude), together with a few adjacent islets (Inaccessible and Nightingale), and the island of Gough (36 mi^2 [93 km^2]), also known as Diego Alvarez, which is located 250 mi [400 km] to the southeast (40° S latitude), and has a weather station. Discovered in 1506, the island of Tristan da Cunha was annexed by the United Kingdom in 1816 and became a dependency of St. Helena in 1938. In 1961 a violent volcanic eruption forced the inhabitants to evacuate the island, to which they were unable to return for two years. The volcano that dominates the island rises to an elevation of 6757 ft [2060 m]. The population, numbering some 300 individuals (Europeans and mulattoes) in 1990, cultivates vegetables and fruit trees, raises livestock, and fishes for crustaceans.

PORTUGUESE AFRICA

With the exception of the island of Madeira, almost nothing remains of Portugal's vast colonial possessions in Africa, that complex of garrisons, trading posts, and military bases that dotted the coast of the continent in the 15th century and opened the way to world domination not only for this Iberian power, but for all of Europe. Under the guidance of Henry the Navigator, the Portuguese were the first explorers to sail south, circumnavigating Africa after Bartolomeo Diaz had rounded the Cape of Good Hope in 1487. The Portuguese commercial empire in reality proved ephemeral; in the next century, the same riches that flowed from it (gold, silver, spices, slaves) attracted the attention of the other European nations who were developing into mercantile powers: the Netherlands, Great Britain, and France. Lisbon continued to hold a few islands (Madeira, Cape Verde, São Tomé, Principe) and several commercial outposts on the coasts of Guinea-Bissau, Angola, and Mozambique. The latter territories assumed renewed importance during the 19th century as bridgeheads for penetration into the African interior, and the creation of real colonies. The borders of these colonies were defined in the last decades of the 19th century, when Portugal gave up the idea of uniting its Atlantic possessions in southern Africa with its territories that faced the Indian Ocean. For Portuguese Africa, the advent of the "Estado Novo" in 1926 marked the beginning of a more comprehensive policy of economic exploitation by the mother country, oppression based on the widespread practice of forced labor, and denial of fundamental human rights. Heavily dependent on the Lisbon government and declared Overseas Provinces, for forty years the Portuguese colonies suffered the brutality of the Salazar regime, and even the reforms of the 1960s could not stem the liberation movement that spread from Cape Verde to Guinea-Bissau, from Angola to Mozambique, and in 1975 brought independence to these nations.

Madeira. On the other hand, Portugal's centuries-long sovereignty over the island of Madeira (total area 306 mi^2 [794 km^2] and population of 274,000 in 1990, together with the nearby islands of Porto Santo, Deserta Grande, etc., and the tiny Ilhas Selvagens 150 mi [250 km] farther south, at 30° N latitude) was never disputed after its initial colonization in 1419, except for brief periods of occupation by the Spanish (1580–1640) and the English.

Entirely volcanic, the island of Madeira has a geological history very similar to that of the Canary Islands, and consists of tufas and basaltic lava that emerged during the Mesozoic and Quaternary eras. Its volcanic relief, now substantially worn down by erosion, reaches a maximum elevation of 6104 ft [1861 m] on Pico Ruivo, which stands to the north of the city of Funchal (pop. 50,000), capital of the island. Administratively, Madeira is an autonomous region of Portugal, like its other Atlantic archipelago, the Azores. The island lies at approximately 33° N latitude, 430 mi [700 km] west of the Moroccan coast and 500 mi [800 km] southwest of Portugal. The shoreline is steep and indented, and difficult to approach by sea. The temperate oceanic climate is characterized by very mild winters (the average temperature in January is 61°F [16°C]) and only moderately hot summers (70°F [21°C] in July). Precipitation is not excessive (annual rainfall at Funchal is about 24 in. [600 mm]), and is more frequent during the period from October to April.

The island's vegetation is extremely lush and agriculture is very well developed: cereals, vegetables, grapes, and many tropical species such as pineapples, sugar cane, and date palms are intensively cultivated. Livestock is also raised, and a fishing fleet supplies a burgeoning canning industry (tuna, sardines, etc.). The most highly developed industries are food processing and especially wine-making: Madeira's famous wines are exported all over the world. Another important economic resource is tourism, which is served by an excellent hospitality infrastructure. The population consists predominantly of ethnic Europeans of Portuguese origin, but Spanish, Italian, and Flemish elements are also present. In addition to Funchal, which has a busy harbor and an international airport, other major centers include Câmara de Lobos, Machico, São Vicente, and Santa Cruz.

SPANISH AFRICA

Engaged in the establishment of a huge colonial empire beyond the Atlantic, in lands of inexhaustible wealth that in the 16th century were found not to be part of the Indies but to belong to a new continent, Spain paid little attention to the possibilities for colonial expansion offered by Africa. Spanish presence on African soil was limited to islands (such as the Canaries) and emplacements of strategic and commercial importance (such as Ceuta) inherited from Portugal after that nation was annexed by Philip II in 1580. Also from Lisbon, two centuries later, Spain acquired Fernando Póo, destined to become independent in 1968 as Equatorial Guinea. Toward the end of the 19th century the Spanish presence in Africa also extended to another originally Portuguese colony, Rio de Oro, which in 1934 expanded into the Spanish Sahara. In recent decades, even before the fall of the Franco regime and the dissolution of the ties binding this territory (renamed Western Sahara) to Madrid, it was the scene of a long battle waged by liberation forces of the Polisario Front against Moroccan occupation troops. The African possessions that remain in Spanish hands today are therefore limited to two enclaves on Moroccan soil, and the Canary archipelago.

Moroccan territories. Spanish sovereignty is still exercised over a few localities on the Mediterranean coast of Morocco, in particular the two cities of Ceuta and Melilla, and over several dependencies and minor islands (Peñón de Vélez de la Gomera, Peñón de Alhucemas, Chafarinas islands). The total area involved is 13 mi^2 [33 km^2], with a population of about 125,000. Spanish control over Ceuta (pop. about 68,000 in 1991) dates back to 1580: located opposite Gibraltar, the city sits within a protected bay, connected by ferry service to the Spanish port of Algeciras.

Spain's sovereignty over Melilla (pop. about 56,000) is older, having begun in 1497. The city lies at the base of Cape Tres Forcas on the same meridian as the island of Alborán, and has a major airport. Administratively, the Spanish territories on Moroccan soil (*plazas de soberanía en el Norte de Africa*, or

North African sovereignty areas) are governed by a delegate based in Cádiz.

Canary Islands. The Canaries archipelago (Islas Canarias), entirely volcanic in origin and located 70 mi [115 km] off the southwest coast of Morocco between 27° and 29° N latitude, constitutes an autonomous Spanish community consisting of the two provinces of Las Palmas and Santa Cruz de Tenerife, with a total area of 2875 mi^2 [7447 km^2] and a population of about 1.5 million in 1990. The closest to the African coast are the islands of Lanzarote (311 mi^2 [806 km^2]) and Fuerteventura (640 mi^2 [1659 km^2]), both elongated and fairly low-lying, with maximum elevations of a few hundred feet. The islands of Gran Canaria (591 mi^2 [1532 km^2]), Tenerife (792 mi^2 [2053 km^2]) and Gomera (146 mi^2 [378 km^2]) lie farther offshore, and the westernmost are the two remaining islands of La Palma (281 mi^2 [728 km^2]) and Hierro (107 mi^2 [278 km^2]). The meridian that passes through the latter (18° west of Greenwich) was once one of the principal reference points for Atlantic navigation. According to geologists, the archipelago was created by the emergence of a submarine volcanic ridge toward the end of the Cretaceous period (Mesozoic era).

Tenerife island contains the highest volcanic structures: Pico de Teide (12,195 ft [3718 m], the highest point in all of Spanish territory), and Pico Viejo (10,280 ft [3134 m]). The other islands also rise to impressive elevations: 7947 ft [2423 m] on La Palma, 4877 ft [1487 m] on Gomera, 4923 ft [1501 m] on Hierro, and 6396 ft [1950 m] on the island of Gran Canaria, which has the highest population (470,000) after Tenerife (pop. 500,000).

Alvise da Ca' da Mosto visited the islands on one of his voyages and left a description of them:

> *These Canary Islands are seven in number: four inhabited by Christians, being Lanzarotta, Forte-Ventura, Gomera, and Ferro; three are inhabited by idolators, being Gran-Canaria, Teneriffe, and La Palma. The governor of those inhabited by Christians is called Ferrera, a gentleman and knight from the city of Sibillia (Seville), and subject to the king of Spain.... These islands are forty or fifty miles distant from one another: all stand in a line one behind the next, from the first to the last, also from east to west.... The other three, inhabited by idolators, are larger and much more populated, especially two, namely Gran Canaria which has some eight to nine thousand souls, and Teneriffe which is the largest of the three, and is said to have fourteen to fifteen thousand souls: La Palma has few people, but it is a most beautiful island. These three islands, being inhabited by many armed men, with their very high mountains and dangerous terrain, are well defended and have never been subjugated by the Christians.*

The climate of the archipelago is maritime tropical, with average monthly temperatures that range, on the coast, from 64°F [18°C] in January to 77°F [25°C] in July. Precipitation is governed by the trade winds, and is relatively abundant only on the highest elevations; in coastal areas it is rather sparse (8–16 in. [200–400 mm] annually), occurring most commonly in the winter months when snow covers the highest mountain peaks. Summers are generally hot and dry, often dominated by Saharan winds. The vegetation therefore varies considerably depending on environmental conditions. Southern and eastern slopes are generally covered by xerophytic species, while plant life on the wetter northern and western exposures is generally more luxuriant and abundant. The fertility of the volcanic soil, together with intensive irrigation using cisterns and numerous artificial reservoirs, allows a wide variety of crops to flourish, including bananas, citrus, vegetables, grapes, and cereals, providing a steady flow of produce for export. Fishing is also very widely practiced; the catch includes cod, sardines, and tuna. The industrial base is still fairly modest, comprising tobacco factories, canning plants, naval shipyards (Las Palmas), and petroleum refineries (Tenerife). But the archipelago's fundamental resource is tourism; visitors are drawn not only by the wonders of nature—the islands' volcanoes, which are still active as demonstrated by the eruptions of Tenerife in 1909 and La Palma in 1971, produce some very interesting features, including numerous hot springs, tunnels several miles long inside the basalt lava fields, lava flows, ash fields, etc.—but also by the mild and healthy climate. In the past their fame was such that they were known as the "Fortunate Isles."

Few descendants remain today of the original population, the Guanci, who lived in extremely primitive conditions. The rest of the population consists of whites, blacks, and mulattoes. The main inhabited centers are the two provincial capitals, Las Palmas (pop. 360,000) on Gran Canaria island, and Santa Cruz de Tenerife (pop. 210,000) on Tenerife. The natural environment is protected by four national parks located on the islands of Tenerife, La Palma, Gomera, and Lanzarote. The islands are connected by daily flights to Spain and the rest of the world, through the two international airports at Santa Cruz and Las Palmas. Santa Cruz also has a well-equipped commercial harbor which serves as a stopover on the major trans-Atlantic shipping routes.

YEMENITE AFRICA

Located about 150 mi [240 km] east of Ras Asir (Somalia) and 225 mi [360 km] southeast of the southern coast of the Arabian peninsula, the island of Socotra, with an area of 1400 mi^2 [3626 km^2] and a population in 1990 of about 15,000 (consisting predominantly of the descendants of Arab and black slaves) constituted an integral part of the former Arab sultanate of Qishn and Socotra before becoming part of the People's Democratic Republic of Yemen (South Yemen) in 1967, before that nation was reunited with the Yemen Arab Republic.

Discovered in antiquity, it was briefly occupied by the Portuguese (1507–1511), who planned to use it as a base for their trade routes to the Indies. Following a treaty (1876) with the sultan of Qishn (a locality on the southern coast of the Arabian peninsula), the island became a British protectorate in 1886, and only 80 years later was incorporated by Yemen.

Elongated in shape (about 60 mi [100 km] long) and consisting of a strip of pre-Paleozoic crystalline rock (a detached fragment of the African continental shield), the island is mountainous, reaching its greatest elevation at 4930 ft [1503 m] on the Hajhir plateau. The climate is tropical—hot and humid—but with little precipitation even though it is subject to the monsoon winds. Its principal economic resources are fish, mother-of-pearl, and date and aloe palms. The capital is Hadiboh (pop. 1500) on the north coast, where there is also an airfield. Socotra also controls several smaller dependent islands to the west, closer to the Somali coast.

GREAT ROUTES AND VOYAGES OF DISCOVERY

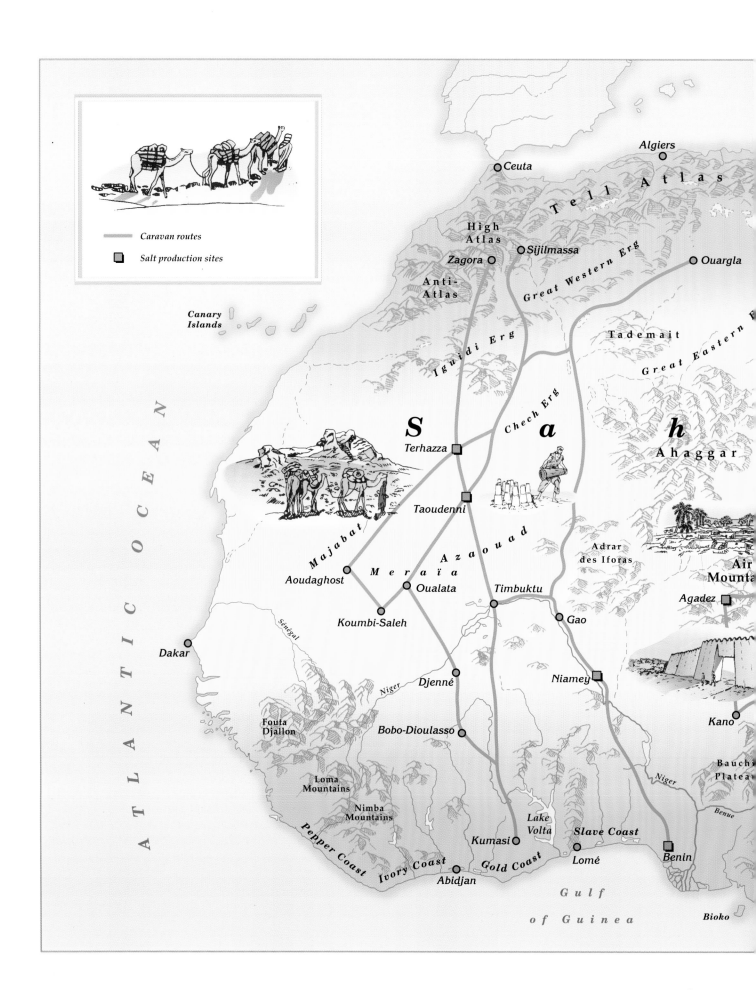

Canary
Islands

A T L A N T I C   O C E A N

Caravan routes

Salt production sites

Ceuta

Algiers

T e l l   A t l a s

**High
Atlas**

Sijilmassa

Zagora

Great Western Erg

Ouargla

**Anti-
Atlas**

Iguidi Erg

Tademait

Great Eastern E

S

Chech Erg

a

h

Terhazza

A h a g g a r

Taoudenni

Majabat

Azaouad

*Adrar
des Iforas*

**Air
Mounta**

Aoudaghost

M e r a ï a

Oualata

Timbuktu

Agadez

Koumbi-Saleh

Gao

Sénégal

Dakar

Niger

Djenné

Niamey

Kano

**Fouta
Djallon**

Bobo-Dioulasso

**Bauchi
Plateau**

Niger

**Loma
Mountains**

Benue

**Nimba
Mountains**

*Lake
Volta*

Slave Coast

Kumasi

Benin

Pepper Coast

Ivory Coast

Gold Coast

Lomé

Abidjan

G u l f

o f   G u i n e a

Bioko

THE SALT ROUTES

*T*rade across the Sahara dates back to at least the 8th century, according to some Arab chronicles. Products from the Mediterranean countries also flowed into the oasis cities in the northern part of the desert, while black Africa supplied primarily gold from the Ashanti realm (Ghana), copper, and cola nuts. "The Sudanese populations had no mines, and as a substitute for salt they used products extracted from the ashes of certain plants; moreover, sea salt … was not sufficiently widespread in the interior…" (J. Ki-Zerbo). Salt was therefore one of the most sought-after and precious consumer goods.

The salt deposits were (and still are) located at Terhazza and Taoudenni in Mali, and at Bilma and Fachi in Niger. They often occupy deep basins not far above sea level. Salt is produced every year by evaporation, and is collected into conical cakes each weighing about 55 lb [25 kg], shaped for easy transportation on dromedaries. At Bilma both salt and dates are produced, and are then sold in the Kano region of northern Nigeria.

Since the 18th century, the salt trade has been managed by the Kel Tamashek Berbers (called in Arabic "Twarigh"), some of whom still understand the art of traveling by camel caravan. There are many routes, both north–south and east–west, the latter being shorter. The journey from Bilma to Agadez involves stages of 16 hours a day, from 6:00 in the morning to 10:00 at night, for six days; only a single well is located on this route, which is extremely strenuous for both humans and animals.

Before the 18th century, however, the salt trade was organized by the sub-Saharan kingdoms and empires, which gradually shifted the customary routes farther west. In the 8th century, the most commonly used route (or darb) ran from Sijilmassa (Morocco) to Aoudaghost and Oualata (Mauritania) and to Tadmekka in Mali; then it was taken over by the former Ghana empire with its capital at Koumbi-Saleh (also in Mauritania), which was conquered by the Almoravids of Morocco in 1076 and later converted to Islam. The even more powerful empire of Mali had its capital first at Niani (Niger), then at Timbuktu (present-day Mali); the capital of the Songhai empire was at Gao in Mali.

These cities gradually inherited the functions of their predecessors, and it was only later that Sijilmassa was replaced by the oases of Fezzan (Libya) and Ouargla (Algeria). The salt deposits of Terhazza and Taoudenni could therefore be reached by the western routes, while those at Bilma lay on the route from Tripoli through Fezzan and Bilma to Lake Chad, site of the kingdom of Kanem with its capital at Nguigmi. Several cities in sub-Saharan Africa prospered on the basis of the trade in salt and gold: Djenné (Mali) and Bobo-Dioulasso (Burkina Faso) are among those still important today, while Salaga in Ghana and Kong in Ivory Coast are no longer significant.

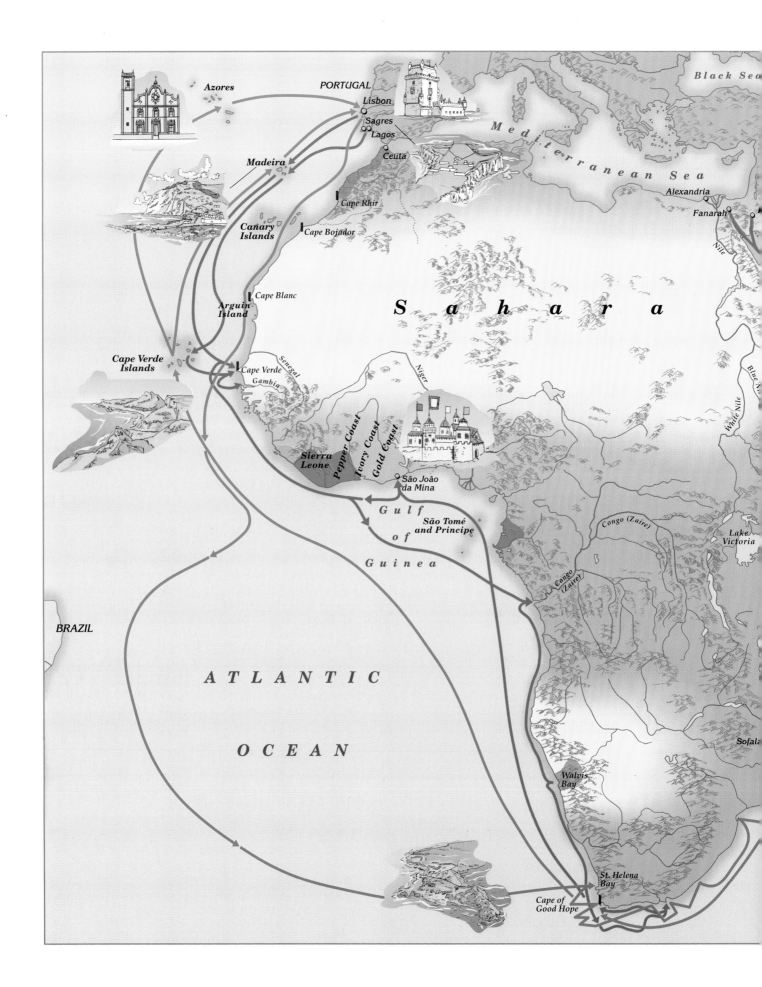

Black Sea

PORTUGAL
Lisbon
Sagres
Lagos
Ceuta

Mediterranean Sea

Alexandria
Fanarah

Azores

Madeira

Cape Rhir

Canary
Islands

Cape Bojador

Cape Blanc

Arguin
Island

S a h a r a

Nile

Cape Verde
Islands

Cape Verde
Gambia

Senegal

Niger

White Nile

Blue Ni

Sierra
Leone

Pepper Coast
Ivory Coast
Gold Coast

São João
da Mina

Gulf

São Tomé
and Principe

of

Congo (Zaire)

Lake
Victoria

Guinea

Congo
(Zaire)

BRAZIL

A T L A N T I C

Sofala

O C E A N

Walvis
Bay

St. Helena
Bay

Cape of
Good Hope

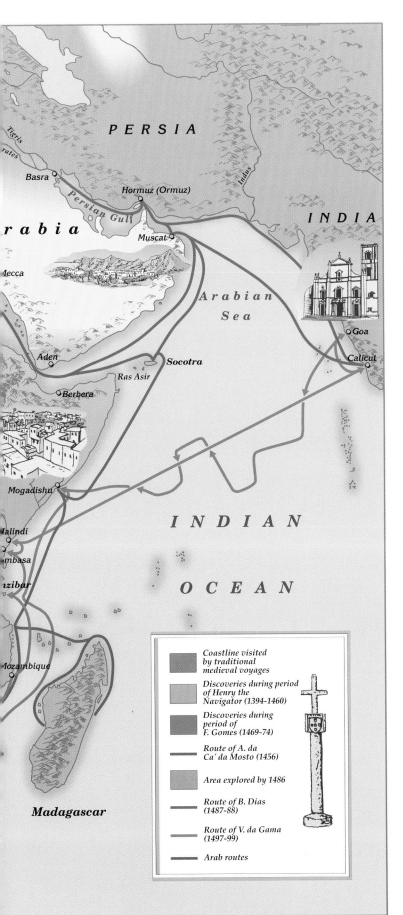

THE PORTUGUESE ROUTES AROUND THE CAPE OF GOOD HOPE

*T*he motives behind Portugal's explorations were both economic (a search for the sources of the gold that was already being received in the West from black Africa by way of the Maghreb) and geopolitical (to encircle the Berber enemy from the south as a continuation of the reconquest of the Iberian peninsula). The capture of Ceuta in 1415 marked the beginning of the African adventure. The first expeditions reached as far as Cape Bojador, frontier of the "Dark Sea." It was not until 1434 that Gil Eanes, sailing in the service of Henry the Navigator, departed from Lagos in Portugal and passed Cape Bojador, opening up the Saharan coastline to exploration. In 1441 Antão Gonsalves reached Cape Blanc, followed shortly by Nuno Tristão.

There was a hiatus from 1444 to 1460 to organize trade in the various commodities—leather, oil, slaves—but this did not prevent the Venetian Alvise da Ca' da Mosto and the Genovese Antoniotto Usodimare from sighting the Cape Verde islands, probably in 1456. From 1469 to 1474, the Portuguese crown entrusted exploration to Fernão Gomes; in 1470, Soerio da Costa, sailing in the service of Gomes, explored some 600 mi [1000 km] of new coastline as far as Cape das Três Pontas. The mouth of the Rio de Sān João was discovered in 1471, and the outpost of São João da Mina was established there between 1482 and 1483. In 1475, the enterprise was placed in the hands of the future king John II (1481); it was under his successor Manuel I (1495) that the route to the Indies was established, marked by padrões (cross-shaped stones) left by the explorers along the coast. In his voyages of 1485–86, Diego Cão reached 22°10' S latitude (Walvis Bay); in July or August 1487, Bartolomeu Dias sailed from the Tagus river with two caravels and a supply ship, and in 1488 arrived at 37–38° S latitude, sailing far offshore around the Cape of Storms, later called the Cape of Good Hope. Dias continued even farther, ascertained that the coastline began to turn towards the northeast, then turned around and arrived back in Portugal in December of 1488.

Organization and planning for the route to the east began in 1488, but it was not until July 8, 1497, that four ships with 150 sailors, commanded by Vasco da Gama, left Lisbon. The fleet made an enormous loop on the high seas, touching the African coast only at 31° S latitude (St. Helena Bay), from which it then sailed against the trade winds to the Cape. On January 25, 1498, it reached the vicinity of the Zambezi river, where the Portuguese found themselves in competition with the Arabs. Da Gama then went ashore at Malindi, where he took on a local pilot who brought the fleet to Calcutta (May 20). On July 10, 1499, Nicolau Coelho, sailing on ahead of da Gama, brought to Lisbon the news that Europe and Asia had been linked by the route to the Indies.

THE SLAVE TRADE

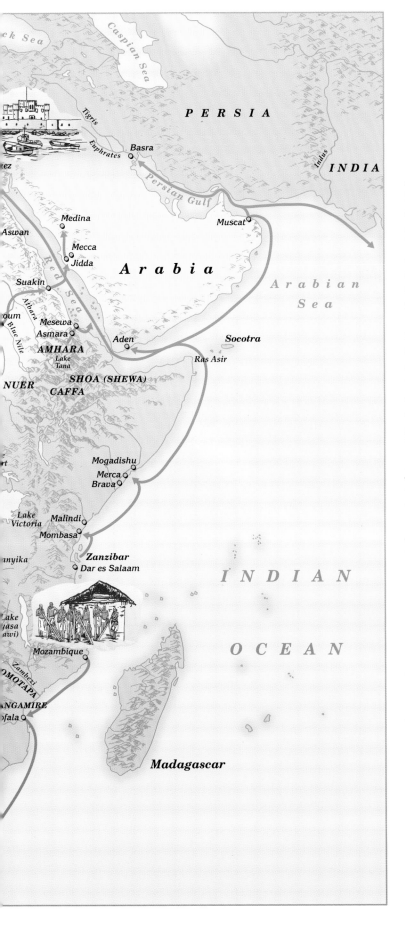

*T*he slave trade was undoubtedly the dominant factor in African history from the 16th to the mid-19th century. Even before this period, slave trading movements existed within the continent's community structures, from black Africa to the Mediterranean coasts across the Sahara or through the Nile Valley, and from the east coast to the Arabian peninsula and India. But the Arabian and Asian trade, although it lasted much longer than the trade with Europe, never assumed the proportions of the trans-Atlantic commerce, which was particularly heavy during the 1800s and persisted into the early 20th century.

The European trade began in 1441, when Antão Gonsalves and Nuno Tristão returned to Lisbon from a voyage of discovery along the coasts of what used to be known as the Spanish Sahara, bringing with them twelve slaves. Other human beings were captured or purchased by the Portuguese and Spanish in the years that followed, but their number remained very limited; the real demand only came later, when labor was needed for the plantations and mines of the Americas, replacing the Indians who (except in Peru and Mexico) had proved to be unsuitable.

After the period of Iberian monopoly, the trade was managed for a short time in the first half of the 17th century by the Dutch, but very soon the large volumes were being controlled by the English, assisted by almost every other European naval power.

The collection areas were located on the coasts, protected by forts. They represented the outlets of states with a well-populated hinterland, capable of supplying the Europeans, in exchange for horses, weapons, liquor, iron, and later textiles, with slaves obtained by wars of conquest or raids. In addition to vanquished enemies, the human cargo consisted of opponents of the particular regimes in power, which used this method to maintain social order.

The departure points for the Atlantic trade consisted of the coastal regions around the Gulf of Guinea, the Congo, and Angola, as well as the southern coast of eastern Africa. The first destination of the slave ships was the Caribbean islands; only later did they come to Brazil, Venezuela, Colombia, Central America, and the southern British colonies in North America. The absolute numbers involved in the trade are extremely uncertain: estimates range from a minimum of 13–14 million slaves taken to the Americas to a maximum of 50 million.

The Atlantic slave trade, declared illegal first in Britain (1807) and then in the United States (1808), did not end until the abolition of slavery in the U.S. (1863), Brazil, and Cuba (1880), when changes in American economic conditions had eliminated the need for the labor of slaves, who were replaced by European wage-earners.

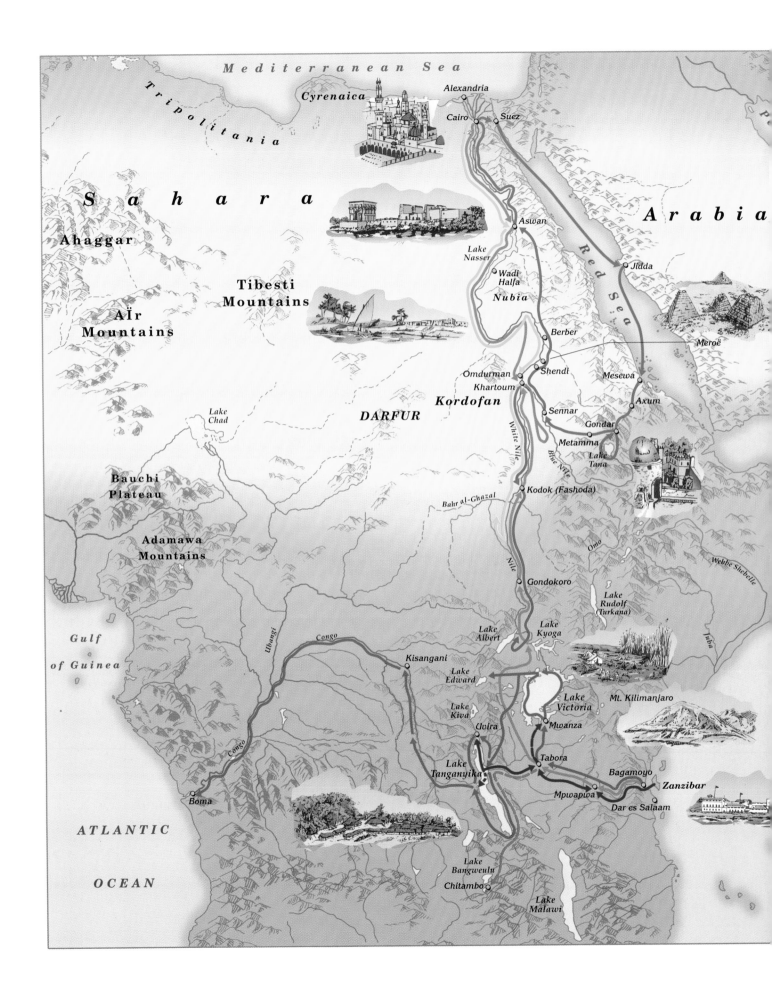

Mediterranean Sea

Cyrenaica

Alexandria

Cairo · Suez

Tripolitania

S a h a r a

Ahaggar

A r a b i a

Aswan

Lake Nasser

Wadi Halfa

Nubia

Jidda

Tibesti Mountains

Red Sea

Aïr Mountains

Berber

Meroë

Shendi

Omdurman

Mesewa

Khartoum

Axum

DARFUR

Kordofan

Sennar

Gondar

Lake Chad

White Nile

Metamma

Lake Tana

Blue Nile

Bauchi Plateau

Bahr al-Ghazal

Kodok (Fashoda)

Adamawa Mountains

Omo

Webbe Shebelle

Nile

Gondokoro

Lake Rudolf (Turkana)

Ubangi

Congo

Lake Albert

Lake Kyoga

Juba

Gulf of Guinea

Kisangani

Lake Edward

Lake Kiva

Mt. Kilimanjaro

Lake Victoria

Uvira

Mwanza

Congo

Lake Tanganyika

Tabora

Bagamoyo

Zanzibar

Boma

Mpwapwa

ATLANTIC

Dar es Salaam

Lake Bangweulu

OCEAN

Chitambo

Lake Malawi

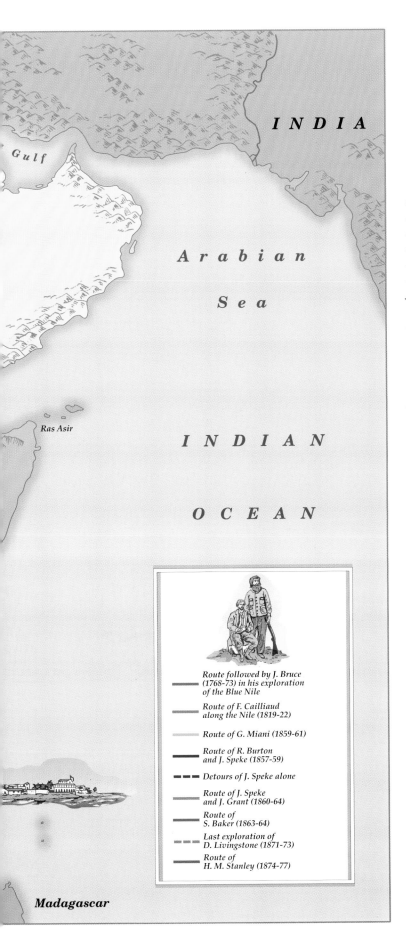

INDIA

Gulf

A r a b i a n

S e a

Ras Asir

I N D I A N

O C E A N

Route followed by J. Bruce
(1768-73) in his exploration
of the Blue Nile

Route of F. Cailliaud
along the Nile (1819-22)

Route of G. Miani (1859-61)

Route of R. Burton
and J. Speke (1857-59)

Detours of J. Speke alone

Route of J. Speke
and J. Grant (1860-64)

Route of
S. Baker (1863-64)

Last exploration of
D. Livingstone (1871-73)

Route of
H. M. Stanley (1874-77)

Madagascar

THE DISCOVERY
OF THE SOURCES
OF THE NILE

*T*he importance of the Nile had been known since antiquity: it is mentioned in the Odyssey; Herodotus sailed up the river, perhaps as far as Elephantine; and the Roman emperor Nero sent an expedition along its course. According to Seneca, the Roman expedition reached perhaps to the confluence of the Bahr el Ghazal and the Blue Nile, a point not reached again by Europeans until the 19th century. On a map of Africa dating from the 2nd century A.D., the Nile rises from two lakes located in the southern hemisphere at the foot of the Mountains of the Moon; along its course it receives two right-bank tributaries, one of which in turn is fed from another lake.

Awareness of the river in Europe then faded, as demonstrated by Giacomo Gastaldi's map of 1564. Not until 1749 did the French geographer J.-B. d'Anville publish a map of Africa purged of all the fantastical information that had slowly crept into African cartography. During a journey into Ethiopia in 1768–73, the Scot James Bruce identified Lake Tana and its effluent the Abbai river as the source of the Nile. The same observations had been published in the Historia de Ethiopia by the Jesuit priest Pedro Xaramillo Páez, who had been sent to Ethiopia in the first half of the 17th century to evangelize the region. Later on, these conclusions were belied by the explorations (1819–22) of the Frenchman Frédéric Cailliaud, who confirmed that the Blue Nile was a tributary of the White Nile.

In 1860, Giovanni Miani came close to the great equatorial lakes, but could get no further than Galuffi in Uganda (3°34' N latitude) some 75 mi [120 km] from Lake Albert, which would be reached from the south by British explorers. It was in fact the expeditions from the south, following the routes of the Arab slave traders, that succeeded in reaching the equatorial lake region, identifying the sources of the White Nile, and confirming that the Nile and Congo rivers occupied two separate basins. The British explorers Richard Burton and John Hanning Speke, sponsored by the Royal Geographic Society, left Zanzibar on June 16, 1857; in 1858 they reached Lake Tanganyika, and on August 3, 1858, Speke alone reached Lake Victoria. A subsequent expedition (1860–64) mounted by Speke and James Augustus Grant explored the area around Lake Victoria, discovering its principal tributary (the Kagera) and effluent (the White Nile). The two ultimately reached Gondokoro in 1863, where their compatriot Samuel Baker, accompanied by his wife, was waiting for them. Following this meeting, and on the basis of information from Speke, the Bakers ascended the White Nile and discovered Lake Albert (1864).

Other expeditions included those of David Livingstone, who looked for the sources of the Nile too far south, and Henry Morton Stanley, who confirmed Speke's discoveries.

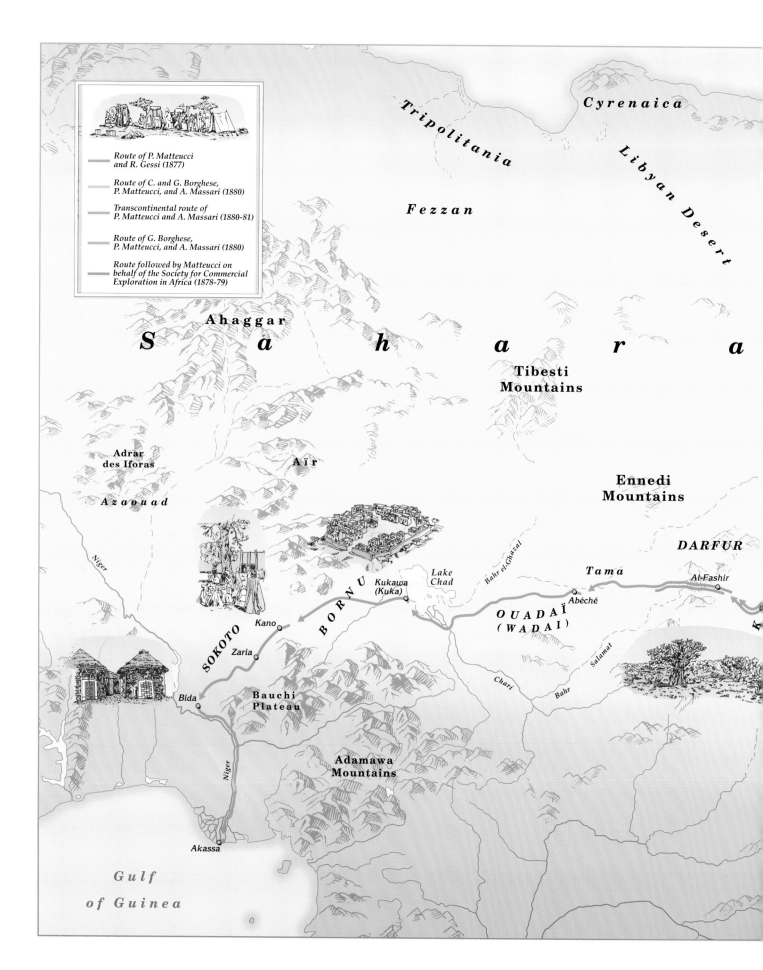

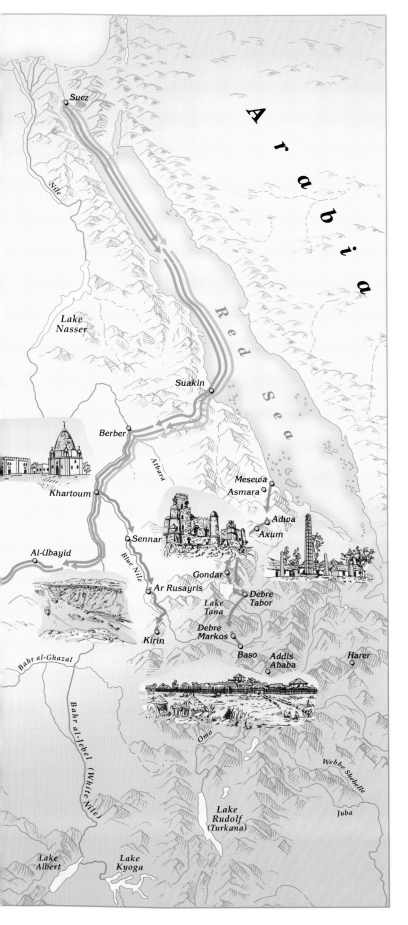

MATTEUCCI'S JOURNEY ACROSS AFRICA

*T*he Association for Promoting the Discovery of Interior Parts of Africa was founded in 1788 in London, initiating a systematic campaign of exploration that regrettably led to the collapse of Africa's environmental and societal structure. Exploration was at first motivated by scientific or humanitarian interests or by the spirit of adventure, but such motives were very soon overwhelmed by political and economic ambition. An important figure in this context was Pellegrino Matteucci (1850–1881), who studied natural history and medicine, and between 1877 and 1879 was a member of the Council of the Italian Geographical Society.

In 1877 Matteucci, together with Romolo Gessi, managed to organize (with the financial support of the Society) an expedition up the Blue Nile to the kingdom of Ghera or Kaffa. The two explorers were halted, however, by hostile natives on the banks of the Jabus. Matteucci's account of the journey was published under the title Sudan and Gallas in 1878. Having thus gained a reputation, he was asked to head a new expedition (1878–79) to Ethiopia, promoted by the Society for Commercial Exploration in Africa. It left from Massawa and reached Debre Tabor, court of the Emperor John; the leader of the expedition then went on alone to Baso, gateway to Galla territory. In his report on the journey entitled In Abyssinia (1880), Matteucci clearly indicated (contradicting most of his contemporaries) that large-scale trade between Italy and Abyssinia was improbable, at least for the near future. Despite the fact that these statements alienated at least some of those with a stake in further exploration, Matteucci succeeded in organizing an expedition to penetrate into the heart of Africa. He was accompanied on the entire journey by Sub-lieutenant Alfonso M. Massari, in charge of recording the itinerary, and for a short stretch by the brothers Giovanni and Camillo Borghese, two of the principal financial backers. The expedition left from Suakin in February 1880, reached Khartoum at the end of March, and from there set out for Ouadaï through Kordofan and Darfur, ultimately arriving at Abéché, capital of Ouadaï. There the sultan refused to allow them to proceed north toward the Mediterranean coast through the Libyan desert, and forced them to continue westward. Matteucci and Massari reached Lake Chad, crossed the kingdom of Bornu, and followed the course of the river Niger; on July 3, 1881, they arrived at the port of Akassa on the Gulf of Guinea, having traveled over 3000 mi [5000 km] and crossed the African continent from east to west. The expedition was unique in terms of its procedure, the enormous distances covered, and the populations who were contacted directly by Europeans for the first time.

Photo Credits

Agenzia Franca Speranza

138-139 (23); R. M. Anzenberger-M. Himmll 45 (8), 50 (15), 58-59 (24); B. Barbier 101 (21); E. Bataille 56 (21); C. Bossu-F. Picat 117 (1); M. Courtny Clarke 190 (9); G. Crivelli 188-189 (6); S. Elbaz 84 (4), 96 (16), 202 (22); V. Gartung 125 (9); A. Heitmann 54 (19); Hoa-Qui/A. Abegg 85 (5); Hoa-Qui/G. Ascani 86-87 (6); Hoa-Qui/J. Boisberranger 322 (12); Hoa-Qui/P. De Wilde 254-255 (6), 323 (14); Hoa-Qui/ G. Gasquet 60 (25); Hoa-Qui/B. Gérard 327 (18), 334 (26); Hoa-Qui/M. Huet 8 (3), 100 (19), 133 (17), 134-135 (19), 136 (21), 195 (14), 264 (16), 268 (21); Hoa-Qui/D. Huot 260 (11); Hoa-Qui/P. Leger 88 (7), 98-99 (18), 100 (20); Hoa-Qui/C. Pavard 88 (8), 96 (15), 97 (17), 261 (13), 266 (18), 270-271 (23), 272 (25), 319 (8), 333 (25); Hoa-Qui/M. Renaudeau 82-83 (2), 90-91 (10), 102-103 (22), 104 (23), 183 (1), 186 (3, 4), 187 (5), 192-193 (11), 196 (5), 197 (17), 200-201 (20), 202 (21), 204 (25), 205 (26), 206 (27), 250-251 (2), 253 (4), 265 (17), 267 (20); Hoa-Qui/X. Richer 128 (12); Hoa-Qui/C. Vaisse 256 (7), 322 (11); Hoa-Qui/E. Valentin 194 (12); Hoa-Qui/ W. Zinder 89 (9); P. Huteau 124 (7); S. Kaufman 41 (5); C. Kutschera 125 (8); D. Laine 206 (28); A. Lanzellotto 249 (1), 316-317 (6); Y. Layma 49 (13), 55 (20); C. Lenars 257 (9); P. Muller 252 (3); Odyssey/R. Frerck 198 (18), 199 (19); C. Poulet 264 (15); U. Vlasak 314 (4), 327 (17);

Archivio fotografico Federico Motta Editore

269 (22); E. Dulevant 93 (13), 190 (8), 191 (10), 203 (23)

L. S. International Cartography

342-351

Guglielmo Mairani

37 (1), 38-39 (2), 40 (3, 4), 42-43 (6), 44 (7), 45 (9), 46-47 (10), 48 (11), 51 (16), 52-53 (17), 54 (18), 56 (22), 57 (23), 197 (16), 204 (24), 253 (5), 272 (24), 311 (1), 320-321 (10), 322 (13), 326 (16), 331 (21, 22), 332 (23, 24)

Marka

B. Maffeis 190 (7)

Laura Ronchi

G. Mairani 120 (4); Photo P. Del Papa 132 (16)

SEF

184-185 (2)

Maria Pia Stradella

L. Caenazzo 125 (10)

Zefa

P. Bauer 315 (5); G. Boutin 324-325 (15); H. R. Bramaz 330 (20); B. Croxford 194 (13); F. Damm 50 (14), 258-259 (10), 260 (12), 262-263 (14), 266 (19); V. Englebert 48 (12), 120 (3), 121 (5), 129 (13), 136 (20); Fabby 81 (1), 92 (11); Geopress 93 (12); K. Honkanen 118-119 (2), 122-123 (6), 126-127 (11), 133 (18); W. D. Horus 256 (8); B. Leidmann 94-95 (14); Masterfile 312-313 (2), 319 (9); NT 130-131 (15), 318 (7); Photofile 314 (3); I. Steinhoff 140 (24); Sunak 129 (14); U. K. 328-329 (19); H. Winter 137 (22)